W9-DDA-970

Molecular Genetics of Hypertension

The HUMAN MOLECULAR GENETICS series

Series Advisors

D.N. Cooper, *Institute of Medical Genetics, University of Wales College of Medicine, Cardiff, UK*

S.E. Humphries, *Division of Cardiovascular Genetics, University College London Medical School, London, UK*

T. Strachan, *Department of Human Genetics, University of Newcastle upon Tyne, Newcastle upon Tyne, UK*

Human Gene Mutation
From Genotype to Phenotype
Functional Analysis of the Human Genome
Molecular Genetics of Cancer
Environmental Mutagenesis
HLA and MHC: Genes, Molecules and Function
Human Genome Evolution
Gene Therapy
Molecular Endocrinology
Venous Thrombosis: from Genes to Clinical Medicine
Protein Dysfunction in Human Genetic Disease
Molecular Genetics of Early Human Development
Neurofibromatosis Type 1: from Genotype to Phenotype
Analysis of Triplet Repeat Disorders
Molecular Genetics of Hypertension

Forthcoming titles
Human Gene Evolution
Transcription Factors
Analysis of Multifactorial Diseases

Molecular Genetics of Hypertension

A.F. Dominiczak
Department of Medicine and Therapeutics, Gardiner Institute, Western Infirmary, Glasgow, UK

J.M.C. Connell
Department of Medicine and Therapeutics, Gardiner Institute, Western Infirmary, Glasgow, UK

F. Soubrier
INSERM U525, Hôpital Saint-Louis, Paris, France

Published in the United States, its dependent territories and Canada by arrangement with BIOS Scientific Publishers Ltd, 9 Newtec Place, Magdalen Road, Oxford OX4 1RE, UK.

First published in 1999

A CIP catalogue record for this book is available from the British Library.

ISBN 0-12-220430-1

Distributed exclusively throughout the United States, its dependent territories and Canada by Academic Press, Inc., A Harcourt Science and Technology Company, 525 B Street, San Diego, CA 92101–4495. www.academicpress.com

Production Editor: Jonathan Gunning.
Typeset by Saxon Graphics Ltd, Derby, UK.
Printed by Biddles Ltd, Guildford, UK.

Contents

Contributors

Barbry, P. Cardiovascular Research Institute, University of California San Francisco, 505 Parnassus Avenue, HSW825, San Francisco, CA 94143, USA

Bianchi, G. University of Milan and S. Raffaele Hospital, via Olgettina 60, I-20132 Milan, Italy

Bonnardeaux, A. Department of Nephrology, Hôpital Maisonneuve-Rosemont, 5415 Boulevard l'assomption, Montreal, Quebec H1T 2M4, Canada

Cambien, F. INSERM U525, 17 rue du Fer à Moulin, 75005 Paris, France

Célérier, J. INSERM U36, Collège de France, 3 rue d'Ulm, F-75005 Paris, France

Corvol, P. INSERM U36, Collège de France, 3 rue d'Ulm, F-75005 Paris, France

Ferrari, P. PRASSIS Research Institute Sigma-Tau, via Forlanini 1/3, I-20019 Settimo Milanese (Milano), Italy

Gimenez-Roqueplo, A.-P. INSERM U36, Collège de France, 3 rue d'Ulm, F-75005 Paris, France

Jacob, H.J. Department of Physiology, Medical College of Wisconsin, 8701 Watertown Plank Road, Milwaukee, WI 53226, USA

Jeunemaitre, X. INSERM U36, Collège de France, 3 rue d'Ulm, F-75005 Paris, France

Kashkoush, S.L. Department of Medicine, Cape Western Reserve University, 2500 MetroHealth Drive – R215, Cleveland, OH 44109–1998, USA

Knoblauch, M. Max Delbrück Centre for Molecular Medicine, Robert-Rössle-Strasse 10, D-13122 Berlin-Buch, Germany

Krege, J.H. Department of Pathology, University of North Carolina at Chapel Hill, Chapel Hill, NC 27599–7525, USA

Kurtz, T.W. Department of Laboratory Medicine, University of California, San Francisco, Box 0134, 505 Parnassus Avenue, Long Hospital 518, San Francisco, CA 94143-1034, USA

Lindpaintner, K. Roche Genetics, F. Hoffman-La Roche AG, CH-4070 Basel, Switzerland

Luft, F.C. Franz Volhard Clinic, Medical Faculty of the Charité, Humboldt University of Berlin, Wiltbergstrasse 50, D-13125 Berlin, Germany

Mellon, S.H. Department of Obstetrics, Gynecology and Reproductive Sciences, University of California, San Francisco, Box 0556, 505 Parnassus Avenue, HSE 1661, San Francisco, CA 94143–0556, USA

Mullins, J.J. Molecular Physiology Laboratory, Wilkie Building, University of Edinburgh Medical School, Teviot Place, Edinburgh EH8 9AG, UK

Mullins, L.J. Molecular Physiology Laboratory, Wilkie Building, University of Edinburgh Medical School, Teviot Place, Edinburgh EH8 9AG, UK

Nabika, T. Department of Laboratory Medicine, Shimane Medical University, Izumo 693, Japan

Ogg, D. Laboratory of Molecular and Cognitive Neuroscience, Department of Neurobiology, The Babraham Institute, Babraham Hall, Babraham, Cambridge CB2 4AT, UK

Payne, C. Veterinary Pathology, Royal (Dick) School of Veterinary Studies, Summerhall, Edinburgh EH9 1QM, UK

Pravenec, M. Institute of Physiology, Czech Academy of Sciences, Videnska 1083, 14220 Prague, Czech Republic

Rubattu, S. Instituto Neuromed, Polo Molisano Universita La Sapienza di Roma, Localita Camerelle – Zona Industriale, I-86077 Pozzilli (Isernia), Italy

Schork, N.J. Department of Epidemiology and Biostatistics, Cape Western Reserve University, 2500 Metro Health Drive – R215, Cleveland, OH 44109–1998, USA

Soubrier, F. INSERM U525, Hôpital Saint-Louis, 1 Avenue Claude Vellefaux, 75475 Paris cedex 10, France

St. Lezin, E. Department of Laboratory Medicine, UCSF/Mount Zion Medical Center 1613, P.O. Box 7921, San Francisco, CA 94120, USA

Stoll, M. Department of Physiology, Medical College of Wisconsin, 8701 Watertown Plank Road, Milwaukee, WI 53226, USA

Struk, B. Laboratory of Molecular Genetics and Genetic Epidemiology, Department of Medicine, Cardiovascular Division, Brigham and Women's Hospital, 20 Shattuck Street, Boston, MA 02115–6195, USA

White, P.C. Department of Pediatrics, University of Texas Southwestern Medical Center, Dallas, TX 75235–9063, USA

Xu, X. Harvard School of Public Health, 865 Huntington Avenue, Boston, MA 02115, USA

Abbreviations

11-HSD	11β-hydroxysteroid dehydrogenase
18OH-DOC	18-hydroxy-11-deoxycorticosterone
ACE	angiotensin converting enzyme
ACTH	corticotropin (adrenocorticotrophic hormone)
AGT	angiotensinogen
AME	apparent mineralocorticoid excess
ANOVA	analysis of variance
Apoe	apolipoprotein e gene
APRE	acute phase responsive element
ASIC	acid sensing ion channel
ASO	allele-specific oligonucleotides
AS-ODN	antisense oligodeoxynucleotide
ANF	atrial natriuretic factor
AVP	arginine vasopressin
BAC	bacterial artificial chromosome
BMI	body mass index
CAH	congenital adrenal hyperplasia
cAMP	cyclic adenosine 3′,5′-monophosphate
CI	confidence interval
CRH	corticotropin releasing hormone
DBP	diastolic blood pressure
DGGE	denaturing gradient gel electrophoresis
D/I	deletion/insertion
DOC	11-deoxycorticosterone
DSE	differentiation-specific element
ECE	endothelin converting enzyme
EDHF	endothelium-derived hyperpolarizing factor
EdinSD	'Edinburgh' Sprague–Dawley
EDRF	endothelium-derived relaxing factor
EIPA	ethylisopropylamiloride
ENaC	epithelial sodium channel
eNOS	endothelial nitric oxide synthase
ES	embryonic stem
ET-1	endothelin-1
GCA	guanylyl cyclase A
GFR	glomerular filtration rate
GH	genetic hypertension
GRA	glucocorticoid-remediable aldosteronism
GRE	glucocorticoid-responsive element
HanSD	'Hannover' Sprague–Dawley
HPLC	high pressure liquid chromatography
HRR	halophyte relative risk

IBD	identical by descent
IGF	insulin-like growth factor
JG	juxtaglomerular
LOD	logarithm of odds ratio
MAP	mean arterial pressure
MH	malignant hypertension
MHS	Milan hypertensive strain of rats
MI	myocardial infarction
MNS	Milan normotensive strain of rats
MR	mineralocorticoid receptor
NEP	neutral endopeptidase
NIDDM	noninsulin-dependent diabetes mellitus
NO	nitric oxide
NOS	NO synthase
OLF	ouabain-like factor
OR	odds ratio
PAC	P1 artificial chromosome
PCR	polymerase chain reaction
PTX	pertussis toxin
PWV	pulse wave velocity
QTL	quantitative trait loci
R	Dahl salt-resistant rat
RAS	renin–angiotensin system
RFLP	restriction fragment length polymorphism
RT	reverse transcribed
S	Dahl salt-sensitive rat
SBP	systolic blood pressure
SF-1	steroidogenic factor-1 (Ad4BP)
SHR	spontaneously hypertensive rat
SHRSP	spontaneously hypertensive stroke-prone rat
SLC	sodium–lithium countertransport
SMG	submaxillary gland
SNP	single nucleotide polymorphism
SSCA	single strand conformation analysis
SSCP	single strand conformation polymorphism
STAR	steroidogenic acute regulatory
TDT	transmission disequilibrium testing
WKY	Wistar–Kyoto rat
YAC	yeast artificial chromosome

Preface

Essential hypertension is a complex quantitative trait under polygenic control. The identification of genes responsible for high blood pressure remains a major challenge because of complex gene–environment interactions, which have proved difficult to dissect with the use of existing genetic and statistical paradigms. However, regardless of the difficulties, identification of the major genes responsible for high blood pressure is important, as this will provide a mechanistic classification of the common hypertensive phenotype, diagnostic markers for individuals and families who are at greatest risk of end-organ complications such as stroke, cardiac ischemia or chronic renal failure and, ultimately, will guide new pharmacogenomic stategies with drug therapies tailored to the underlying primary genetic abnormalities.

In this book experts in their respective fields explore a variety of strategies and concepts which have transported us from the very limited understanding of earlier decades to a much better, albeit incomplete, appreciation of the molecular genetic determinants of this important disorder.

To provide as broad coverage as possible, we have included experimental genetic strategies with a special emphasis on successful translational approaches. Two examples of these with potential clinical relevance are the functional mutations in rat and human α-adducin genes and the use of the major rat QTL on chromosome 10 to investigate the homologous region on human chromosome 17 in familial essential hypertension. Both strategies demonstrate how rodent genetics can help to develop diagnostic and pharmacogenomic tools for human essential hypertension. Furthermore, novel transgenic, 'knock-out' and 'knock-in' mice models illustrate how alterations of a single gene might produce a hypertensive phenotype, thus providing mechanistic explanations to guide human investigation. Several chapters discuss rare Mendelian forms of hypertension and try to address the clinically relevant controversy of whether more subtle mutations within the same genes might be responsible for specific sub-types of essential hypertension. Moreover, the analogy between these rare Mendelian syndromes and the commonly used paradigm of candidate gene strategies is described in detail. These are compared and contrasted with genome wide scanning strategies, and their future potential for discovery of causal genes with the new panels of thousands of single nucleotide polymorphisms. Lastly, current and future developments, including pharmacogenomic approaches are discussed, again with clinically relevant examples.

The last decade has seen significant progress in the understanding of molecular genetics of hypertension but this effort has to continue through translational approaches, functional genomics and collaborative efforts to achieve positional cloning of the major susceptibility and severity genes for high blood pressure and its major complications.

A.F. Dominiczak
J.M.C. Connell
F. Soubrier

Foreword

We live in a time of breathtaking transitions in the biological and informational sciences which are beginning to have profound global effects on biomedical research, education and industry. The mapping and sequencing of genomes ranging from yeast to man has currently advanced to the point that a variety of experimental approaches have now been developed which enable us to link genes to their complex functional pathways and inherited diseases. The successful application of genetics for the diagnosis, prevention and treatment of cardiovascular diseases now lies in our ability to develop novel approaches which can effectively utilize the wealth of tools emerging from these genome projects. The 12 chapters of this book provide an exciting glimpse of many of the new paradigms which are paving the way in the hunt for genes of essential hypertension.

Essential hypertension is the quintessential prototype of a highly prevalent cardiovascular disorder in which both genes and environment play an important role. Considerable advances have been made in our understanding of the complex physiological systems which regulate arterial pressure. There have been useful therapeutic agents developed which lower arterial pressure. However, while progress has been made in identifying some rare Mendelian syndromes associated with severe hypertension, the genes responsible for susceptibility to hypertension and the wide variation in phenotypical expression of this disease are mostly unknown. Clearly, current approaches to understanding the cause, and approaches for the prevention and treatment of essential hypertension have been insufficient.

Despite the scientific challenges before us, it is evident that exciting new tools and approaches are rapidly emerging to explore the relationship between gene(s), function(s), and disease(s). One cannot read this book without a sense of hope and optimism. The research presented here represents the elements of a creative and powerful new science which is being driven by the unparalleled opportunities which exist for those who combine the exploding knowledge of the genome with that of functional biology. If there remain any doubts that we are entering a new age of exploration in our quest for the underlying causes of essential hypertension, the chapters in this book should dispel such thoughts. It seems inevitable that this new science will lead us to a deeper understanding of complex biological functions and discoveries which will one day result in the improved diagnosis, health care, and quality of life for the millions of individuals afflicted with hypertension.

A.W. Cowley, Jr.

1

Hypertension as a complex trait amenable to genetic analysis: basic strategies and integrative approaches

Nicholas J. Schork, Silvia L. Kashkoush and Xiping Xu

1. Introduction

Hypertension and blood pressure regulation are a related disease and physiological endpoint which have received considerable attention from every conceivable medical research angle, including genetics, physiology, pharmacology and epidemiology. There are numerous reasons for this, not the least of which include the fact that hypertension is an extremely prevalent condition in industrialized countries accounting for a great deal of morbidity and mortality, and blood pressure is an integrated physiological parameter with numerous correlates and determinants. Of these research angles, genetics has been receiving the most pronounced attention. In this review we consider contemporary research investigating the genetics of hypertension and blood pressure and argue that, despite the fact that both hypertension and blood pressure regulation are complex and multifactorial in nature, they are amenable to genetic dissection provided that a few fundamental themes are acknowledged. These themes center around the notion that many of the factors influencing hypertension and blood pressure regulation are 'context dependent' in the sense that they interact and form synergistic bonds of influence. This context dependency tends to force researchers to focus on either the effects of single factors within isolated settings or consider costly and inferentially and interpretatively problematic studies which seek to assess the effects of multiple factors simultaneously. To explore these themes we first outline six areas of genetic research from which one can study the genetic basis of a complex disease or trait like hypertension and/or blood pressure regulation. We then

Molecular Genetics of Hypertension, edited by A.F. Dominiczak, J.M.C. Connell and F. Soubrier.

briefly describe studies which attest to the complexity of hypertension and blood pressure regulation, and document studies of relevance to hypertension and blood pressure within the six identified research areas. Finally, we consider the possibility of integrating results of studies from within these areas and further describe potential study designs which might facilitate this integration. We end with a brief discussion and a few concluding comments.

2. Modern genetics: the six-fold way

Although there are many ways in which one could partition modern research paradigms in genetic analysis, we offer the following as a heuristic device which might promote greater understanding of the need for integration rather than as dogmatic categorization which might promote greater division.

2.1 Gene discovery

Gene discovery research seeks to identify the genes (i.e. their chromosomal locations, signature sequences, intron/exon boundaries, etc.) which influence traits and diseases when no *a priori* evidence for their location exists. Gene discovery research can be pursued in two ways. First, one can attempt to determine a gene with a single phenotypic endpoint in mind, such as meiotic or linkage mapping studies. Second, one can attempt to determine the existence, effects and possible locations of genes *en masse* without focus on a specific genetic effect, such as in random cDNA or expression monitoring assays (Hwang *et al.*, 1997).

2.2 Structural genomics

Structural genomics considers the construction and patterning of the nucleotide and amino acid sequences associated with genes in an effort to predict function and relatedness of genes. Structural genomic inquiry has become one of the flagship research areas in the burgeoning field of *bioinformatics* (Boguski, 1994) and plays an important role in settling modern evolutionary questions as well as contributing to medical research.

2.3 Clinical genetics

The identification of important phenotypic endpoints is crucial to deciphering the genetic basis of any trait or disease. The reason for this is that most complex traits and diseases can be subdivided and categorized into component diseases, each with slightly different characteristics and phenotypic manifestations that are likely caused by subtle differences in their genetic determinants as well as the interaction of these genes with environmental factors. The identification of unique phenotypes in clinical settings has produced some of the most prominent and insightful research findings in contemporary hypertension research (Lifton, 1996).

2.4 Integrative physiology

The goal of integrative physiology is to determine and characterize low-level (i.e. molecular and cellular) physiological phenomena and processes that probably reflect the immediate correlates and outcomes of gene activity. These processes include gene expression, protein interactions and the flow of metabolites across integrated pathways. Although not traditionally associated with *genetic* analysis, integrative physiology will become an increasingly important avenue for tying genetic effects together and for the development and assessment of therapies based on gene effects.

2.5 Functional genomics

The goal of functional genomics is to correlate physiological endpoints with genes. Obviously, the mere association between a genetic variant and a disease gives tremendous insight into the ultimate physiological endpoints associated with the gene in question, but such association does not reveal what has been referred to as the 'mechanism' or pathway within which the gene operates to induce its effects. Such knowledge can only be obtained through sophisticated assays and model development. Ultimately, functional genomics research can be seen as attempting to connect newly 'discovered' genes with the integrative physiological endpoints that connect gene effects with more remote clinically relevant phenotypes.

2.6 Population genetics and genetic epidemiology

Population genetics and genetic epidemiology seek to identify and characterize genetic, social and environmental processes that allow disease to emerge and be maintained *in the population at large*. It is worth emphasizing that although a great deal of epidemiologic research has focused on the discovery of disease genes via linkage mapping strategies in families (Lander and Schork, 1994), this is not the only, nor necessarily the most important, research that epidemiologists should undertake with respect to genetics research. Rather, genetic epidemiologists should focus on phenomena such as gene × environment interaction, effective population size, migration rates, and so on, in an effort to characterize factors that contribute to the sustenance of disease within and across populations (Schork *et al.*, 1988).

3. Blood pressure regulation as a genetically complex phenomenon

Blood pressure has, for the most part, always been seen as a *multifactoral* phenotype that, like respiration and immune system function, is extremely well integrated physiologically (i.e. there are many physiological processes and phenomena which influence and respond to changes in blood pressure). This physiological complexity is ultimately mediated by a great genetic complexity, whereby a number of genes, when upset due to mutation or other changes such as gross deletion, can detrimentally affect the physiological components of blood pressure regulation and thereby potentially lead to hypertension. The genetic and physiological complexity underlying blood pressure is further compounded by the fact that many genes and physiological parameters influencing blood pressure react in pronounced ways to various environmental stimuli (see Schork, 1997, for a discussion, and *Table 1* for a list of environmental factors thought to influence human blood pressure and hypertension susceptibility).

The fact that many factors influence blood pressure and hypertension susceptibility creates situations in which the effect of a single factor may be more or less pronounced in the presence of others. Thus, the effects of factors that influence blood pressure and hypertension susceptibility are 'context dependent' (see Sing *et al.*, 1996 for a discussion of context dependency and other concepts and their relation to cardiovascular disease). This context dependency manifests itself in the laboratory as well as in 'natural' surroundings and has important consequences in the design and interpretation of studies meant to disclose factors that influence blood pressure and hypertension susceptibility.

Three well-studied phenomena corroborate this theme. Firstly, elegant studies of the renin gene by John Rapp and colleagues among inbred rat strains have shown conclusively that certain rat renin gene variants that impact blood pressure in one rat strain have virtually no, or even an opposite, effect when introduced into other strains (Rapp *et al.*, 1990). Thus, gene variants at other rat loci may compensate for, nullify, or in other ways interact with, the renin gene. Sorting out biochemical and physiological pathways implicated in such compensatory activities will be difficult at best, simply because the specification and implementation of any particular study to sort out features of this context dependency (e.g. knock-out, transgene, *ex vivo* experiments involving immortalized cells, pharmacological probes, etc.) may themselves introduce a very context-specific outcome. Secondly, one of the most widely studied hypertension-inducing phenomena is salt-sensitivity (Svetkey *et al.*, 1996). The fact that some individuals manifest hypertension when challenged with salt whereas others do not suggests that some hypertension predisposing factors (i.e. genes) are very *environmentally* context dependent, in that salt-challenge provides a trigger for their detrimental effects. This salt-sensitive context dependence manifests itself both in molecular studies and population-based studies (Lopez *et al.*, 1995). Thirdly, theoretical and empirical studies of vascular remodelling among prehypertensive and hypertensive individuals suggests that the very examination of elements of the pathophysiology or etiology of hypertension may itself be context dependent. Thus, it has been argued that many hypertensive individuals have undergone a transition from a state in which their elevated blood pressure was caused by mechanisms inducing a high-cardiac output to a state in which their elevated blood pressure was caused by mechanisms inducing an increased total peripheral resistance (Eich *et al.*, 1996; Folkow, 1978; Iyengar *et al.*, 1998; Sivertsson, 1970). The link between these two states involves a physiological evolutionary process whereby the blood vessels of these individuals adapt to, or attempt to compensate for, a high blood flow by slowly losing their distensibility and, in fact, developing a structure which would, consequently, *promote* this lack of distensibility and therefore greater resistance to blood flow (see Iyengar *et al.*, 1988 for a review). These studies suggest that the mechanisms

Table 1. Example environmental factors thought to influence hypertension susceptibility and blood pressure level

Ref	Year	Factor	Population	Outcome
1	1997	Information-processing stress	32 healthy US volunteer subjects	Increased mental stress correlated with blood pressure response
2	1997	Cigarette smoking	429 men and 661 women in Framingham heart study	Smoking was associated with higher blood pressure
3	1996	Mental stress	132 volunteer men and women aged 30–40	SBP negatively related to the expression of anger from mental stress
4	1996	Caffeine	48 healthy men aged 20–35	Borderline hypertensive subjects exhibited a blood pressure response 2–3 times larger than controls from caffeine ingestion
5	1996	Trace elements	Review of the literature	Some evidence linking blood pressure to trace elements, but it is vague and limited
6	1994	Recreational exercise	857 subjects – Tecumseh, Michigan	Positive correlation between recreational exercise and reduction in blood pressure
7	1993	Anxiety levels	1123 men in the Framingham study	Among middle aged men anxiety levels are predictive of future hypertension
8	1987	Suppressed anger	171 white and 279 black female high-school sophomores	Suppressed anger levels correlated with higher blood pressure
9	1983	Vegetarian diet	59 Western Australians	Decrease in DBP and SBP for those on vegetarian diet
10	1983	Saturated fat	8479 Eastern Finlandians	Increase in mean arterial pressure with increased intake of fat
11	1983	Potassium intake	Untreated hypertensive patients	A decrease of 4% in mean blood pressure with an increase intake of potassium
12	1982	Life events	Construction employees in Stockholm	Higher prevalence – life change and high rank of life discord showed correlation to higher blood pressure
12	1982	Noise level	Adults living in Munich, Germany	Exposure to noise increased blood pressure response while solving mental arithmetic

continued overleaf

Table 1. *continued*

Ref	Year	Factor	Population	Outcome
13	1982	Contraceptive pill use	23 000 pill-users and study group	Demonstrated 2–6 times more hypertension in those using contraceptive pills
14	1980	Marital status	Framingham study participants	Widows and widowers had higher blood pressures than those who stay married
15	1996	Calcium intake	Hypertensive rats	High calcium intake lowered the blood pressure in hypertensive rats
16	1978	Employment type	British civil servants	Higher mean blood pressures among men in 'lower' grades of employment
17	1977	Socio-economic stress	2305 interviewed persons in Detroit derived from census sampling of families	Economic deprivation, family instability, density and crime rates correlated with higher blood pressures
18	1976	Body weight	1 million Americans	Strong association between obesity and higher blood pressure
19	1976	Education level	40 000 employed Chicagoans	Blood pressure level correlates with education level
20	1980	Alcohol consumption	5000 middle age men and women	Blood pressure twice as high in drinkers of 60 oz+ per month
21	1974	Social class	(Not determined)	Social status correlated with blood pressure
22	1973	Water hardness	(Not determined)	Higher blood pressure found in areas with soft water (in plumbing system)
23	1973	Sodium intake	Adults 50–59 years old	A positive correlation between sodium intake and elevated blood pressure
24	1958	Physical activity	Males in South Wales	Reduction in blood pressure when physically active

References: 1. Deter *et al.* (1997); 2. Wilson *et al.* (1997); 3. Steptoe *et al.* (1997); 4. Pincomb *et al.* (1996); 5 Houtman (1996); 6. Gudbrandsson *et al.* (1994); 7. Markovitz *et al.* (1993); 8. Johnson *et al.* (1987); 9. Rouse *et al.* (1983); 10. Salonen *et al.* (1983); 11. MacGregor *et al.* (1983); 12. Von Eiff *et al.* (1982); 13. Weir (1982); 14. Ayachi (1979); 15. Kannel (1996); 16. Marmot *et al.* (1978); 17. Harburg *et al.* (1973); 18. Stamler *et al.* (1976); 19. Dyer *et al.* (1976); 20. Kannel *et al.* (1980); 21. Syme *et al.* (1974); 22. Stitt *et al.* (1981); 23. Gleibermann (1973); 24. Miall and Oldham (1958).
DBP = diastolic blood pressure; SBP = systolic blood pressure.

initiating hypertensive disease may be different from those that actually sustain it!

The above examples and arguments do not imply that there do not exist factors that elevate or reduce blood pressure uniformly or in all contexts, but it does emphasize the fact that the bulk of the evidence linking certain genes and environmental factors to blood pressure and hypertension susceptibility indicates that most of these factors have a somewhat small and seemingly context-dependent effect. A good question to ask is then: how should hypertension researchers proceed? The answer to this question may reside in studying two seemingly antithetical themes in future research: the distinction (or lack thereof) between the characterization of factors that allow disease to emerge and be maintained *within individuals* as opposed to *within populations*, and placing an emphasis on the design of studies which promote the integration and synthesis of results from disparate experimental settings (to the degree that this is possible).

3.1 Gene discovery

The discovery of genes that influence particular traits can proceed in a very directed manner, as in the assessment of the linkage, association, unique expression pattern or metabolic pathway involvement of a *particular* gene, or as a part of a large screening effort of a number of genes or an entire genome. Most modern gene discovery programs are pursued in one of four ways:

(i) meiotic or 'linkage' mapping;
(ii) candidate polymorphism and association analysis;
(iii) model-organism gene mapping studies with subsequent homology search;
(iv) expression analyses.

Meiotic or linkage mapping strategies attempt to trace transmission, cosegregation and recombination phenomena between alleles at landmark positions along the genome, known as 'marker' loci, and hypothetical or putative trait-influencing alleles, across and within generations and members of families. Statistical methods are used to draw inferences about such possible cosegregation as well as the likely location of a trait-influencing locus (Schork and Chakravarti, 1996). These efforts can either focus on marker loci near or within a gene whose relationship to blood pressure and hypertension is unknown or in need of corroboration, or on marker loci dispersed throughout the genome, as in 'genome-scan' paradigms (Lander and Schork, 1994; Schork and Chakravarti, 1996). Genome scan designs for complex traits such as hypertension, and linkage analysis in general, are known to lack power and can be heavily resource intensive (Risch and Botstein, 1996). Although linkage analysis has been pursued to identify hypertension susceptibility genes (see *Table 2* for some recent examples) the results of these studies have been difficult to reconcile. One recent notable exception is the study by Lathrop *et al.*, which provides reasonably compelling evidence for an hypertension susceptibility locus on chromosome 17 (Julier *et al.*, 1997). Schork *et al.* (1998) have argued that meiotic mapping approaches to gene discovery have dominated the thoughts of many geneticists and epidemiologists – a fact which has probably diverted attention from more 'holistic' approaches to understanding the emergence and maintenance of hypertension in populations and individuals.

Table 2. Examples of recent candidate gene studies examining blood pressure and/or hypertension in humans

Ref	Year	Gene or Locus	#P	Analysis	Population	Outcome
1	1997	Angiotensinogen	1	ASL, MG	46 San Antonio families	+ HT linkage
2	1997	Adducin	3	ASL, R	282 HTs, 151 NTs (Italy) 137 French sibpairs	+ HT linkage; + HT association
3	1997	Neuronal nitric oxide synthase	1	CT	131 HTs, 147 NTs (Japanese)	– HT association
4	1997	Renin	2	CT	86 Hts, 107 Nts (Chinese)	+ HT association
5	1996	Angiotensinogen	4	R	76 black/139 white Indiana Children	± BP association
6	1996	11β-hydroxysteroid dehydrogenase		CT	79 Black HTs with ESRD, 67 Black NTs (Alabama)	+ HT association
7	1996	Adrenergic receptor	2	ASL	109 North Carolina black sib pairs	+ DBP response to salt association
8	1995	Angiotensinogen	1	ASL, CT	63 African sibpairs	+ HT linkage; + HT association
9	1995	Nitric oxide synthase	1	CT, ASL	269 French sibpairs, 198 HTs, 106 NTs (French)	– HT linkage; – HT association
10	1994	Angiotensin II type I receptor	3	CT, ASL	267 French sibpairs, 206 HTs, 298 NTs (French)	– HT linkage; + HT association
11	1994	Angiotensinogen	2	CT, ASL	63 White European families	+ HT linkage; +HT association;
12	1994	SA	2	CT, PL, ASL	306 HT French sibpairs, 206 HTs, 298 NTs (French)	– HT association; – HT linkage
13	1993	SA	1	CT	89 HTs, 81 NTs (Japanese)	+ HT association
14	1992	Renin, haptoglobin, cardiac myosin, neuropeptide Y	5	CT	120 HTs, 115 NTs (Austrialia)	– HT association
15	1992	ACE, human growth hormone	9	ASL	237 Utah sibpairs	– HT linkage
16	1991	Antithrombin III	1	PL	9 Australian pedigrees	± HT linkage
17	1991	Insulin receptor, insulin gene	2	CT	67 HTs, 75 NTs (Austrialia)	+ HT association with insulin receptor
18	1990	Na^+–H^+ antiporter	1	PL, R, ASL	14 Utah pedigrees	– HT linkage

#P = Number of polymorphisms studied. Analysis: R = regression; CT = contingency table; ASL = allele sharing linkage; PL = parametric linkage; MG = measured genotype. Population: HT = hypertension, HTs = hypertensives; NT = normotension; NTs = normotensives; BP = blood Pressure; SBP = systolic blood pressure; DBP = diastolic blood pressure; ESRD = end stage renal disease. Outcome: – = Negative finding; ± = Marginal finding; + = Positive finding.
References: 1. Atwood *et al.* (1997); 2. Cusi *et al.* (1997); 3. Takahashi *et al.* (1997); 4. Chiang *et al.* (1997); 5. Bloem *et al.* (1996); 6. Watson *et al.* (1996); 7. Svetkey *et al.* (1996); 8. Caulfield *et al.* (1995); 9. Bonnardeaux *et al.* (1995); 10. Bonnardeaux *et al.* (1994); 11. Caulfield *et al.* (1994); 12. Nabika *et al.* (1994); 13. Iwai *et al.* (1993); 14. West *et al.* (1992); 15. Jeunemaitre *et al.* (1992); 16. Griffiths *et al.* (1991); 17. Ying *et al.* (1991); 18. Lifton *et al.* (1990).

In other words, gene discovery, and more importantly, gene discovery via meiotic mapping approaches, are simply *not* capable of promoting our understanding of hypertension and blood pressure regulation by themselves and hence should not be overemphasized.

Unlike linkage analysis, which focuses on genomic regions that are likely to harbor trait-influencing genes, candidate polymorphism analysis tests specific variants within or around genes of interest for association with a trait or disease in a sample of individuals. As the number of identified polymorphisms increases, testing of this sort will become more and more pronounced, possibly to the point of motivating 'whole genome association studies' in which polymorphisms over the entire genome are tested with a trait or disease in an effort to identify trait-influencing loci (Risch and Merikangas, 1996). *Table 2* offers examples of recent candidate polymorphism studies for hypertension and blood pressure regulation. Three important points should be emphasized in the context of candidate polymorphism analysis. Firstly, the 'candidacy' of a test polymorphism for a particular trait or disease is only as good as the biological motivation for testing it. Thus, motivation for testing a specific polymorphism or set of polymorphisms can derive from, for example, knowledge that the gene in which that polymorphism is located is uniquely expressed in a tissue known to be upset in the pathogenesis of a disease, previous tests have suggested association, it is in a gene family studied in model organisms, and so on. The important point here is that the candidacy of a polymorphism relates to *ancillary* or additional information about the polymorphism and not just that it exists. Secondly, associations between polymorphisms and diseases can arise for at least three reasons: (i) the polymorphism is causally associated with the disease; (ii) the polymorphism is in linkage disequilibrium (or essentially associated) with another polymorphism which is causally associated with the disease; and (iii) the polymorphism is spuriously associated with the disease (due to, e.g., a type I error or population stratification). Determining which of these three reasons is behind a specific study result can be difficult. Thirdly, a particular polymorphism's effect may interact with other factors, thus obscuring its influence on a trait or disease if ignored. This fact complicates comparison of results across studies involving different populations, designs and additional data (e.g. environmental, dietary, lifestyle, etc.) collections.

Genome-scan approaches to trait-influencing locus identification are greatly eased in model organism contexts, since breeding strategies and environmental controls can be exploited (Frankel, 1995). Once a gene has been identified that is known to impact a particular phenotype in a model organism, the human homolog of that gene can be studied to see if variation in the human homolog influences the human analog to the phenotype originally studied in the model organism. Schork *et al.* (1996) have reviewed model-organism based gene mapping studies of blood pressure regulation and hypertension and offer a table describing the results of these studies. What is worth emphasizing from these studies, as a collection of studies, is that they have revealed a great variety of genes, gene defects and unique strain effects, which can only corroborate the great potential for genetic heterogeneity in *human* hypertension.

Analysis of gene expression has become the focal point of a great deal of contemporary medical and biological research. Technologies are now at hand that will allow researchers to monitor the expression of hundreds or thousands of genes simultaneously, possibly in an effort to identify genes that influence or participate in disease pathogenesis (DeRisi *et al.*, 1997; Velculescu *et al.*, 1995). In order to identify specific disease genes, however, contrasts in gene expression need to made. Thus, one could contrast gene expression profiles in the blood vessels of hypertensives and normotensives. Those genes showing differential expression may be genes that influence hypertensive susceptibility. *Table 3* lists recent gene expression studies of relevance to hypertension and blood pressure research and suggests that gene expression studies can shed enormous light on physiological phenomena mediating blood pressure and hypertension. Although gene expression studies will probably replace meiotic mapping strategies as the method of choice for disease gene discovery, there are a number of problems plaguing the large scale use of expression assays for screening a number of genes. For example, the mere identification of a gene that is differentially expressed between diseased and nondiseased tissues does not mean that the gene in question is *causally* related to the disease; rather, the gene may be merely secondarily affected as a result of disease pathogenesis.

It should be emphasized that unless an attempt at identifying a disease gene is planned purely for its diagnostic potential, disease gene discovery programs are typically pursued in an effort to gain a foothold or 'entry point' into biochemical and physiological processes involving those genes, which can in turn shed light on the pathogenic processes that mediate the disease (see *Figure 1*). Thus, meiotic mapping efforts seek to link genes to gross clinical endpoints such as hypertension, whereas gene expression-based disease gene discovery paradigms seek to identify genes that are upset in, and thereby associated with, pathological tissues (*Figure 1*). Each of these strategies must initially ignore processes that go on between, or connect, the deleterious genes with relevant clinical endpoints or diseased tissues. This is important when considering the potential for integrating blood pressure and hypertension-susceptibility gene discovery with other avenues of research, since it suggests that a great many factors, systems, and stimuli might need to be studied if understanding of the nature of hypertension and blood pressure regulation is to be anywhere near 'complete' in the sense that connections between genes and clinical endpoints have been identified.

3.2 Structural genomics

Structural genomics studies for hypertension are very much in their infancy in terms of their sophistication and yield. Two good examples of structural genomic investigations are studies by Jeunemaitre and colleagues on the angiotensinogen gene (Jeunemaitre *et al.*, 1992a,b) and mutations responsible for glucocorticoid-remediable aldosteronism (GRA) (Lifton *et al.*, 1992). The study of the angiotensinogen gene not only revealed aspects of its intron/exon structure, but the authors were able to identify a number of polymorphisms that could either be of functional significance or be used to test the gene in association and/or linkage analysis settings (Jeunamaitre *et al.*, 1992a,b). The study of GRA led to the finding that the responsible mutation involved a chimeric gene duplication arising from an unequal crossing over between genes encoding aldosterone synthase and

Table 3. Examples of recent gene expression studies of relevance to hypertension pathophysiology

Ref	Year	Gene	Organism and design	Outcome
1	1997	*c-myc, B-actin*	Expression responses to shear stress in mesenteric arteries of Wisar rats	Wall stress significantly correlated with *c-myc* m gene expression
2	1997	Angiotensinogen	Gene expression levels at multiple time points in Wistar fatty (WFR) and lean (WLR) rats	Gene regulated differently in WFR and WLR
3	1997	Atrial natriuretic peptide (ANP)	Expression differences in Sprague–Dawley rats fed/not fed a high salt diet	High salt diet did not affect expression
4	1996	Proto-oncogene	Expression responses to shear stress in mesenteric arteries of Wistar rats	Increased pressure stimulates proto-oncogene expression
5	1996	Kallikrein, kininogen, kallikrein-binding protein	Expression differences in Sprague–Dawley rats treated/not treated with L-NAME	Treatment with L-NAME influenced expression of all genes
6	1996	Endothelin-1	Expression differences in SHR treated/not treated with L-NAME	Increase of endothelin-1 expression in L-NAME treated SHR.
7	1996	Angiotensinogen	Expression comparison between WKY and SHR rats	Expression differences in angiotensinogen between WKY and SHR
8	1996	ANP, collagen types I and II, myosin	Expression changes after ACE-inhibitor and beta blocker treatment in SHR	ACE inhibitor affected gene expression profiles Beta blocker affected collogen expression
9	1996	Renin	Affect of Renin gene transfer from Dahl S rats to Dahl R rats	Renin gene expression modified by gene transfer
10	1996	Angiotensin II, plasminogen	Comparison of WKY and SHR gene expression in the brain after Ang II stimulation	Ang II stimulation causes 10-fold increase in plasminogen gene expression
11	1996	NPC receptor	Comparison of SHRSP and WKY expression after Ang II antogonist administration	Ang II administration caused down regulation of NP-C receptor in SHRSP
12	1996	*c-fos, c-jun*	Male Wistar rats either administered phenylephrine/Ang II or restrained	*c-fos* and *c-jun* expression changes after acute elevation of BP
13	1995	Heat shock protein 70 (hsp70)	Male Fischer 344 rats	hsp70 induction occurs as a physiological response to acute hypertension

References: 1. Allen *et al.* (1997); 2. Tamura *et al.* (1997); 3. Wolf and Kurtz, (1997); 4. Allen *et al.* (1996); 5. Chao *et al.* (1996); 6. Sventek *et al.* (1996); 7. Tamura *et al.* (1996); 8. Ohta *et al.* (1996); 9. St. Lezin *et al.* (1996); 10. Yu *et al.* (1996); 11. Yohimoto *et al.* (1996); 12. Xu *et al.* (1996); 13. Xu *et al.* (1995).

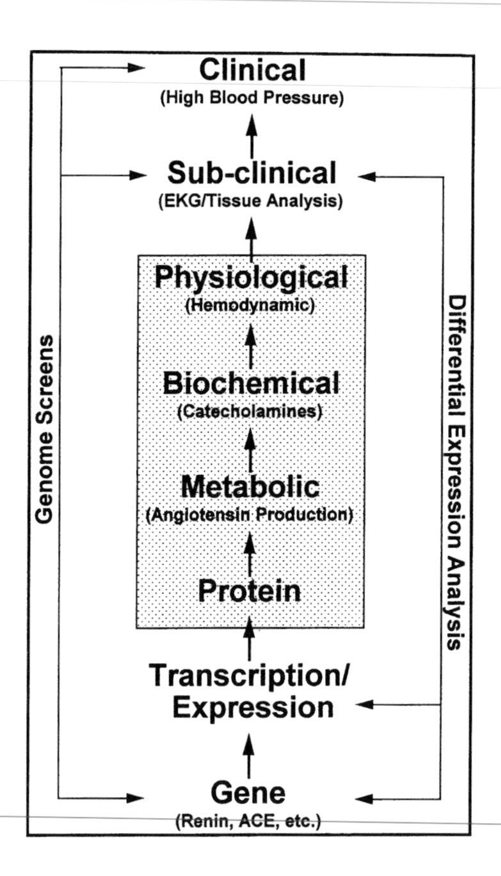

Figure 1. Diagrammatic representation of a 'phenotypic hierarchy'. Although an obvious oversimplification, the diagram is meant to characterize the various levels implicated in disease pathogenesis that may serve as targets for pharmacological intervention and manipulation. Example phenotypes at each level that may be associated with hypertension are also presented. The shaded box represents phenomena initially ignored in most gene discovery paradigms, but knowledge of which is absolutely crucial for proper development and assessment of pharmacotherapies and other disease interventions. The lines denoted 'genome screens' and 'differential expression analysis' depict the levels which are assessed or exploited in these two modern disease gene discovery strategies.

steroid 11β-hydroxylase (Lifton, 1996). Structural genomic studies of this kind can not only shed light on the nature of the putative mutations causing disease, but also their relative conservation over an evolutionary time period. Conservation of function is crucial for the proper assessment of homology and synteny studies between humans and other species, especially as these studies relate to something as complex hypertension and blood pressure regulation. In addition, conservation of function and structure arguments are also crucial when evolutionary arguments are invoked as to the nature of the emergence and maintenance of HT (Julius, 1995; Schork *et al.*, 1998; Weder and Schork, 1994).

3.3 Clinical genetics

A number of unique clinically recognizable syndromes and phenotypes of which HT is a component have been identified. Disclosure of the genetic basis of these phenotypes can yield tremendous insight into genes and physiological structures which influence garden-variety hypertension and elevated blood pressure in the population at large (Lifton, 1996; Thibbonier and Schork, 1995). *Table 4* describes studies that have successfully identified genes or chromosomal regions thought to influence unique and often very rare clinical forms of hypertension. The use of such phenotypes for understanding the genetic basis of hypertension is often

Table 4. Recent examples of unique clinical phenotypes used to identify hypertension and/or blood pressure regulation genes

Ref	Year	Phenotype	Gene or locus	Outcome
1	1997	Pheochromocytoma	von-Hippel Lindau gene (*vhl*)	A point mutation in *vhl* was identified in all patients with pheochromocytoma in a large kindred
2	1997	Familial hyperkalaemia (PHAII)	1q31–42 and 17p11-q21	Linkage heterogeneity identified in 8 PHAII kindreds
3	1997	Familial primary pulmonary hypertension (PPH)	2q31–32	Linkage analysis revealed a 25cM region on chromosome 2 likely to harbor a PPH gene
4	1997	Pseudohypoaldosteronism (PHA-1)	Epithelial sodium channel (ENaC)	Missense mutation identified as likely to be causally related to PHA-1
5	1996	Hypokalemic alkalosis	Na–Cl cotransporter and Na–K–2Cl cotransporter	Bartter's and Gitelman's syndromes influenced by unique mutations in two different genes causing hypotension
6	1996	Brachydactyly and hypertension (HT) syndrome	12p	Linkage analysis of a large Turkish pedigree identified a region on chromosome 12 likely harboring an HT locus
7	1995	Pre-eclampsia	Angiotensinogen (AGT)	Functional studies revealed that a specific mutation in the AGT gene cause pre-eclampsia in a single patient
8	1995	Liddle's syndrome	Epithelial sodium channel (ENaC)	A specific mutation in the γ subunit of ENaC is found to be responsible for some cases of Liddle's syndrome
9	1994	Liddle's syndrome	Epithelial sodium channel (ENaC)	A specific mutation in the β subunit of ENaC is found to be responsible for some cases of Liddle's syndrome
10	1992	Glucocorticoid-remediable aldosteronism (GRA)	11 β-hydroxylase and aldosterone synthase	GRA is found to be caused by a chimerism in the 11 β-hydroxylase and aldosterone synthase genes on 8q

References: 1. Garcia *et al.* (1997); 2. Mansfield *et al.* (1997); 3. Nichols *et al.* (1997); 4. Grunder *et al.* (1997); 5. Simon and Lifton (1996); 6. Schuster *et al.* (1996); 7. Inoue *et al.* (1995); 8. Hansson *et al.* (1995); 9. Shimkets *et al.* (1994); 10. Lifton *et al.* (1992).

obscured through the somewhat artificial distinction between 'primary' and 'secondary' forms of hypertension.

3.4 Integrative physiology

As noted in the section on gene discovery, the identification of a particular disease gene can lead to insights into metabolic or physiological pathways that, when upset or dysfunctional due to mutations in genes participating in them, contribute to the pathophysiological cascade that leads to disease. Determining these pathways is thus an important element in understanding the genetic basis of diseases such as hypertension. Although the search for such pathways and processes may not be considered a component of *genetic* research by many, the definition of a 'gene' and 'genetics' often includes information on the 'function' of genes in order to distinguish one gene from another. In addition, the expression of a gene or its encoded protein are intimately related to a gene and hence capture information about that gene that is not available from sequence information alone. A distinction may help clarify things. Consider the study of gross physiological phenomena, such as hemodynamic (i.e. pressure-resistance-flow) relationships affecting blood pressure. These relationships can be studied via tilt tables, plethysmography, and so on, and are the product of a coordinated series of high-level physiological events and processes (Folkow, 1978, 1993, 1995). Now consider studying the expression pattern of genes, the behavior of proteins and the flow of metabolites across biochemical networks. The latter phenomena are less removed from the genetic milieu and typify what can be coined 'molecular physiology'. These phenomena and processes, although they impact the higher level physiological endpoints such as hemodynamic relationships, are the typical targets for drug intervention and provide vital links for relating genes to diseases and disease processes. *Table 5* lists some examples of cellular or molecular processes thought to influence blood pressure regulation and hypertension susceptibility. Of those listed, the renin–angiotensin system has been studied most extensively, as knowledge of a number of genes (e.g. renin, angiotensin converting enzyme, angiotensinogen, etc.) implicated in that system has been obtained.

3.5 Functional genomics

The study of the function of genes, or 'functional genomics', has become one of the most pronounced research areas in genetics research (Fields, 1997; Miklos and Rubin, 1996). There are at least three ways in which one could determine the function of a gene. Firstly, *gene targeting* approaches, such as knock-outs, knock-ins, gene titration, directed mutagenesis and transgene experiments, focus on a particular gene and typically involve manipulating that gene in experimental organism settings. *Table 6* lists knock-out and transgene experiments of relevance to hypertension and blood pressure research. For further discussion of gene targeting strategies see also Chapter 6. Random mutagenesis approaches, in which a genome is exposed to mutagenic stimuli and any resulting mutation is examined for its phenotypic effect, are related strategies. Secondly, *simple association studies* investigating the effect that the presence or absence of a particular

Table 5. Example cellular, molecular, and low-level physiological factors thought to influence blood pressure regulation

Ref	Year	Defective mechanism	Defective system	Comments
1	1997	Contractile proteins	Smooth muscle cell activity	Greater contractility may underlie HT susceptibility
2	1996	Endothelial cells	Vessel lining	High blood pressure may upset endothelium and hence cause exacerbate development of HT
3	1996	Angiotensin II levels	Renin–angiotensin system	Elevated pulse pressure is associated with elevations in angiotensin II
4	1996	Angiotensin-converting enzyme	Renin–angiotensin system	Correlation between BP and ACE levels
5	1996	Catecholamine levels	Sympathetic reflexes	Elevated BP is correlated with elevated norepinephrine levels
6	1996	Cellular calcium	Exchange across cell membrane	Increased Ca^{2+} cellular load or enhanced external Ca^{2+} associated with HT
7	1992	Renin secretion	Renin–angiotensin system	High renin levels found to be associated with infarction among hypertensives
8	1986	Prostaglandins	Arteriolar vasoconstriction	Diminished endogenous synthesis of prostaglandins may contribute to elevated vasoconstriction leading to HT
9	1983	Mineralocorticoid production	Renal Na/K levels	Mineralocorticoid pathway is suppressed in prehypertensive individuals
10	1980	Na^+/K^+ cotransport	Cellular transport	Decrease in Na^+/K^+ exchange associated with HT
11	1980	Na^+/Li^+ countertransport	Cellular transport	Increase in Na^+/Li^+ exchange associated with HT

HT=hypertension; BP=blood pressure.
References: 1. Savineau and Marthan (1997); 2. Haller (1996); 3. Harrap *et al.* (1996); 4. Schunkert *et al.* (1996); 5. Ward *et al.* (1996); 6. Aviv (1996); 7. Laragh (1992); 8. Stoff (1986); 9. Guthrie *et al.* (1983); 10. Garay *et al.* (1980); 11. Canessa *et al.* (1980).

Table 6. Recent examples of knock-out and transgene experiments performed in an effort to determine gene function in the context of blood pressure regulation and hypertension susceptibility

Ref	Year	Gene	Study	Organism and strain	Outcome
1	1997	Angiotensin 1b receptor (Agtr1b)	Knock-out	C57BL/6 mice	No abnormal phenotypes observed among homozygous and heterozygous deleted mice
2	1997	Bradykinin-B_2 receptor	Knock-out	129Sv mice	Bradykinin-B_2 receptor knock-out significantly influenced salt-sensitive hypertension
3	1997	Human angiotensinogen and renin	Transgenic	Male Sprague–Dawley rats	Rats transgenic for human angiotensinogen and renin genes show significantly altered physiology relative to control rats
4	1996a	α_2 Adrenergic receptors *2b* and *2c*	Knock-out	129Sv/J × FVB/N and 129Sv/J and *C57BL/6J mice*	Stimulation of *2b* receptors resulted in hypertension
					No hemodynamic effect produced by disruption of *2c* subtype
5	1996	Nitric oxide (eNOS)	Knock-out	129 × C57 BL/6J mice	eNOS essential for maintenance of normal blood pressure and heart rate
6	1996	Duplicated renin (Ren-2)	Knock-out	129 mice	No clear effects of Ren-2 knock-out
7	1995	Angiotensin type-2 receptor (AT_2)	Knock-out	129/SV × FVB/N mice	AT2 deficient mice exhibit behavioral changes and increased response to Ang II administration
8	1995	Angiotensin type-2 receptor (AT_2)	Knock-out	129/OLa × C57BL/6 mice	AT2 deficient mice exhibit behavioral changes and increased response to Ang II administration
9	1995	Mouse Ren-2	Transgenic	Sprague–Dawley rats	Transgenic rats exhibited hemodymanic effects consistent with their elevated blood pressure
10	1995	Guanylyl cyclase A receptor (GC-A)	Knock-out	129/SVJ × C57BL/6J mice	Knock-out mice exhibited a sodium-resistant increase in blood pressure
11	1995	Rabbit Na +/– proton exchanger	Transgenic	C3H × C57B46 mice	Transgenic mice have increase blood pressure response to salt challenges

12	1995	Angiotensinogen (AOGEN)	Gene Titration	129 × C57BL/6J mice	Blood pressure increases observed as a function of gene copy number
13	1995	Atrial natriuretic Factor	Knock-out	129 × C57BL/6J mice	Knock-out mice show increased blood pressure to salt intake
14	1995	ACE	Knock-out	129 × C57BL/6J mice	Knock-out mice show reduced blood pressure
15	1994	Angiotensinogen (AOGEN)	Knock-out	C57BL6J/CBA × ICR	Knock-out mice exhibit reduced blood pressure
16	1994	Endothelin-1	Knock-out	129/SVJ × C57BL/6J or ICR	Knock-out mice exhibited elevated blood pressure and craniofacial defects
17	1994	Mouse brain natriuretic peptide	Transgenic	C57BL/6J	Transgenic mice exhibit reduced blood pressure

References: 1. Chen *et al.* (1997); 2. Alfie *et al.* (1997); 3. Bohlender *et al.* (1997); 4. Link *et al.* (1996); 5. Shesely *et al.* (1996); 6. Sharp *et al.* (1996); 7. Hein *et al.* (1995); 8. Ichiki *et al.* (1995); 9. Gardiner *et al.* (1995); 10. Lopez *et al.* (1995); 11. Smithies and Maeda (1995); 12. Kim *et al.* (1995); 13. John *et al.* (1995); 14. Krege *et al.* (1995); 15. Tanimoto *et al.* (1994); 16. Kurlhara *et al.* (1994); 17. Villarreal *et al.* (1995).

polymorphism has on a (possibly molecular) phenotype is another way of assigning function (e.g. *Table 2*). Thirdly, *bioinformatics approaches*, in which one tries to determine the function of a gene by searching for elements in its sequence that are known to induce certain outcomes (e.g. binding properties of a protein) or by relating the sequence to known sequences in other species whose gene functions have been identified, are being used more and more often , but have not found unique and specific application in hypertension and blood pressure research as yet. Unfortunately, as emphasized throughout this chapter, the experimental conditions, the collection of additional information, the population or model organism strain used, and so on, can all impact on the interpretability and generalizability of an outcome obtained from a functional genomics study and as such can contribute to the context dependency that is the hallmark feature of factors influencing hypertension and blood pressure as well as other complex phenotypes.

3.6 Population genetics and genetic epidemiology

The identification of genes and the characterization of their functions, particularly with respect to certain pathologies, is an important and prominent activity in modern medical genetics. Knowledge obtained from relevant experiments in medical genetics can lead to the development of diagnostic, preventive, and therapeutic strategies for the disease. The focus of this research is on the development and treatment of disease *within individuals* from *clinical* vantage points. Equally important, however, is understanding of the processes that contribute to the emergence and maintenance of disease *within populations* (Schork *et al.*, 1998). The concepts, experimental strategies and outcomes used to examine disease for the sake of individuals and populations are not necessarily the same, although there is tremendous and obvious overlap in terms of the information they produce. Consider *Table 1*, which lists gross environmental factors that influence blood pressure and hypertension susceptibility. Each of these factors exists to varying degrees in different populations. Hence, the 'health', in terms of hypertension frequency, of these populations will vary considerably. In fact, studies have shown that when individuals migrate from communities possessing a lesser amount of these (and other) environmental stimuli to communities with greater amounts, their blood pressures change (*Table 7*). There are other factors that contribute to the emergence and maintenance of hypertension in disparate populations. For example, founder effects, migration and immigration rates, mutation rates, and drift, can all contribute to differential allele frequencies across different populations (Cavalli-Sforza *et al.*, 1994). Thus, it is entirely possible that some populations have a lesser degree of hypertension in them, not necessarily because of the lesser presence of toxic environmental stimuli or better health care, but because deleterious mutations are simply not as frequent. This and other phenomena will greatly affect public health and epidemiological campaigns aimed at hypertension if not accounted for properly. Greater attention to population-level processes and events are thus called for (Schork *et al.*, 1998).

Table 7. Examples of migration studies investigating the gross effects of environmental change on blood pressure and hypertension prevalence

Ref	Year	Population	Design	Outcome
1	1997	316 Kuna Indians	Kuna migration to Panama City	Increase in BP, likely changes in salt intake
2	1995	313 Yi farmers (male), 265 Yi migrants, 253 Urban Han	Rural villagers migrate to Urban areas; Contrast of migrants and nonmigrants	Yi farmers on average had lower BP than Yi migrants Urinary sodium and calcium positively related to BP Urinary potassium and magnesium inversely related to BP Increased adiposity could not explain increased BP
3	1995	25 Normotensive blacks	Migration from Somalia to Italy	After 6 months SBP and DBP increased (likely diet-induced) Urinary sodium excretion increased Plasma renin activity significantly reduced
4	1991	8241 Yi farmers, 2575 urban Yi migrants, 3689 Han urban residents	Rural villagers migrate to Urban areas	Yi migrant men and Han men had higher BP Yi farm women's age-related rise in BP and HT prevalence lower than in other two groups.
5	1989	654 Tokelua Islanders	Migration to urban areas	Increase in body mass SBP and DBP of migrant men significantly higher than nonmigrants, but not for women
6	1988	90 Kenyan males	Migration to Nairobi	Migrants had higher BP than rural-based controls
7	1987	617 European and 155 Bengalis	Contrast of Bengali immigrants with native Europeans	Both groups showed rise in BP with increasing age and BM. Increased IHD risk in Benagli group not explained by an BP elevations.
8	1985	532 Tokelau adults (nonmigrants), 280 Tokelau adults (migrants)	Migration from Tokelau Island to New Zealand	Significant difference between migrant and nonmigrants in rates of change for both SBP and DBP

continued overleaf

Table 7. *continued*

Ref	Year	Population	Design	Outcome
				Age, body mass,and pulse pressure-corrected rates of change were greater in migrants than in nonmigrants
9	1983	784 Tokelau (nonmigrants), 1119 New Zealanders (migrants)	Migration from Tokelau Island to New Zealand	Elevation of BP associated with migration (more apparent in males than females)
				Both migrant and nonmigrant BP in females rises more steeply than that of males DBP of migrants responds more sharply to migration than does SBP
10	1979	2284 Samoan male and females American Samoa (nonmigrants) and Hawaii (migrants)	Migration from Samoa to Hawaii	Nonmigrants had higher BP Migrants from more traditional areas of Samoa had higher BP than nonmigrants
				Body fat positively related to BP
11	1971	3102 subjects from Evans County Georgia	Migration from Evans County to Urban areas	Migrants had lower levels of BP and serum cholesterol.

BP = blood pressure; HT = hypertension; BM = body mass; IHD = ischemic heart disease.
References: 1. Hollenberg *et al*. (1997); 2. Klag *et al*. (1995); 3. Modesti *et al*. (1995); 4. He *et al*. (1991); 5. Salmond *et al*. (1989); 6. Poulter *et al*. (1988); 7. Silman *et al*. (1987); 8. Salmond *et al*. (1985); 9. Joseph *et al*. (1983); 10. McGarvey and Baker (1979); 11. Wetherbee and Tyroler (1971).

4. Synthesis and integration

Given the variety of approaches used to unlock aspects of the genetic basis of hypertension and other complex, multifactorial diseases, a good question concerns just how one might be able to reconcile or integrate the results of each approach in an effort to construct a comprehensive understanding of relevant pathophysiological and population-level processes that cause and sustain these diseases. We outline four very simple approaches that demand cross-talk among geneticists interested in hypertension and blood pressure and then consider two more sophisticated strategies in greater detail.

4.1 Theoretical frameworks

The development of theoretical frameworks that describe the origin of hypertension and the underpinnings of blood pressure regulation can motivate a series of studies whose ultimate outcomes should, if the framework is correct, converge on a common theme. Evolutionary angles provide one such framework and can easily lead to insights into the *genetic* basis of disease (Schork *et al.*, 1998). Two of the most notable evolutionary arguments on the origin and maintenance of hypertension are the 'flight or fight' response argument by Julius and Jamerson (Julius, 1995) and the developmental hypothesis of Weder and Schork (Iyengar *et al.*, 1998; Weder and Schork, 1994).

4.2 Candidate polymorphism characterization

Obviously, genes known to be expressed in certain tissues of relevance to hypertension, for example, should be studied in other contexts, such as in knockout experiments and association studies. Thus, the identification of genes worth testing in population-based samples can easily proceed by polling researchers (or the literature) that have identified genes possibly involved in hypertension and blood pressure regulation through other means.

4.3 Environmental factor control in gene analysis

Epidemiological studies which have identified environmental stimuli that induce or exacerbate hypertension susceptibility provide information that should be accounted for in, for example, meiotic mapping or candidate polymorphism association studies. Although not unprecedented, sophisticated ways of accommodating gene $\times$ environment interactions must be developed for these purposes.

4.4 Monogenic hypertension dissection

Lifton and colleagues (see Lifton, 1996, for a review) have clearly shown that one can determine genes, phenotypes, pathways and potential therapies for hypertension by studying rare monogenic forms of hypertension. Although the actual strategies used in such research will not necessarily carry over to the study of essential or 'garden-variety' hypertension, the genes, pathways, and so on, implicated in these rare hypertensions can be scrutinized for variants that influence common hypertension.

Each of the above strategies is simple to execute, but may not be comprehensive enough, in terms of guiding actual experiments, to confront and characterize the context dependency behind the factors that influence hypertension and blood pressure. We outline two strategies below, one molecular and physiological in orientation, and the other epidemiological or population-based in orientation, that may provide comprehensiveness of the desired type.

4.5 Model organism phenotype dissection

To facilitate the integration of research involving the genetics, physiology and clinical pathology of hypertension, one could design a study involving model organism inbred strain cross hybrids whose biomaterials are made available to researchers, and for which a database of results is created. Consider, as an example, the conduct of a large intercross experiment between a normotensive rat (or mouse) strain and a hypertensive rat (or mouse) strain in which all the individual rats are genotyped. Consider further that tissues, blood samples, cell lines, and so on, are preserved from these mice and rats as well. These materials could then be made available to interested researchers who have developed, or can implement, different biochemical or clinical chemistry tests. The results of these tests could then be compiled in a database and used to link particular genotypic information with phenotypic information. This database could then be queried by interested scientists to either confirm their own physiological or genetic hypotheses or motivate further experiments with the knowledge that the same animals were assessed in each experiment. This design would allow linkages between a number of phenotypic, physiological and genetic processes to be made in a manner that would be more uniform in orientation and scope than the current method for linking information, which is basically based on literature review and *ad hoc* comparison and synthesis of results obtained from widely different study designs.

One could imagine a number of associated designs, such as the dissemination of transgenic or mutagenized animals from a larger stock, or the tracking of animals as in a longitudinal project organized to examine features of the *development* of a chronic disease. Each design would have its own obvious weaknesses, not the least of which would be the assurance of the reliability of the data put into the database, as well as manner in which the data would be stored for intuitive and easy access. Although the dissemination of animals for research purposes is hardly unprecedented, the coordination of research results in, for example, a comprehensive database, and the global motivation for the studies would be.

4.6 Uniform cross-population comparisons

The development of a 'global' genetic epidemiological network which would facilitate the contrast and comparison of a number of different human populations in terms of genetic backgrounds, social factors and environmental factors influencing hypertension susceptibility and blood pressure regulation, could lead to tremendous insights into the emergence and maintenance of hypertension within and across populations. The hallmark feature of the implementation of such studies would, of necessity, have to be centered around *uniformity* in data collection. This study design could easily embrace candidate polymorphism testing in which the contribution to hypertension by environmental factors (that may interact with the polymorphism) is accounted for and characterized. Such studies could also shed enormous light on the utility of various public health campaigns (e.g. the encouragement of non salty diets, the prescription of a certain type of drug or therapy, etc.) and the economic repercussions of interventions.

Although each of the above two strategies may seem to be an expensive undertaking, it is a simple fact that many of the component studies that could be implicated in them are being pursued, but in an isolated and independent manner. Consider the fact that many efforts to examine candidate polymorphisms and environmental factors in different populations have been completed or are being pursued (e.g. the studies listed in *Table*s *1, 2* and *3*) and many molecular geneticists are studying phenotypes and specific genes of all types (e.g. the studies listed in *Table*s *3, 5* and *6*). What is lacking in these studies is coordination, oversight and comprehensive design – features which may simply be impossible to implement in the current, very competitive, global intellectual and research funding climate.

5. Conclusions

The genetic dissection of complex diseases and traits such as hypertension and blood pressure regulation poses one of the greatest challenges facing modern medical researchers and biologists. The wide variety of factors influencing such traits makes complete isolation and appropriate characterization of each factor difficult, if not impossible. What is needed at this point in our understanding of hypertension and blood pressure is not necessarily new technologies or experimental strategies for determining potential factors, but rather ways of synthesizing and making use of the collection of information that is already at hand (Schork, 1997; Thibbonier and Schork, 1995). Although there are ways of synthesizing such information, it remains to be seen if one can actually simultaneously embrace or acknowledge the fact that 'context-dependency' is the rule rather than the exception for factors influencing hypertension and blood pressure *and* determine, in as unequivocal way as possible, the myriad ways in which hypertension can emerge and be maintained in individuals as well as populations.

The various chapters of this book will make clear the variety of approaches taken to dissect the genetic basis of hypertension as well as the enormous problems confronting the integration of their results within a more comprehensive framework. In this regard they stand as a testament to the belief that progress in understanding the genetic basis of hypertension can be made despite its complexity.

Acknowledgments

The construction of ideas and arguments discussed in this chapter were supported, in part, by United States National Institutes of Health grants:

HL94-011 (NHLBI), HL54998-01 (NHLBI) and RR03655-11 (NCRR), of which Nicholas J. Schork is a participating researcher.

References

Alfie, M.E., Sigmon, D.H., Pomposiello, S.I. and Carretero O.A. (1997) Effect of high salt intake in mutant mice lacking bradykinin-B2 receptors. *Hypertension* **29**: 483–487.

Allen, S.P., Liang, H.M., Hill, M.A. and Prewitt, R.L. (1996) Elevated pressure stimulates protooncogene expression in isolated mesenteric arteries. *Am. J. Physiol.* **271**: H1517–H1523.

Allen, S.P., Wade, S.S. and Prewitt, R.L. (1997) Myogenic tone attenuates pressure-induced gene expression in isolated small arteries. *Hypertension* **30**: 203–208.

Atwood, L.D., Kammerer, C.M., Samollow, P.B., Hixson, J.E., Shade, R.E. and MacCluer, J.W. (1997) Linkage of essential hypertension to the angiotensinogen locus in Mexican Americans. *Hypertension* **30**: 326–330.

Aviv, A. (1996) The links between cellular Ca^{2+} and Na^+/H^+ exchange in the pathophysiology of essential hypertension. *Am. J. Hypertension* **9**: 703–707.

Ayachi, S. (1979) Increased dietary calcium lowers blood pressure in the spontaneously hypertensive rat. *Metabolism* **27**: 1234.

Bloem, L.J., Foroud, T.M., Ambrosius, W.T., Hanna, M.P., Tewksbury, D.A. and Pratt, J.H. (1996) Association of the angiotensinogen gene to serum angiotensinogen in blacks and whites. *Hypertension* **29**: 1078–1082.

Boguski, M.S. (1994) Bioinformatics. *Curr. Opin. Gen. Dev.* **4**: 383–388.

Bohlender, J., Fukamizu, A., Lippoldt, A., *et al.* (1997) High human renin hypertension in transgenic rats. *Hypertension* **29**: 428–434.

Bonnardeaux, A., Davies, E., Jeunemaitre, X., *et al.* (1994) Angiotensin II Type 1 receptor gene polymorphisms in human essential hypertension. *Hypertension* **24**: 63–69.

Bonnardeaux, A., Nadaud, S., Charru, A., Jeunemaitre, X., Corvol, P. and Soubrier, F. (1995) Lack of evidence for linkage of the endothelial cell nitric oxide synthase gene to essential hypertension. *Circulation* **91**: 96–102.

Canessa, M., Adragna, N., Solomon, H., Connolly, T. and Tosteson, D. (1980) Increased sodium–lithium countertransport in red cells of patients with essential hypertension. *N. Engl. J. Med.* **302**: 772–776.

Caulfield, M., Lavender, P., Farrall, M., Munroe, P., Lawson, M., Turner, P. and Clark, A. (1994) Linkage of the angiotensinogen gene to essential hypertension. *N. Engl. J. Med.* **330**: 1629–1633.

Caulfield, M., Lavender, P., Newell-Price, J., *et al.* (1995) Linkage of the angiotensinogen gene locus to human essential hypertension in African Caribbeans. *J. Clin. Invest.* **96**: 687–692.

Cavalli-Sforza, L.L., Menozzi, P. and Piazza, A. (1994) *The History and Geography of Human Genes*. Princeton University Press, Princeton.

Chao, C., Maddeddu, P., Wang, C., Liang, Y., Chao, L. and Chao, J. (1996) Differential regulation of kallikrein, kininogen, and kallikrein-binding protein in arterial hypertensive rats. *Am. J. Physiol.* **271**: F78–F86.

Chen, X., Li, W., Yoshida, H., Tsuchida, S., *et al.* (1997) Targeting deletion of angiotensin type 1B receptor gene in the mouse. *Am. J. Physiol.* **272**: F299–F304.

Chiang, F.T., Hsu, K.L., Tseng, C.D., Lo, H.M., Chern, T.H. and Tseng, Y.Z. (1997) Association of the renin gene polymorphism with essential hypertension in a Chinese population. *Clin. Genet.* **51**: 370–374.

Cusi, D., Barlassina, C., Azzani, T., *et al.* (1997) Polymorphism of alpha adducin and salt sensitivity in patients with essential hypertension. *Lancet* **349**: 1353–1357.

DeRisi, J.L., Iyer, V.R. and Brown, P.O. (1997) Exploring the metabolic and genetic control of gene expression on a genomic scale. *Science* **278**: 680–686.

Deter, H.C., Buchholz, K., Schorr, U., *et al.*, (1997) Psychophysiological reactivity of salt-sensitive normotensive subjects. *J. Hypertension* **15**: 839–844.

Dyer, A., Stamler, J., Shekell, R. and Schoenberger, J. (1976) The relationship of education to blood pressure: findings on 40,000 employed Chicagoans. *Circulation* **54**: 987.

Eich, R.H., Cuddy, R.P., Smulyan, H. and Lyons, R.H. (1996) Hemodynamics in labile hypertension: a follow-up study. *Circulation* **34**: 299–307.

Fields, S. (1997) The future is function. *Nature Genet.* **15**: 325–327.

Folkow, B. (1978) Cardiovascular structural adaptations: its role in the initiation and maintenance of primary hypertension. *Clin. Sci. Mol. Med.* **55**: 3–22.

Folkow, B. (1993) Physiological organization of neurohormonal responses to psychosocial stimuli: implications for health and disease. *Ann. Behav. Med.* **15**: 236–244.

Folkow, B. (1995) Integration of hypertension research in the era of molecular biology: G.W. Pickering memorial lecture. *J. Hypertension* **13**: 5–18.

Frankel, W.N. (1995) Taking stock of complex trait genetics in mice. *Trends in Genetics* **11**: 471–477.

Garay, R., Dagher, G., Pernollet, M., Devynck, M. and Meyer, P. (1980) Inherited defect in a Na^+, K^+-co-transport system in erythrocytes from essential hypertensive patients. *Nature* **284**: 282–283.

Garcia, A., Matias-Guiu, X., Cabezas, R., Chico, A., Prat, J., Baiget, M. and DeLeiva, A. (1997) Molecular diagnosis of von Hippel disease in a kindred with a predominance of familial phaeochromocytoma. *Clin. Endocrinol.* **46**: 359–363.

Gardiner, S.M., March, J.E., Kemp, P.A., Mullins, J.J. and Bennett, T. (1995) Haemodynamic effects of losartan and the endothelin antagonist, SB 209670, in conscious, transgenic ((*mRen-2*)27), hypertensive rats. *Br. J. Pharmacol.* **116**: 2237–2244.

Gleibermann, L. (1973) Blood pressure and dietary salt in human populations. *Ecol. Food Nutrition* **2**: 143.

Griffiths, L., Zee, R., Ying, L. and Morris, B. (1991) A locus on the long arm of chromosome 1 as a possible cause of essential hypertension. *Clin. Exp. Pharmacol. Physiol.* **18**: 363–366.

Grunder, S., Firsov, D., Chang, S.S., *et al.* (1997) A mutation causing pseudohypoaldosteronism type 1 identifies a conserved glycine that is involved in the gating of the epithelial sodium channel. *EMBO J.* 5: 899–907.

Gudbrandsson, T., Julius, S., Jamerson, K., Smith, S., Krause, L. and Schork, N. (1994) Recreational exercise and cardiovascular status in the rural community of Tecumseh, Michigan. *Blood Pressure* **3**: 174–184.

Guthrie, G., Kotchen, T. and Kotchen, J. (1983) Suppression of adrenal mineralocorticoid production in prehypertensive young adult men. *J. Clin. Endocrinol. Metab.* **56**: 87–92.

Haller, H. (1996) Hypertension, the endothelium and the pathogenesis of chronic vascular disease. *Kidney Blood Press. Res.* **19**: 166–171.

Hansson, J.H., Nelson-Williams, C., Suzuki, H., *et al.* (1995) Hypertension caused by a truncated epithelial socium channel Y subunit: genetic heterogeneity of Liddle syndrome. *Nature Genet.* **11**: 76–82.

Harburg, E., Erfurt, J., Chape, C., Hauenstein, L., Schull, W. and Schork, M. (1973) Socio-ecological stressor areas and black–white blood pressure. *J. Chron. Dis.* **26**: 595–611.

Harrap, S., Dominiczak, A., Fraser, R., Lever, A., Morton, J., Foy, C. and Watt, G. (1996) Plasma angiotensin II, predisposition to hypertension, and left ventricular size in healthy young adults. *Circulation* **93**: 1148–1154.

He, J., Klag, M.J.,Whelton, P.K., Chen, J.Y., *et al.* (1991) Migration, blood pressure pattern, and hypertension: The Yi migrant study. *Am. J. Epidemiol.* **134**: 1085–1101.

Hein, L., Barsh, G.S., Pratt, R.E., Dzau, V.J. and Koblika, B.K. (1995) Behavioural and cardiovascular effects of disrupting the angiotensin II type-2 receptor gene in mice. *Nature* **377**: 744–747.

Hollenberg, N.K., Martinez, G., McCullough, M., *et al.* (1997) Aging, acculturation, salt intake, and hypertension in the Kuna of Panama. *Hypertension* **29**: 171–176.

Houtman, J. (1996) Trace elements and cardiovascular diseases. *J. Cardiovascular Risk* **3**: 18–25.

Hwang, D.M., Dempsey, A.A., Wanh, R.X., *et al.* (1997) A genome-based resource for molecular cardiovascular medicine: toward a compendium of cardiovascular genes. *J. Clin. Invest.* **96**: 4146–4203.

Ichiki, T., Labosky, P, Shiota, C., *et al.* (1995) Effects on blood pressure and exploratory behavior of mice lacking angiotensin II type-2 receptor. *Nature* **377**: 748–750.

Inoue, I., Rohrwasser, A., Helin, C., *et al.* (1995) A mutation of angiotensinogen in a patient with preeclampsia leads to altered kinetics of the renin–angiotensin system. *J. Biol. Chem.* **270**: 11430–11436.

Iwai, N., Ohmichi, N., Hanai, K., Makamura, Y. and Kinoshita, M. (1993) Human SA gene locus as a candidate locus for essential hypertension. *Hypertension* **23**: 375–380.

Iyengar, S.K., Weder, A.B., Koike, G., Jokelainen, P., Jacob, H.J. and Schork, N.J. (1998) The developmental evolution of hypertension: evidence, genetic models, and mechanisms. In: *Development of the Hypertensive Phenotype: Basic and Clinical Studies: Handbook of Hypertension* (eds R. McCarty, D.A. Blizard and R.L. Chevalier). Elsevier, Amsterdam.

Jeunemaitre, X., Lifton, R., Hunt, S.C., Williams, R.R. and Lalouel, J.M. (1992a) Absence of linkage between the angiotensin converting enzyme locus and human essential hypertension. *Nature Genet.* **1**: 72–75.

Jeunemaitre, X., Soubrier, F., Kotelevtsev, Y.V., *et al.* (1992b) Molecular basis of human hypertension: role of angiotensinogen. *Cell* **71**: 169–180.

John, S., Krege, J., Oliver, P., *et al.* (1995) Genetic decreases in atrial natriuretic peptide and salt-sensitive hypertension. *Science* **267**: 679–681.

Johnson, E., Schork, N. and Spielberger, C. (1987) Emotional and familial determinants of elevated blood pressure in black and white adolescent females. *J. Psychosomatic Res.* **31**: 731–741.

Joseph, J.G., Prior, I.A., Salmond, C.E. and Stanley, D. (1983) Elevation of systolic and diastolic blood pressure associated with migration: the Tokelau Island migrant study. *J. Chron. Dis.* **36**: 507–516.

Julier, C., Delephine, M., Keavney, B., *et al.* (1997) Genetic susceptibility for human familial essential hypertension in a region of homology with blood pressure linkage on rat chromosome 10. *Hum. Mol. Genet.* **6**: 2077–2085.

Julius, S. (1995) The defense reaction: a common denominator of coronary risk and blood pressure in neurogenic hypertension? *Clin. Exp. Hypertension* **17**: 375–386.

Kannel, W.B. (1996) Blood pressure as a cardiovascular risk factor. *JAMA* **275**: 1571–1578.

Kannel, W.B., Sorlie, P. and Gordon, T. (1980) Labile hypertension: a faulty concept? The Framingham Study. *Circulation* **61**: 1183–1187.

Kim, H., Krege, J., Kluckman, K., *et al.* (1995) Genetic control of blood pressure and the angiotensinogen locus. *Proc. Natl Acad. Sci.* **92**: 2735–2739.

Klag, M.J., He, J., Coresh, J., *et al.* (1995) The contribution of urinary cations to the blood pressure differences associated with migration. *Am. J. Epidemiol.* **142**: 295–303.

Krege, J., John, S., Langenback, L., *et al.* (1995) Male-female differences in fertility and blood pressure in ACE-deficient mice. *Nature* **375**: 146–148.

Kurihara, Y., Kurihara, H., Suzuki, H., *et al.* (1994) Elevated blood pressure and craniofacial abnormalities in mice deficient in endothelin-1. *Nature* **368**: 703–710.

Lander, E.S. and Schork, N.J. (1994) Genetic dissection of complex traits. *Science* **265**: 2037–2048.

Laragh, J. (1992) Role of renin secretion and kidney function in hypertension and attendant heart attack and stroke. *Clin. Exp. Hypertension* **14**: 285–305.

Lifton, R.P. (1996) Molecular genetics of human blood pressure variation. *Science* **272**: 676–680.

Lifton, R., Hunt, S., Williams, R., Pouyssegur, J. and Lalouel, J.M. (1990) Exclusion of the Na^+–H^+ antiporter as a candidate gene in human essential hypertension. *Hypertension* **17**: 8–14.

Lifton, R.P., Dluhy, R.G., Powers, M., Rich, G.M., Cook, S., Ulick, S. and Lalouel, J.M. (1992) A chimeric 11B-hydroxylase/adosterone synthase gene causes glucocorticoid-remediable aldostronism and human hypertension. *Nature* **355**: 262–265.

Link, R.E., Desai, K., Hein, L., *et al.* (1996) Cardiovascular regulation in mice lacking -2 - adrenergic receptor subtypes b and c. *Science* **273**: 803–805.

Lopez, M.J., Wong, S.K., Kishimoto, I., *et al.* (1995) Salt-resistant hypertension in mice lacking the guanylyl cyclase-A receptor for atrial natriuretic peptide. *Nature* **378**: 65–68.

MacGregor, G., Smith, S. and Markandu, N. (1983) Moderate potassium supplementation in essential hypertension. *Lancet* **2**: 567.

Mansfield, T., Simon, D., Farfel, Z., *et al.* (1997) Multilocus linkage of familial hyperkalaemia and hypertension, pseudohypoaldosteronism type II, to chomosomes 1q31-42 and 17p11-q21. *Nature Genet.* **16**: 202–205.

Markovitz, J., Matthews, K., Kannel, W., Cobb, J. and D'Agostino, R. (1993) Psychological predictors of hypertension in the Framingham Study – is there tension in hypertension? *J. Am. Med. Assoc.* **270**: 2439–2443.

Marmot, M., Rose, G., Shipley, M. and Hamilton, P. (1978) Employment grade and coronary heart disease in British civil servants. *J. Epidemiol. Commun. Health* **32**: 244.

McGarvey, S.T. and Baker, P.T. (1979) The effects of modernization and migration on Samoan blood pressure. *Hum. Biol.* **51**: 461–479.

Miall, W. and Oldham, P. (1958) Factors influencing arterial blood pressure in the general population. *Clin. Sci.* **17**: 409.

Miklos, G. and Rubin, G. (1996) The role of the genome project in determining gene function: insights from model organisms. *Cell* **86**: 521–529.

Modesti, P.A., Tamburini, C., Hagi, M.I., Cecioni, I., Migliorini, A. and Serneri, G.G. (1995) Twenty-four-hour blood pressure changes in young Somalian blacks after migration to Italy. *Am. J. Hypertension* **8**: 201–205.

Nabika, T., Bonnardeaux, A., James, M., *et al.* (1994) Evaluation of the SA locus in human hypertension. *Hypertension* **25**: 6–13.

Nichols, W., Koller, D., Slovis, B. *et al.* (1997) Localization of the gene for familial primary pulmonary hypertension to chromosome 2q31-32. *Nature Genet.* **15**: 277–281.

Ohta, K., Kim, S. and Iwao, H. (1996) Role of angiotensin-converting enzyme, adrenergic receptors, and blood pressure in cardiac gene expression of spontaneously hypertensive rats during development. *Hypertension* **28**: 627–634.

Pincomb, G., Lovallo, W.R., McKey, B.S., *et al.* (1996) Acute blood pressure elevations with caffeine in men with borderline systemic hypertension. *Am. J. Cardiol.* **77**: 270–274.

Poulter, N.R., Khaw, K.T. and Sever, P.S. (1988) Higher blood pressures of urban migrants from an African low-blood pressure population are not due to selective migration. *Am. J. Hypertension* **1**: 143S–145S.

Rapp, J.P., Wang, S.M. and Dene, H. (1990) Effect of genetic background on cosegregation of renin alleles and blood pressure in Dahl rats. *Am. J. Hypertension* **3**: 391–396.

Risch, N. and Botstein, D. (1996) A manic depressive history. *Nature Genet.* **12**: 351–353.

Risch, N. and Merikangas, K. (1996) The future of genetic studies of complex human diseases. *Science* **273**: 1516–1517.

Rouse, I., Beilin, L., Armstrong, B. and Vandongen, R. (1983) Blood pressure lowering effect of a vegetarian diet: a controlled trial in normotensive subjects. *Lancet* **1**: 5–10.

Salmond, C.E., Joseph, J.G., Prior, I.A., Stanley, D.G. and Wessen, A.F. (1985) Longitudinal analysis of the relationship between blood pressure and migration: the Tokelau Island migrant study. *Am. J. Epidemiol.* **122**: 291–301.

Salmond, C.E., Prior, I.A. and Wessen, A.F. (1989) Blood pressure patterns and migration: a 14-year cohort study of adult Tokelauans. *Am. J. Epidemiol.* **130**: 37–52.

Salonen, J., Tuomilheto, J. and Nissinen, A. (1983) Relation of blood pressure to reported intake of salt, saturated fats, and alcohol in a healthy middle-aged population. *J. Epidemiol. Community Health* **37**: 32.

Savineau, J. and Marthan, R. (1997) Modulation of the calcium sensitivity of the smooth muscle contractile apparatus: molecular mechanisms, pharmacological and pathophysiological implications. *Fundam. Clin. Pharmacol.* **11**: 289–299.

Schork, N.J. (1997) Genetically complex cardiovascular traits: origins, problems, and potential solutions. *Hypertension* **29**: 145–149.

Schork, N. and Chakravarti, A. (1996) A nonmathematical overview of modern gene mapping techniques applied to human diseases. In: *Molecular Genetics and Gene Therapy of Cardiovascular Disease* (ed. S. Mockrin). Marcel Dekker, New York, pp. 79–109.

Schork, N.J., Weder, A.B., Trevisan, M. and Laurenzi, M. (1994) The contribution of pleiotropy to blood pressure and body-mass index variation: The Gubbio Study. *Am. J. Hum. Genet.* **54**: 361–373.

Schork, N.J., Nath, S.P., Lindpaintner, K. and Jacob, H.J. (1996) Extensions of quantitative trait locus mapping in experimental organisms. *Hypertension* **28**: 1104–1111.

Schork, N.J., Cardon, L.R. and Xu, X. (1998) The future of genetic epidemiology. *Trends Genet.* **14**: 266–271.

Schunkert, H., Hense, H., Muscholl, M., Luchner, A. and Riegger, G. (1996) Association of angiotensin converting enzyme activity and arterial blood pressure in a population-based sample. *J. Hypertension* **14**: 571–575.

Schuster, H., Wienker, T., Bahring, S., *et al.* (1996) Severe autosomal dominant hypertension and brachydactyly in a unique Turkish kindred maps to human chromosome 12. *Nature Genet.* **13**: 98–100.

Sharp, M.G., Fettes, D., Brooker, G., Clark, A.F., Peters, J., Fleming, S. and Mullins, J.J. (1996) Targeted inactivation of the *Ren-2* gene in mice. *Hypertension* **28**: 1126–1131.

Shesely, E.G., Maeda, N., Kim, H.S., *et al.* (1996) Elevated blood pressure in mice lacking endothelial nitric oxide synthase. *Proc. Natl Acad. Sci. USA* **93**: 13176–13181.

Shimkets, R.A., Warnock, D.G., Bositis, C.M., *et al.* (1994) Liddle's Syndrome: heritable human hypertension caused by mutations in the B subunit of the epithelial sodium channel. *Cell* **79**: 407–414.

Silman, A.J., Evans, S.J. and Loysen, E. (1987) Blood pressure and migration: a study of Bengali immigrants in East London. *J. Epidemiol. Community Health* **41**: 152–155.

Simon, D. and Lifton, R. (1996) The molecular basis of inherited hypokalemic alkalosis: Bartter's and Gitelman's syndromes. *Am. J. Physiol.* **271**: F-961–F966.

Sing, C.F., Haviland, M.B. and Reilly, S.L. (1996) Genetic architecture of common multifactorial diseases. In: *Variation in the Human Genome*, Vol. 197 (ed. C.F. Sing). Ciba Foundation, New York, pp. 211–232.

Sivertsson, R. (1970) The hemodynamic importance of structural vascular changes in essential hypertension. *Acta Physiol. Scand.* **79**: 3–56.

Smithies, O. and Maeda, N. (1995) Gene targeting approaches to complex genetic diseases: atherosclerosis and essential hypertension. *Proc. Natl Acad. Sci. USA* **92**: 5266–5272.

St. Lezin, E.M., Pravenec, M., Wong, A.L., *et al.* (1996) Effects of renin gene transfer on blood pressure and renin gene expression in a congenic strain of Dahl salt-resistant rats. *J. Clin. Invest.* **97**: 522–527.

Stamler, J., Stamler, R., Riedliner, W.F., *et al.* (1976) Hypertension screening of 1 million Americans: Community Hypertension Evaluation Clinic Program. *J. Am. Med. Assoc.* **235**: 2299.

Steptoe, A., Fieldman, G., Evans, O. and Perry, L. (1996) Cardiovascular risk and responsivity to mental stress: the influence of age, gender and risk factors. *J. Cardiovasc. Res.* **3**: 83–93.

Stitt, F., Crawford, M., Clayton, E. and Morris, J. (1981) Clinical and biochemical indicators of cardiovascular disease among men living in hard and soft water areas. *Lancet* **1**: 122.

Stoff, J. (1986) Prostaglandins and hypertension. *Am. J. Med.* **80**: 56–61.

Sventek, P., Li, J., Grove, K., Deschepper, C.F. and Schiffrin, E.L. (1996) Vascular structure and expression of endothelin-1 gene in L-NAME-treated spontaneously hypertensive rats. *Hypertension* **27**: 49–55.

Svetkey, L., Chen, Y.T., McKeown, S., Preis, L. and Wilson, A. (1996) Preliminary evidence of linkage of salt sensitivity in black americans at the α-adrenergic receptor locus. *Hypertension* **29**: 918–922.

Syme, S., Oakes, T., Friedman, G. (1974) Social class and differences in blood pressure. *Am. J. Publ. Health* **64**: 619.

Takahashi, Y., Nakayama, T., Soma, M., Uwabo, J., Izumi, Y. and Kanmatsuse, K. (1997) Association analysis of TG repeat polymorphism of the neuronal nitric oxide synthase gene with essential hypertension. *Clin. Genet.* **52**: 83–85.

Tamura, K., Umemura, S., Nyui, N., *et al.* (1996) Tissue-specific regulation of angiotensinogen gene expression in spontaneously hypertensive rats. *Hypertension* **27**: 1216–1223.

Tamura, T., Umemura, S., Yamakaw, T., *et al.* (1997) Modulation of tissue angiotensinogen gene expression in genetically obese hypertensive rats. *Am. J. Physiol.* **272**: R1704–R1711.

Tanimoto, K., Sugiyama, F., Goto, Y., *et al.* (1994) Angiotensinogen-deficient mice with hypotension. *J. Biol. Chem.* **269**: 31334–31337.

Thibbonier, M. and Schork, N.J. (1995) The genetics of hypertension. *Curr. Opin. Genet. Devel.* **5**: 362–370.

Velculescu, V.E., Zhang, L., Vogelstein, B. and Kinzler, B. (1995) Serial analysis of gene expression. *Am. J. Hum. Genet.* **270**: 484–487.

Villarreal, F.J., MacKenna, D.A., Omens, J.H. and Dillmann, W.H. (1995) Myocardial remodeling in hypertensive *Ren-2* transgenic rats. *Hypertension* **25**: 98–104.

Von Eiff, A., Friedrick, G. and Neus, H. (1982) Traffic noise, a factor in the pathogenesis of essential hypertension. *Contrib. Nephrol.* **30**: 82.

Ward, K., Sparrow, D., Landsberg, L., Young, J., Vokonas, P. and Weiss, S. (1996) Influence of insulin, sympathetic nervous system activity, and obesity on blood pressure: the Normative Aging Study. *J. Hypertension* **14**: 301–308.

Watson, B., Bergman, S., Myracle, A., Callen, D., Acton, R. and Warnock, D. (1996) Genetic association of 11 β-hydroxysteroid dehydrogenase Type 2 (HSD11B2) flanking microsatellites with essential hypertension in blacks. *Hypertension* **28**: 478–482.

Weder, A.B. and Schork, N.J. (1994) Adaptation, allometry and hypertension. *Hypertension* **24**: 145–156.

Weir, R. (1982) Effect on blood pressure on changing from high to low steroid preparations in women with oral contraceptive induced hypertension. *Scott. Med. J.* **27**: 212.

West, M.J., Summers, K.M. and Huggard, P.R. (1992) Polymorphisms of candidate genes in essential hypertension. *Clin. Exp. Pharmacol. Physiol.* **19**: 315–318.

Wetherbee, H. and Tyroler, H.A. (1971) The relation of migration to coronoary heart disease risk factors. *Arch. Intern. Med.* **128**: 976–981.

Wilson, P.W., Hoeg, J.M., D'Agastino, R.B., *et al.* (1997) Cumulative effects of high cholesterol levels, high blood pressure and cigarette smoke on carotid stenosis. *New Engl. J. Med.* **337**: 516–522.

Wolf, K. and Kurtz, A. (1997) Influence of salt intake on atrial natriuretic peptide gene expression in rats. *Pflugers Arch.* **433**: 809–816.

Xu, Q., Li, D.G., Holbrook, N.J. and Udelsman, R. (1995) Acute hypertension induces heat-shock protein 70 gene expression in rat aorta. *Circulation* **92**: 1223–1229.

Xu, Q., Liu, Y., Gorospe, M., Udelsman, R. and Holbrook, N. (1996) Acute hypertension activates mitogen-activated protein kinases in arterial wall. *J. Clin. Invest.* **97**: 508–514.

Ying, L., Zee, R., Griffiths, L. and Morris, B. (1991) Association of a RFLP for the insulin receptor gene, but not insulin, with essential hypertension. *Biochem. Biophys. Res. Commun.* **181**: 486–492.

Yohimoto, T., Naruse, M., Naruse, K., *et al.* (1996) Angiotensin II-dependent down-regulation of vascular natriuretic peptide Type C receptor gene expression in hypertensive rats. *Endocrinology* **137**: 1102–1107.

Yu, K., Lu, D., Paddy, M.R., Lenk, S.E. and Raizada, M.K. (1996) Angiotensin II regulation of plasminogen activator inhibitor-1 gene expression in neurons of normotensive and spontaneously hypertensive rat brains. *Endocrinology* **137**: 2503–2513.

2

Improved strategies for the mapping of quantitative trait loci in the rat model

Monika Stoll and Howard J. Jacob

1. Introduction

Over the past three decades, various genetic models of hypertension have been used to study the etiology of hypertension and the regulation of blood pressure. Prior to the development of genetic models of hypertension, animal models of high blood pressure were induced by producing some type of damage to the kidney by surgical removal of kidneys, or by ligation or clipping of renal arteries (Barger, 1979; Goldblatt *et al.*, 1934). Although these experimentally induced models of hypertension provided insight into the regulation of blood pressure, they were not always consistent with respect to producing equivalent levels of hypertension, nor were they very similar to human essential hypertension. Genetic models were developed to provide reliable and reproducible forms of high blood pressure, as well as to be more representative of the clinical phenotype. For instance, the genetic hypertensive (GH) rat, the first genetic model of hypertension (Smirk and Hall, 1958), develops spontaneous hypertension that is relatively consistent from rat to rat in the severity and age of onset of the hypertension. The development of the spontaneously hypertensive rat (SHR; Okamoto and Aoki, 1963) and the Dahl salt-susceptible strain of rat (Dahl *et al.*, 1962) provided additional reproducible models of hypertension. Over the past 30 years at least nine different genetically hypertensive rat strains have been developed. Importantly, eight of the nine strains were developed by selecting the breeders for the next generation based on blood pressure measurements using a tail-cuff protocol, yet each strain is slightly different in its etiology of hypertension (Yamori, 1984) and has different degrees of susceptibility to end-organ damage associated with hypertension. Since each strain was developed from a different founder colony, we can infer that the etiological differences are most probably due to different and potentially unique gene combinations being fixed

Molecular Genetics of Hypertension, edited by A.F. Dominiczak, J.M.C. Connell and F. Soubrier.

within each strain, as well as their interaction with environmental factors. Physiologists, pharmacologists and other basic science investigators have known about these etiological differences and have tried to exploit them to find the underlying causes of hypertension. Unfortunately, the study designs employed required the use of 'control' strains such as the WKY (Wistar–Kyoto), once thought to only differ from the SHR at 'hypertensive' genes, which in retrospect do not exist. Strains used as 'control' strains do have lower blood pressure, but also many other phenotypic differences very likely due to strain differences. Consequently, although investigators have found many differences between the hypertensive strains and the 'control' strains they have made little headway in defining the initial cause of high blood pressure.

Genetic analyses provide means to distinguish primary factors causing hypertension in inbred strains (isogenic) of the rat. Genes playing a role in causing the trait will cosegregate (be linked) with the trait in a cross; whereas, the genes responsible for secondary responses or unrelated strain differences will not. The nine different genetically hypertensive inbred strains thus provide the unique opportunity to investigate the genetic basis of different 'patterns' of hypertension. The etiological differences among the strains should enable molecular geneticists to make comparisons, and by this means to identify major genes which differ among strains; this information may help us to model the effect of genetic heterogeneity found in human hypertension, while minimizing the loss of power in these studies. Furthermore, the regions of the rat genome that cosegregate with blood pressure can become candidate regions for genetic studies in man.

At issue is how helpful are experimental animal models in defining human disease etiology. Given that the infrastructure for the human genome is far superior to that of the rat genome, one may ask, 'how useful will experimental animal models be for hypertension or any other human disease?' Typically the debate about the utility of animal models hinges upon the assumption that a model system must reflect the clinical picture. It must be remembered that in the case of hypertension, even two people with essential hypertension do not necessarily have the same clinical picture. Consequently, it is not surprising that the rat genetic models have differences from the clinical condition. First, there are obvious species differences between humans and rats. Second, the genetic models were developed by simply selecting for breeding pairs with the highest blood pressure, irrespective of other characteristics or phenotypes.

Despite these differences, the nine different strains provide a controlled framework that can be used to identify gene combinations and interactions that result in hypertension. Specifically, one can ask, 'are there genomic regions in common amongst all nine or the majority of the strains?' If so, these genomic regions would be ideal candidates for the human disease. This assumption supposes that gene defects will have similar effects between these very divergent species.

We are not overly concerned about a direct cross-species translation of gene defects. Our view has been that in the worst case scenario we will identify important pathways where defects result in hypertension. We also believe that studies using these genetic models will provide essential information about the use of intermediate phenotypes to understand disease etiology.

2. Study design

As discussed in Chapter 3, genetic studies begin by selection of the cross design and strains to be used to study a disorder, such as hypertension. Unfortunately, in many cases considerably less attention is paid to the phenotyping. For example, investigators (including ourselves) frequently write and speak about using molecular genetics to identify the genes responsible for hypertension. Yet, hypertension by definition represents the upper end of the blood pressure distribution curve that is defined by a clinically relevant threshold; whereby intervention has been determined to modify the rates of morbidity and mortality (Group HDaF-UPC, 1977). We, and we presume our colleagues, are not using this definition, but rather studying the genes responsible for 'higher' blood pressure.

Blood pressure is a continuous trait or quantitative trait that exhibits a relatively normal distribution in the population (Group HDaF-UPC, 1977). This continuous distribution of a quantitative trait is a hallmark of multifactorial disorders (multiple genes interacting with the environment). However, it should be noted that a qualitative trait may still be polygenic and display a dichotomous distribution because of the definition of the phenotype used. For example, insulin-dependent diabetes mellitus is a disease with multifactorial inheritance that is typically studied as a qualitative trait.

In our studies to determine the genetic components of blood pressure, we have used two different types of rat crosses. By mating a hypertensive with a normotensive inbred animal a first filial (F_1) progeny is generated derived from two inbred animals with very different levels of blood pressure and with a very different genetic predisposition to high blood pressure. However, these animals are still genetically identical, since they share one chromosome from each parent. In most of the crosses, the F_1 animals were near to the average of the two progenitor strains, suggesting that the responsible genes act together in an additive manner. A second generation cross is required to perform linkage analysis. For an additive mode of inheritance, an F_2 intercross (F_1 progeny is mated to each other) is best. In some cases, the phenotypes from the F_1 progeny are closer to one of the parental strains; in this case a backcross (in which F_1 animals are mated with one of the parental strains) is preferred. As a result of genetic segregation and environmental effects, the progeny from the second generation displays a greater variation. The ability to separate the total variance into genetic and environmental components is the major advantage of using inbred animals over outbred animals or humans. With the genetic component defined, a search can be conducted to identify the genes producing the effects. It is important to realize that genetic studies using animal models can only distinguish genetic differences between the two strains studied. Since the selection of the strains determines the numbers of genes segregating within the cross, it is meaningless to speak about the 'number of genes responsible for high blood pressure' without specifying the strains, cross structure and, as will be outlined later, phenotypes that were studied. For instance, in 1970, Tanase *et al.* used estimates of heritability (a heritability score of 1 means the trait is entirely genetic; whereas, a score of 0 means the trait is entirely environmental) to perform a genetic analysis of hypertension in three crosses, where SHR rats were crossed with different normotensive rat strains. It was evident in these studies that crosses between SHR and different normotensive strains yielded a different estimate of the number of loci responsible for the high blood pressure. These data demonstrated that the genetic background of the normotensive strain determines how many genes are segregating with high blood pressure in a given cross. Obviously, the selection of a cross markedly influences the number of loci for a distinct phenotype, and, therefore, limits the identification of the genes responsible for traits such as high blood pressure.

Genetic analysis in its simplest form is assigning a component of the phenotypic variance, in our case blood pressure, to the three possible genotypes: homozygous for allele 1, heterozygous and homozygous for allele 2 (in an F_2 intercross of inbred strains). If allele 1 is from the SHR, then we would expect to find the animals with the highest blood pressure to be homozygous more often at allele 1 and less often at allele 2 than the 1:2:1 Mendelian segregation ratio predicted. When a locus (either a gene-marker or anonymous DNA-marker) shows a significant deviation from the Mendelian segregation ratio, the locus is 'linked' to the phenotype or the phenotype has been 'mapped'. For most multifactorial disorders phenotypes are linked to quantitative trait loci (QTLs). The appropriate level for significance remains controversial; however, the traditional view is a logarithm of the odds ratio (LOD score) greater or equal to 3.0 or a *p*-value greater or equal to 0.001 for an analysis of variance (ANOVA) define the threshold of significance. Interested readers in the statistical genetics are encouraged to start with a review by Lander and Schork (1994), followed by Lander and Kruglyak (1995) and Elston (1997).

3. Phenotyping

The method and protocol used to collect a given phenotype will affect the environmental component and, therefore, the ability to carry out genetic dissection. The age at which an animal is phenotyped influences the number of genes contributing to the expression of this phenotype, what we call a 'phenotyping window effect' (Schork *et al.*, 1996; Wright, 1968). The data reported by Tanase *et al.* (1970) suggest such a 'phenotyping window effect' also for hypertension, in a F_2 progeny of a SHR × WKY cross. These findings rise an important issue. Are phenotypes collected at the right time point?

The phenotyping method represents a second pitfall in studies of the genetics of complex diseases. The selection scheme used to develop all inbred strains of hypertensive rats was based on the tail-cuff method (sphygomanometer). This may have resulted in the selection for a form of hypertension which is associated with stress since this method requires anesthesia and heating of the animal and tail-cuff itself. In this case, reducing the stress by using radio-telemetry may decrease one's ability to map genes responsible for high blood pressure associated with stress genetically. In addition, however, radio-telemetry increases the accuracy in the measurement of blood pressure, minimizing the environmental component to the blood pressure variance, and therefore, may increase one's ability to determine the contribution of genetic factors. The trade-off between using more accurate phenotyping protocols (less stress) versus tail-cuff is not easy, particularly since longitudinal data may improve the genetic dissection of complex traits (Schork *et al.*, 1996). Unfortunately, the collection of longitudinal data may necessitate the use of the tail-cuff procedure for measuring blood pressure. However, the use of (two) different methods to measure blood pressure offers the possibility of distinguishing between genes responsible for blood pressure regulation in general, as well as those being important in blood pressure increases in response to stress.

In a recent study we performed with our collegues in New Zealand, Drs Grigor and Harris, a total genome scan (~ 200 genetic markers covering the entire genome are genotyped in each animal) in a F_2 progeny of a GH × BN intercross, independently analyzing the data for linkage of QTLs to both blood pressure measured by direct arterial catheter and by the tail-cuff method. Interestingly, depending on the phenotyping method used, we identified some QTLs cosegregating with blood pressure (Koike *et al.*, 1997) dependent on the phenotyping method, while other QTLs were identical for blood pressure and therefore, independent of the phenotyping method used. This observation underlines how different phenotyping protocols can modify the outcome of a study using molecular genetic techniques. It is quite interesting to speculate that environmental factors such as stress can reveal different genes responsible for blood pressure regulation or maintenance. We have used pharmacological agents and different physiological stressors to characterize further the genes responsible for blood pressure.

4. Summary of genome scanning results

Several groups have used our genetic markers (Jacob *et al.*, 1995) and those developed by Levan, Szpirer and colleagues (Levan *et al.*, 1991; 1992), by Remmers and colleagues (Goldmuntz *et al.*, 1993; Zha *et al.*, 1993), by Serikawa and colleagues (Serikawa *et al.*, 1992; Yamada *et al.*, 1994) and Bihoreau and colleagues (1997) to initiate genome scans of crosses between hypertensive and normotensive rats. However, many of these studies have not been *complete* genome scans and have largely focused on a single QTL at a time. The results of these studies are summarized in *Table 1*.

Several points emerge from this table. First, it appears that hypertension-related genes are located on nearly every chromosome. This identification of a large number of loci for the trait high blood pressure is not surprising, since the blood pressure regulatory pathway is rather complex and, therefore, involves a large number of genes. Second, only very few studies have identified QTLs associated with basal blood pressure. This might be due to various reasons: at baseline all animals are at the genetically predetermined blood pressure, that means they are in homoeostasis. There may be a large number of genes involved in the maintenance of this homoeostasis, of which each one contributes to a very small degree of baseline blood pressure.

If this hypothesis is correct, a very large number of animals would need to be studied before QTLs can be detected accounting for basal blood pressure. To reveal the responsible genes more readily, it may be necessary to 'challenge' the cardiovascular system in order to detect where the regulatory pathway can no longer compensate for a given stimulus. These 'challenges' may include drugs interfering with the renin–angiotensin system such as angiotensin converting

Table 1. QTLs identified in rat models of genetic hypertension

Cross	Hypertensive locus	Phenotype	Reference
SHR × WKY	Chr. 1 (SA)	SBP, DBP	Samani *et al.*, 1993
SHR × WKY	Chr. 1 (MT1PA)	SBP	Samani *et al.*, 1993
SHRSP-H x WKY-H	Chr. 1 (SA)	SBP after salt	Lindpaintner *et al.*, 1993
*RI (SHR × BN)	Chr. 1 (Kallikrein)	SBP	Pravenec *et al.*, 1991
SS/JR × Lew	Chr. 1 (SA)	SBP after salt	Pravenec *et al.*, 1991
LEW × SS/JR	Chr.1 (SA)	SBP	Gu *et al.*, 1996
WKY × SS/JR	Chr.1 (SA)	SBP	Gu *et al.*, 1996
LEW × SS/JR	Chr.1 (D1Mco1)	SBP	Gu *et al.*, 1996
SHRSP × WKY	Chr.1 (CYP2A3)	SBP	Gu *et al.*, 1996
SHRSP × WKY	Chr.1 (SA)	SBP	Kreutz *et al.*, 1997
SBH × SBN	Chr.1 (?)	SBP	Yagil *et al.*, 1998
SHR × BB/OK	Chr.1 (IGF2)	SBP	Kovacs *et al.*, 1987
SHR × WKY	Chr. 2 (?), Chr. 4 (SPR)	BP	Nakajima *et al.*, 1994
*RI (SHR × BN)	Chr. 2 (?)	DBP	Pravenec *et al.*, 1995
LH × LN	Chr. 2 (?)	PP, SBP	Dubay *et al.*, 1993
SS/JR × MNS	Chr. 2 (GCA)	SBP after salt	Deng *et al.*, 1994
	Chr. 10 (ACE)	SBP after salt	Deng *et al.*, 1994
SHRSP × WKY	Chr2(CPB)	SBP, DBP	Clark *et al.*, 1996
SS/JR × WKY	Chr. 2 (GCA)	SBP after salt	Cicila *et al.*, 1994
SS/JR × WKY	Chr. 2 (NaK(1)	SBP after salt	Cicila *et al.*, 1994
SHR-SP × SS/JR	Chr. 2 (GCA)	MAP	Dubay *et al.*, 1993
GH × BN	Chr. 2 (?), Chr. 10 (ACE)	SBP	Deng *et al.*, 1994
MNS × SS/JR	Chr. 2 (NaK(1)	SBP after salt	Deng *et al.*, 1994
SHR × BN	Chr.2 (GCA)	SBP	Schork *et al.*, 1995
SS/JR × F1 (SS/JR x SR/JR)	Chr. 3 (ET-3)	SBP after salt	Cicila *et al.*, 1984
SHRSP × WKY	Chr.3	PP, PP after salt	Clark *et al.*, 1996
SHRSP × WKY	Chr.3	SBP and DBP after salt	Clark *et al.*, 1996
SHRSP × WKY	Chr.3 (?)	SBP	Matsumoto *et al.*, 1996
*RI (SHR × BN)	Chr. 4 (IL-6)	MAP	Pravenec *et al.*, 1995
SHR × WKY	Chr. 4 (NPY, SPR)	SBP, DBP, MAP	Katsuya *et al.*, 1993
SHR × BN	Chr.4 (NPY)	SBP	Schork *et al.*, 1995
SHR × BB/OK	Chr.4 (SPR, IL6)	SBP	Kovacs *et al.*, 1997
SS/JR × Lew	Chr. 5 (GLUTB,ET-2)	SBP after salt	Deng *et al.*, 1994
SS/JR × Lew	Chr. 5 (ET-2)	SBP after salt	Deng *et al.*, 1994
SHRSP-H × WKY-H	Chr. 10 (ACE), Chr. 18 (?)	SBP after salt	Jacob *et al.*, 1991
SHRSP-H × WKY-H	Chr. 10 (ACE), Chr. (?)	SBP after salt	Hilbert *et al.*, 1991
SHR × DRY	Chr. 10 (NGF)	MAP	Kapusczinski *et al.*, 1994
SHRSP-I × WKY-I	Chr. 10 (ACE)	SBP after salt	Nara *et al.*, 1991
SHRSP-H × WKY-H	Chr. 10 (BP/SP-1a)	basal BP	Kreutz *et al.*, 1995a
SHRSP-H × WKY-H	Chr. 10 (BP/SP-1b)	BP after salt	Kreutz *et al.*, 1995a
SHRSP-H × WKY-H	Chr. 10 (ACE)	ACE activity after salt	Kreutz *et al.*, 1995b
		ACE activity basal	Kreutz *et al.*, 1995b
SS/JR × MNS	Chr. 10 (ACE, ANPR)	SBP after salt	Deng and Rapp, 1992
SS/JR × MNS	Chr. 10 (NOS-2)	SBP after salt	Deng and Rapp, 1992
SS/JR × MNS	Chr. 10 (NOS)	SBP after salt	Deng and Rapp, 1992
SHRSP × WKY	Chr. 10 (ACE)	SBP, DBP after salt	Deng and Rapp, 1992

Table 1. *continued*

Cross	Hypertensive locus	Phenotype	Reference
SHR × WKY	Chr. 13 (renin)	SBP, DBP	Sun *et al.*, 1993
LH × LN	Chr. 13 (renin)	DBP, SBP	Dubay *et al.*, 1993
*RI (SHR x BN)	Chr. 13 (renin)	SBP	Yagil *et al.*, 1998
SS/JR × SR/JR	Chr. 13 (renin)	SBP after salt	Rapp *et al.*, 1989
SHR × LEW	Chr. 13 (renin)	MAP	Kurtz *et al.*, 1990
SHRXBB/OK	Chr.13 (ATP1A2)	SBP	Kovacs *et al.*, 1997
SS/JR × MNS	Chr. 17 (AT1a)	SBP after salt	Deng *et al.*, 1994
SHRXBB/OK	Chr.18 (?)	SBP	Kovacs *et al.*, 1997
*RI (SHR × BN)	Chr. 19 (?)	SBP	Pravenec *et al.*, 1995
*RI (SHR × BN)	Chr. 20 (HSP70)	SBP	Hamet *et al.*, 1992
SS/JR x Lew	Chr. 17 (HITH)	SBP after salt	Pravenec *et al.*, 1995
SHR × WKY	Y-Chr. (?)	SBP	Ely and Turner, 1990; Ely *et al.*, 1993

Candidate genes: angiotensin II AT_{1a} receptor (AT_{1a}), angiotensin converting enzyme (ACE), ATPase-1A2 subunit (ATP1A2), cytochrome p450–2a3 subunit (CYP2A3), endothelin-2 (ET-2), guanylylcyclase A/atrial natriuretic peptide receptor (GCA), heat-shock protein-70 (HSP-70), sodium potassium channel α_1 subunit ($NaK_{\alpha1}$), nitric oxide synthase (NOS), nerve growth factor (NGF), neuropeptide Y (NPY), substance P receptor (SPR). *Rat cross*: stroke prone-spontaneously hypertensive rat–Heidelberg (SHRSP–H), Wistar–Kyoto rat–Heidelberg (WKY–H), spontaneously hypertensive rat (SHR), Brown Norway (BN), salt sensitive-rat/John Rapp (SS/JR), Lyon hypertensive rat (LH), Lyon normotensive rat (LN), Lewis rat (Lew), recombinant inbred cross (RI (SHR×BN)), Milan normotensive rat (MNS), sabra hypertensive rat (SBH), sabra normotensive rat (SBN), biobreeding rat (BB/OK). *Phenotypes*: blood pressure (BP), systolic blood pressure (SBP), diastolic blood pressure (DBP), mean arterial pressure (MAP), left ventricular hypertrophy (LVH).
*Recombinant inbred strains.

enzyme (ACE) inhibitors and angiotensin II receptor antagonists or infusion of endocrine hormones such as angiotensin, norepinephrine or arginine vasopressin known to alter the cardiovascular response via different pathways.

A third point emerging from *Table 1* is that particular QTLs segregate with blood pressure only in particular crosses. The fact that different QTLs are seen in different crosses is not unexpected, since it must be remembered that molecular genetic studies can only determine genes which are different between two strains. Therefore, the selection of the cross studied, and the number of informative markers between the parental strains, limits the detection of genes which potentially contribute to blood pressure.

These published studies illustrate the power of using molecular genetic techniques to identify regions of the genome that segregate with blood pressure. While the data reduction is immense, from the whole genome down to a few regions of the genome, the challenge of identifying the genes response is also immense. One solution to accelerating gene discovery and etiology definition is to use intermediate phenotypes.

5. Intermediate phenotypes

An intermediate phenotype represents a measurable trait regulated by a single gene within at least one of the pathways involved in the development of hypertension.

This definition has two pitfalls. First, there are few examples of intermediate phenotypes. Enzyme levels, such as ACE, and substrate levels, such as angiotensinogen have been linked to specific genotypes. However, the role these genes play in high blood pressure is not clear, illustrating the second problem with using intermediate phenotypes; the pathways involved in the initiation of this disease are not known. Consequently, it would appear that there are no clear intermediate phenotypes for high blood pressure. However, the regulation of blood pressure is fairly well known, and some of the assays or test systems used to study blood pressure regulation could be considered as intermediate phenotypes (*Figure 1*). We refer to these as likely determinant phenotypes. Likely determinant phenotypes represent a measurable trait which defines a specific element of at least one of the pathways controlling blood pressure that *may* contribute to high blood pressure.

In a recent study using a Dahl salt-sensitive × Brown Norway F_2 intercross we, with our collegues Drs Greene, Roman and Cowley, set out to investigate the genes underlying salt-dependent hypertension (Simon *et al.*, 1995), as well as to assess if we could convert likely determinant phenotypes to intermediate phenotypes. We performed a total genomic search (273 genetic markers covering 99% of the rat genome) in an F_2 progeny which were derived from a cross between inbred Dahl salt-sensitive rats (SS/MCW, which are a different substrain than SS/JR) and inbred normotensive Brown Norway rats (BN/SSN/Hsd). The SS/MCW consistently develop hypertension and when placed on a high salt (8% NaCl) diet develop hypertension at a faster rate. After extensive phenotyping of the animals (67 direct

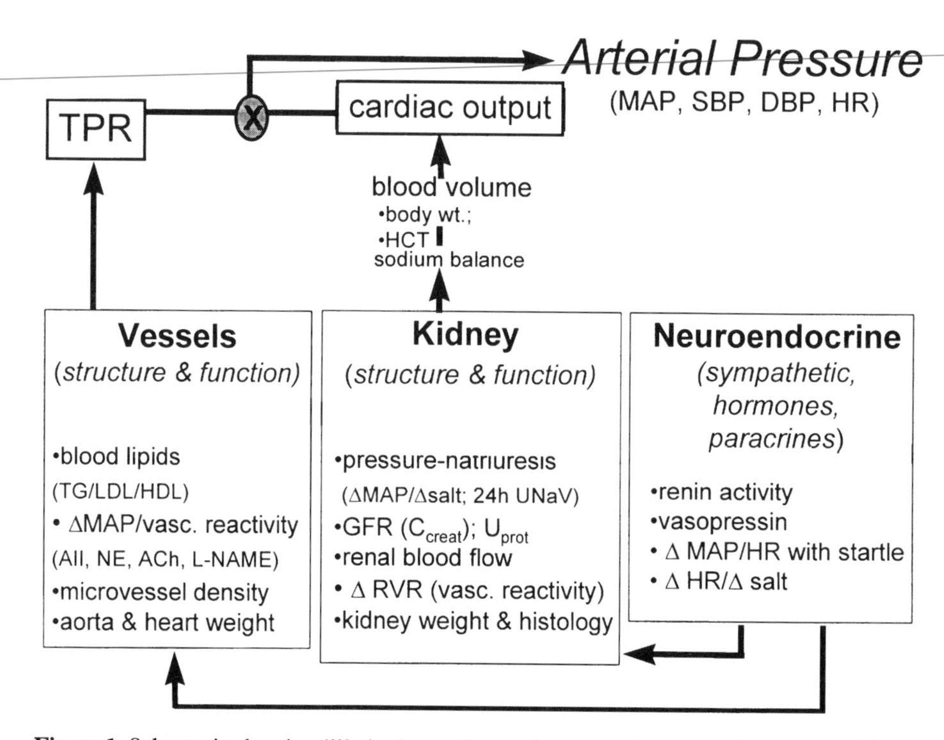

Figure 1. Schematic showing 'likely determinant phenotypes' representing measurable traits that contribute to blood pressure regulation.

measurements: 40 cardiovascular, 10 renal, nine plasma and eight miscellaneous plus 24 derived phenotypes totaling a number of 91 phenotypes) we performed a total genomic search. Among these were genetic markers for a number of genes known or speculated to be involved in blood pressure regulation such as ACE, angiotensin AT_1 receptors, arginine vasopressin (AVP), and the alpha2b adrenergic receptor. The genome was then scanned for QTLs for the 91 phenotypes studies. Linkage analysis resulted in the identification of two major QTLs for high blood pressure on chromosomes 3 and 18. Interestingly, the genes for AVP and the alpha2b adrenergic receptor mapped within the 95% confidence interval of the QTL (a 1.6 LOD drop on either side of the peak LOD score) on chromosome 3. Therefore, both AVP and the alpha2b adrenergic receptor represent positional candidate genes for the differences in systolic blood pressure after a high salt diet. To account for the polygenic nature of blood pressure, the genome was screened for QTLs linked to a variety of other traits. Analysis of these 'likely determinant phenotypes' for blood pressure/hypertension resulted in the identification of a number of QTLs (data not shown), none of which were located on chromosome 3 near the likely positional candidate genes such as AVP or the alpha2b adrenergic receptor. However, it was evident that most of the likely determinant phenotypes we selected were themselves quantitative traits. These data illustrate two essential points – firstly, intermediate phenotypes will not be easily defined and secondly, high blood pressure is likely to be determined by other complex traits. Hence the complete genetic dissection of high blood pressure is likely to remain a major challenge for some time. Unfortunately, searching for quantitative traits that are determined by other quantitative traits or differences in phenotyping protocols are not the only challenges for studying high blood pressure.

6. Influence of gender and age

The gender of the animal also appears to influence the number of genes responsible for the increase in blood pressure, as well as the overall genetic contribution. Unfortunately, very little has been reported about these gender differences. Most of the studies have used a single-cross design in which epigenetic factors such as mitochondrial or imprinting effects and sex chromosome effects cannot be studied and, therefore, lacks the possibility of distinguishing between genetic and epigenetic determinants or sex-linked effects. For most genetic studies, in the first generation (G_0) a male hypertensive animal is crossed with a female normotensive animal. Consequently, one cannot determine whether a phenotypical difference is the result of a sex-linked gene (carried on X- or Y-chromosome) or due to a gender-specific effect (that means the results of hormonal differences or other factors which distinguish between the genders). Sex-linked effects and epigenetic factors can be studied by using a reciprocal-cross design, in which a male hypertensive is crossed with a female normotensive *and* a male normotensive is crossed with a female hypertensive animal. A comparison is then made between the male and the female F_1 progeny of the two crosses. If males and females in the two crosses have the same blood pressures but differ between the genders, then the difference is sex-specific. On the other hand, if animals of a given sex from the reciprocal crosses are different from each other, then a sex-linked gene, mitochondrial effect or imprinting effect may be involved. Therefore, in order to cover these gender-dependent effects, investigators may need to adjust the number of animals by gender in each cross to locate 'hypertensive' genes. This is of particular importance in view of the fact that the age of onset for a given phenotype may differ between genders. For example, recently we and our colleague Dr Provoost identified two genes, *Rf-1* and *Rf-2*, which both contribute to the development of renal failure, estimated by proteinuria, in a backcross between the Fawn-hooded hypertensive rat (FHH/EUR) and a low proteinuric strain the ACI (Brown *et al.*, 1996). In this study, only male animals were used, because male animals develop renal failure at the age of 6–9 months, whereas females do not. However, after 2 years of age, a comparable state of renal failure was also detectable in female rats, which again could be attributable to *Rf-1* – indicating that in this case the age of onset of the disease was completely different (unpublished observation). This different onset of a disease in males and females causes a phenomenon, which may be described as the 'phenotyping window effect' that may give rise to inconsistent results when studying gender mixed populations using an age-matched study design. Harris *et al.* (1995) have found that genetic loci for high blood pressure were different between males and females, suggesting gender differences that maybe the result of the 'phenotyping window effect'. Recently, Samani *et al.* (1996) reported age-related effects for loci responsible for high blood pressure similar to those predicted by Schork *et al.* (1996).

In view of these phenotyping window effects and gender difference, the question arises whether 'gender-matched, age-matched study designs' are still appropriate or if gender-mixed studies should use epidemiologically defined ages rather than simply making sure the mean age is not different between the genders.

One means to overcome this problem may be to set up large crosses involving both genders but only to study males or females at the same time. Unfortunately, this is not always possible for many reasons. Consequently, investigators need to take into consideration the questions they seek to address with a molecular genetic analysis. Failure to think through the issue outlined above will result in limited study power and the necessity to repeat the study. While reproducibility of data is a hallmark of 'good' science, it can be particularly problematic for genetic studies, as the molecular genetic studies are sensitive to environmental influences and/or genetic background. For example, Kreutz *et al.* (1995b) generated a second cross and were unable to reproduce the original linkage to the ACE gene due to two substrains of WKY within their colony.

7. Improving the utility of total genome searches

When studying the genetics of hypertension using the rat model, it becomes evident that different QTLs are identified using crosses between different strains, underscoring the impact of genetic heterogeneity on the expression of hypertension (*Table 1*). The choice of the strains used in a given study determines the number of loci which may be identified since each inbred strain has been genetically 'fixed' for a few high blood pressure genes. On the other hand, such strain-specific founder-alleles and/or mutations actually drive the creation and use of strains with slightly different features (e.g. SHR vs. SHR-SP). The various rat strains created for genetic hypertension allow investigators to model 'heterogeneity' by utilizing the results of QTL total genome scans across different crosses. As yet, formal tests for heterogeneity and differential effects of a QTL among different strain crosses are still lacking.

Most of the studies reported in *Table 1* were primarily candidate gene studies that did not take into consideration the effects of the polygenic nature of high blood pressure. Failure to use a polygenic model results in inflated estimates of a QTL contribution to the genetic variance, and the failure to identify other gene regions. Many recent articles have emerged that define statistical models and methods for detection and simultaneous testing of multiple QTLs (Haley and Knott, 1992; Lander and Botstein 1989). Most of the proposed models are based on linear regression models and exploit the so called 'conditioning' principle. This principle suggests that once a QTL has been identified, one can statistically condition the model, by 'fixing' the effect of that QTL and then search the genome for additional QTLs. 'Fixing' the QTL simply means that the portion of variance explained by the locus is removed from any subsequent analysis, thereby minimizing the noise, and allowing the identification of additional loci contributing to the trait. Polygenic modeling that exploits that principle improves estimates of gene contribution to the trait and most importantly improves the map localization of the QTL and will, therefore, be absolutely essential if greater progress and insight are to be gleaned from QTL mapping studies of blood pressure.

To improve the accuracy and detection abilities of total genome scans, we have developed an 'integrated' map for the five genetically hypertensive strains, of which each shares similarities and differs etiologically, but all having one feature in common: high blood pressure using a polygenic model for the statistical analyses. This 'integrated map' offers the opportunity to look for a common gene, as well as for important gene combinations and clusters where blood pressure and likely determinant phenotypes overlap; thereby suggesting potential mechanisms. We have analyzed the progenies from seven different rat intercrosses (FHH × ACI, GH × BN, LH × LN, SHR × BN, SHR × DNY, SHR × WKY, SS/MCW × BN, totaling 1250 animals) to dissect the genes for hypertension and associated end-organ diseases such as renal failure and left ventricular hypertrophy. Summing the number of phenotypes from the seven crosses, 262 phenotypes were measured, of which 67 additional traits were derived, totaling 329 phenotypes. With an average of ~200 markers for genotyping of the progenies most informative for the phenotypes of interest, approximately 210 000 genotypes were determined. We then analyzed the patterns of inheritance in all rat strains in order to find common genomic regions cosegregating (linkage) with a given phenotype (blood pressure related and 'likely determinant phenotypes'). By integrating these data and the QTLs reported by others generated in the various genetically hypertensive models into one map, we identified several common regions cosegregating with high blood pressure within the various strains. However, we did not find even one region in common in all seven crosses, illustrating the complexity of the phenotype and the consequences of heterogeneity, as well as gender and age-related consequences. As outlined before, genes linked with the phenotype 'high blood pressure' may be identified on practically every chromosome depending on the rat strains and the phenotyping protocol used. On the other hand, a cluster of QTLs within the same confidence interval suggests that the same gene might be causal for hypertension and a 'likely determinant phenotype' (pleiotrophy – one gene influencing more than one phenotype) or a tight cluster of genes influencing the same phenotype (*Figure 2*). We may assume the existence of multiple alleles for a given interval among the various rat strains harboring polymorphisms/mutations within the coding sequence. In another study, not presented here, we have used the genetic linkage map and alleles at 1000 genetic markers to determine phylogenic relationships between the different strains as a tool to predict where multiple alleles are likely to be found within the rat genome. Where multiple alleles exist, a comparison of these sequences between hypertensive and normotensive strains should enable investigators to identify mutations in the causal gene, an approach which is comparable to the approaches used in human genetic studies for the identification of the causal gene and its mutations leading to Bartter's syndrome (Simon, D.B. *et al.*, 1996), GRA (Lifton *et al.*, 1992), Liddle's syndrome (Shimkets *et al.*, 1994).

An example of the integrated map for chromosome 2 which carries at least four such 'phenotypic clusters' is shown in *Figure 2*. The clustering of loci that cosegregate with the phenotypes mean arterial pressure (MAP), diastolic blood pressure (DBP), systolic blood pressure (SBP) before and after salt-load, as well as other hypertension-related phenotypes on rat chromosome 2. It is obvious that a variety of phenotypes cluster in distinct regions of this chromosome, of which some are direct blood pressure estimates, others are kidney- or growth-related – but again showing the complex nature of hypertension. Some of the identified clusters of QTLs harbour 'positional candidate genes', such as the Na^+,K^+-$ATPase_{\alpha1}$ isoform, the calmodulin-dependent protein kinase II-delta loci ($NaK_{\alpha1}$ – CAMK)

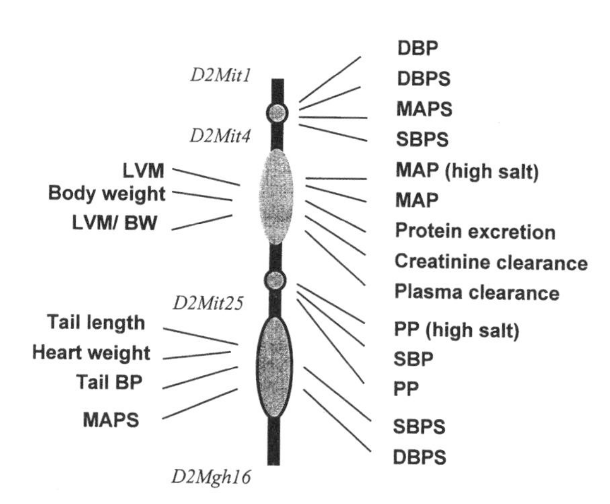

Figure 2. 'Clustering' of hypertension-related quantitative trait loci (QTLs) on rat chromosome 2. Abbreviations used in this figure: LVM (left ventricular mass), SBP (systolic blood pressure), DBP (diastolic blood pressure), BW (body weight), BP (blood pressure), MAPS, SBPS, DBPS (blood pressures after 2 weeks high salt diet (1% NaCl), PP (pulse pressure).

and the guanylyl cyclase A/atrial natriuretic peptide receptor A (GCA), which have been shown multiple times to cosegregate with high blood pressure after salt-load in various rat strains (*Table 1*). In addition to these, the genes for neutral endopeptidase/enkephalinase (NEP) and the angiotensin II AT_{1B} receptor have been mapped to this interval (Deng and Rapp, 1994). The latter genes have also been investigated for their cosegregation with blood pressure phenotypes in various rat crosses using the candidate gene approach, but both have failed to reach the threshold of significance (Deng and Rapp, 1994). Other possible candidate genes mapping to the various QTL clusters on rat chromosome 2 are fibrinogen gamma, carboxy-peptidase B and the prolactin receptor. However, the intervals spanning the QTLs are rather large and may contain a variety of as yet unknown genes, which may contribute to the regulation of blood pressure. The clusters identified do not represent absolute genetic distances because they are based on calculation of recombination in several crosses and from this an estimated distance between different loci. This estimated distance, however, differs depending on the coverage of the genome due to the availability of informative markers in the cross investigated and the chance of recombination, and, therefore, differs within the various strains. Furthermore, the data reported in the literature have been largely limited to candidate gene searches, focusing on a specific subset of genes which were previously associated with hypertension.

Leaving these problems aside, the integrated map may provide a tool to focus on specific positional candidates for hypertension for further investigation using congenic animals and positional cloning, and may provide further insight in the pathology of this disease, as well as the underlying etiology.

Given the number (~4000) of genetic markers for the rat, and a two or more standard deviations in the trait of interest between the parental strains; enough animals of the appropriate cross design; and linkage is virtually assured (Lander and Botstein, 1986, 1989). However, linkage is merely the entry point into determining the causal gene. Most of the studies published to date have focussed on selected genomic regions using candidate genes. When linkage is found to the candidate gene, the investigator next must determine if the gene is causal. As we and others have found, proving the gene is causal is a very time consuming (Koike *et al.*, 1995a,b; Simon, J.S. *et al.*, 1996) effort that more often than not turns up without the gene. A total genome search provides the investigator with positional candidate genes, which are defined as a genes within an interval that has been linked to a certain trait. Once linkage is found to a region, the next step is to look for positional candidate genes residing within the genetic locus, which may then serve as a starting point for future investigations. Therefore, the candidate gene approach and the total genomic search strategies do not exclude but complement each other. Genetic mapping studies are not capable of confirming or refuting that a gene is causal.

8. From linkage to gene identification

Cloning mammalian genes by position remains to be a challenge even for genes controlling simple Mendelian traits. Less than a hundred genes have been cloned by position, and none of these for a quantitative trait. However, in anticipation of the continuous improvement in genomic mapping tools, the task of cloning by position is becoming realistic. In the following section, we briefly outline an approach to cloning a QTL by position.

Once linkage has been confirmed in a second cross using the original parental strains and phenotyping protocol, congenic strains may be created. In a congenic animal a relatively small region (10–25 cM) of a genome from another strain has been introgressed onto its isogenic (inbred) background using a successive series of backcrosses. The advantage of the generating congenic animals is that the multifactorial trait is reduced to a simple Mendelian trait (there is only one region different). By definition it takes 10 successive backcrosses to generate an isogenic congenic animal. The time, money and effort behind making congenic animals would seem to dissuade investigators from pursuing this strategy. In mouse genetics, congenics have become a relatively common tool. Even in hypertension research using the rat, several congenic animals have been reported by Kurtz and colleagues, by Rapp and colleagues and by ourselves (Cicila *et al.*, 1997; Deng *et al.*, 1997; Frantz *et al.*, 1998; Iwai *et al.*, 1998; Jiang *et al.*, 1997; Kren *et al.*, 1997; Rapp *et al.*, 1998; St. Lezin *et al.*, 1996; St. Lezin *et al.*, 1997; Zhang *et al.*, 1997). Before discussing how congenic animals can be generated more rapidly, it is worth discussing the strategy.

Within the hypertension field there is a debate about how congenics should be made. Some researchers feel that introgression of the gene region from the normotensive strain onto the hypertensive background is all that is required. Furthermore, the belief is that this approach is most likely to result in seeing a change in baseline blood pressure (in this case a decrease). We and other researchers feel that two congenics should be made for every region – placing the normotensive interval into the hypertensive background and the hypertensive region into the normotensive background. This approach allows the investigation of how a single locus may influence blood pressure on a permissive and resistant genetic background, and facilitating the dissection of mechanisms underlying complex diseases. The generation of two congenic strains also provides a classic 2 × 2 study design; whereby for the first time physiological experiments have appropriate controls. Failure to generate both lines removes an important control and leaves open questions about the effect of genetic background. For example, Rapp *et al.* (1990) have reported the effects of different backgrounds on the expression of the renin gene. While the notion of making congenic animals is logical, how likely is it that changing one gene region in a multifactorial disease results in an observable phenotypic difference? By and large investigators are looking for a significant change in blood pressure at baseline. Our data suggests that this definition is likely to be too narrow, given the effects of gender, variations in phenotyping protocols and age-related differences in expression. Our use of many different 'likely determinant phenotypes' provides us with additional information within a given interval. Furthermore, we have found that challenging the cardiovascular system can be more powerful in illustrating defects in it.

With the advent of molecular genetics and the complete linkage map for the rat, it is possible to accelerate this process using marker assisted selection reducing the number of backcrosses to three to five generations for the construction of a congenic strain (Lounde and Thompson, 1990). In these 'speed congenics', marker assisted selection is used to increase the rate of introgression of one genetic locus onto the selected background by successive backcrosses. Genetic markers flanking the region are used to ensure that the genetic locus is not lost during introgression and, since recombination is a chance event, to determine which animals have more of the background desired. This approach has been used by many groups. The major criticism is that with the speed congenic approach is that there is an increased risk of passenger loci (the wrong background between two markers showing the correct genotypes could not be determined) being carried. While it is a formal possibility that the passenger loci could interact with the introgressed region, we do not believe the risk is so great to offset the 3-fold reduction in time and money required to build these essential reagents.

Generation of congenic rat strains will also facilitate *cloning by position*, since they reduce a complex trait to a Mendelian model and these animals provide a region containing a rather small genomic interval from a unique strain. This interval may be further narrowed by arranging additional crosses and looking for recombinant animals (animals that have exhibited a crossover between the two closest flanking markers, yet still exhibit the phenotype), which points the investigator towards the gene. If the recombinant animal loses the phenotype, then the investigator knows he/she has gone past the gene. Once the region has been adequately narrowed (~1–2 cM) a physical map (a region of the genome cloned in an ordered array) may be constructed to identify the genes in the region and to analyse its sequences for putative causal mutations. Until now there have been very few physical mapping reagents for the rat. However, thanks to European, German and U.S. Rat Genome Projects there are two 10-fold coverages of rat genome with yeast artificial chromosomes (YACs) (Coi *et al.*, 1997; Haldi *et al.*, 1997); a radiation hybrid panel and a PAC library (Woon *et al.*, 1998).

Unfortunately, even with the new physical mapping reagents, identification of the gene responsible for high blood pressure will not be easy. The major challenge is knowing when one has identified the gene in question, since it is likely that the mutations are rather subtle and, therefore, difficult to distinguish from sequence variation within the population. Recall that the congenic rats are comparing only two regions of the genome and within a 2–4 million bp region there will be thousands of sequence differences. So how does one determine which sequence variant is important? Obviously if there is a marked change in primary structure due to a mutation such as a deletion, insertion or a point mutation, identification will be easy, but what if it is a point mutation in a novel gene sequence. In human genetics, the problem is more tractable at this level because one can look at more alleles and look for a clustering of mutations within a single gene.

We have tried to capitalize on allele variants in the rat in two ways. First, we have studied seven different crosses between hypertensive and normotensive strains looking for regions in common, which would suggest the same allele has been fixed in two different strains of hypertensive rat, or mutations have occurred independently in the different strains, or that there is more than one gene capable of modifying blood pressure within the same interval. Second, we have determined the allele sizes of 48 different strains of rats for more than 2500 genetic markers. This information can be used to determine the likelihood that alleles are independent by using phylogenic analyses to determine how similar the rat strains are to one another. We believe these tools will help us to determine which regions of the rat genome will be the best targets for positional cloning.

Proving a gene is responsible is another matter. To date only one gene in the rat is likely to be causal for high blood pressure – 11β-hydroxylase in the Dahl salt-sensitive rat (SS/JR). However, this gene was not identified using a positional cloning approach, but rather a candidate gene approach combined with using information from a human genetic study (Cicila *et al.*, 1993). Yet, even in this case the final proof is lacking – transfer of a resistant or susceptible gene with the phenotype. For genes with a recessive mode of inheritance and in some case an additive mode of inheritance, transgenic models will allow the investigation of whether the transfer of a 'disease-free' allele reduces blood pressure or a 'disease-causing' allele increases blood pressure. What about genes with dominant or additive mode of inheritance? The first experiment many investigators would wish to perform is knock-out or knock-in, once this technique is available for the rat. However, a knock-out or knock-in may have difficulties with differential genetic background effects. At least four studies have reported the loss of a knock-out phenotype when moved to an isogenic background by successive backcrosses. Extensive phenotyping of the congenic animals including challenges of the cardiovascular system and the determination of 'likely determinant phenotypes', may help to identify defects in cardiovascular regulation, which are not primarily accompanied by a change in blood pressure. The availability of these phenotypes provides additional bioassays to be tested during positional cloning. Since 'likely determinant phenotypes' represent a measurable trait which defines a specific element of at least one of the pathways controlling blood pressure, mapping data from these phenotypes can be compared to that of blood pressure. When blood pressure and these other phenotypes map to the same genomic region, they may provide potential insight to gene function and therefore gene identity. In some cases the only solution may be to construct congenics with and without the candidate gene.

With the background and caveats associated with the production of congenic animals outlined above, let us review some literature. Several congenic strains have been reported that do have changes in blood pressure. Starting with a plausible candidate gene may accelerate the process, but does not necessarily lead to an immediate success in proving that this particular gene is causal for a given phenotype. The most recent publications on congenic animals carrying the renin gene from a hypertensive strain on a normotensive background (and vice versa) (Jiang *et al.*, 1997; St. Lezin *et al.*, 1996; Zhang *et al.*, 1987) confirm the foreseen difficulties in proving a candidate gene to correspond to a given QTL and to be causal for a given trait. Due to the fact that renin was repeatedly reported to cosegregate with an increased systolic blood pressure in various rat strains (Deng and Rapp, 1992; Sun *et al.*, 1993), several investigators started to construct congenic animals for the renin gene. St. Lezin *et al.* (1996) reported that in congenic Dahl R rats carrying the Dahl S renin gene and fed on a 8% high salt diet, the systolic blood pressure was significantly lower than in the progenitor Dahl R rats; the opposite of what was predicted from the original linkage data (Rapp, 1989). The lower systolic blood pressure in the congenic animals was, furthermore, accompanied by decreased kidney mRNA and decreased plasma renin levels. While these results were confusing, they were substantiated by Jiang *et al.* (1997) who constructed the other congenic line for the same region and reported that, in his studies, the transfer of an R-allele on a Dahl S background increased systolic blood pressure, as well as renin activity in their congenic animals (again the opposite of the original linkage data). Jiang *et al.* used pharmacological agents to investigate their congenics. They found that administration of captopril and losartan both reduced blood pressure after a salt load in the congenic but not SS/MCW parental line, suggesting a role of the renin–angiotensin system. Yet, Zhang *et al.* (1997) observed the predicted effect, when transferring the R-allele of the renin gene on a susceptible SS/JR background. In their study, transfer of the 'normotensive' R-allele significantly lowered blood pressure in the congenic animals compared to the progenitor susceptible SS/JR rats. Alam *et al.* (1993) have reported that the renin genes in the SR/JR and SS/JR are not different at the sequence level. The discrepancies in these studies may be due to various reasons, and give rise to the speculation that the renin gene by itself is not the causal gene for the phenotype 'high blood pressure' in these particular strains, but that there exists an as yet unknown gene or molecular variants residing near or in the renin gene in the SR/JR rat. This hypothesis is supported by the fact that in the three studies, different intervals flanking the renin gene were transferred (*Figure 3*). While in

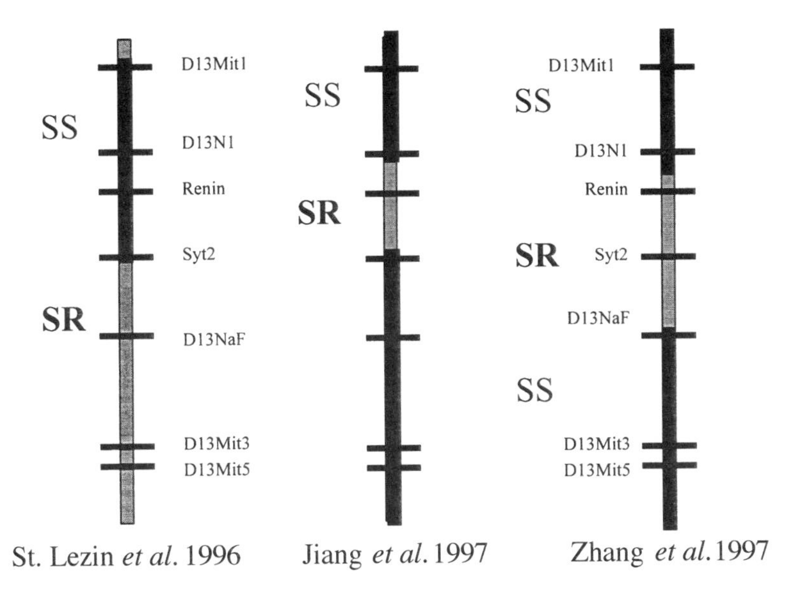

Figure 3. Maps of rat chromosome 13 in three different congenic strains developed for regions harboring the renin gene. This plot illustrates the different regions transferred on the genetic background of a different strain. St. Lezin (1996) reported a congenic strain, where the locus from a SS/JR rat was moved onto a SR/JR background, while Jiang *et al.* (1997) and Zhang *et al.* (1997) performed the opposite approach, moving the locus from a SR/JR rat on the background of a SS/JR rat. Maps are adapted from their previous publications and markers given only represent an estimate of the genetic distance being transferred.

the studies performed by St. Lezin *et al.* (1996) and Jiang *et al.* (1987), an interval containing the renin gene and 5′ flanking regions were transferred, in the study by Zhang *et al.* (1997) it was the renin gene plus predominantly the 3′ flanking region that was transferred. In all cases, it is most likely that the intervals transferred contain various additional genes, since the intervals span between 10 and ~20 million bp. The genetic length of the rat genome is ~1600 cM with 3×10^9 bp; therefore 1 cM = ~2 Mbp and 25 cM = ~50 Mbp of rat chromosome 13. The differential transfer of additional genes or regulatory elements residing in the 5′ or 3′ flanking region of the renin gene may account for the discrepant results observed in the above mentioned congenic strains. How likely is it that there is more than one gene responsible for blood pressure regulation within the confidence interval containing renin? Furthermore the phenotyping protocols used to measure blood pressure were not identical to the phenotyping protocol used to determine the initial linkage of renin to high blood pressure. Thus, it is obvious that further studies are needed to prove the identity of a given candidate gene with a given QTL involving knock-out approaches using the mouse, transgenic models and cloning by position.

9. Summary

In this chapter we have outlined the premise of molecular genetic dissection of high blood pressure. However, the approaches presented here could equally be applied to other multifactorial disorders. The major emphasis in our laboratory has been to attach various phenotypes to the genome that are viewed as being important in the regulation of blood pressure. It is perhaps surprising that in the age of high-throughput genotyping and genomics, disease discovery appears to be limited by the phenotyping both at the experimental animal model level and clinically. The production of congenic animals for many different genetic regions, positional cloning and combining congenic rats in controlled ways are likely to have far reaching consequences for biomedical research and clinical applications well into the 21st century.

Acknowledgments

This work was supported by grants to M.S. from DFG; to H.J.J. from NCRR (5R01RR08888-02), NCHGR, NHLBI (1RO1HL56284-01, 5R01HL55729-02, 1P50,HL 54998-01), NIDDK (5 RO1 DK46612-04) and sponsored research from Bristol–Myers Squibb. We are indebted to the work of many people from around the world for their contributions to this overview chapter by providing phenotyped crosses: FHH × ACI (Abraham P. Provoost and his lab in the Netherlands), GH × BN (Murray Grigor and Eugenie Harris in New Zealand), LH × LN (Nilesh Samani in the UK, Jean Sassard in France and their respective labs), SHR × BN (Jose E. Krieger and Eduardo M. Krieger and their laboratories in Brazil), SHR × DNY (Stephen B. Harrap in Australia and his lab), SHR × WKY (Morton P. Printz and Darrell Farnestil in San Diego and their respective labs) in SS/MCW

× BN (Allen W. Cowley, Jr, Andrew Greene, Richard J. Roman in Milwaukee and their respective labs) for their contribution to the SCOR project. The genotyping work and analytical work has been largely done by members of the Jacob lab (past and present): Donna Brown, Michelle Brown, Alec Goodman, Mary Granados, Jo Handley, Brendan Innes, Jian Jiang, George Koike, Anne Kwitek-Black, Rebecca Majewski, Michael McLaughlin, Vishwanathan Nadig, Marcelo Nobrega, Carole Roberts, Masahide Shiozawa, Chang Sim, Jason Simon, Maria Trolliet, and Eric Winer.

References

Alam, K.Y., Wang, Y., Dene, H. and Rapp, J.P. (1993) Renin gene nucleotide sequence of coding and regulatory regions in Dahl rats. *Clin. Exp. Hypertension* **15**: 599–614.

Barger, A.C. (1979) The Goldblatt Memorial Lecture. Part I. Experimental renovascular hypertension. *Hypertension* **1**: 447–455.

Bihoreau, M.T., Gauguier, D., Kato, N., Hyne, G., Lindpaintner, K., Rapp, J.P., *et al.* (1997) A linkage map of the rat derived from three F_2 crosses. *Genome Res.* **7**: 434–440.

Brown, D.M., Provoost, A.P., Daly, M.J., *et al.* (1996) Renal disease susceptibility and hypertension are under independent genetic control in the fawn-hooded rat. *Nature Genet.* **12**: 44–52.

Cai, L., Schalkwyk, L.C., Schoeberlein-Stehli, A., Zee, R.Y., Smith, A., Haaf, T., Georges, M., Lehrach, H. and Lindpaintner, K. (1997) Construction and characterization of a 10-genome equivalent yeast artificial chromosome library for the laboratory rat, *Rattus norvegicus*. *Genomics* **39**: 385–392.

Cicila, G.T., Rapp, J.P. and Wang, J. *et al.* (1993) Linkage of 11(-hydroxylase mutations with altered steroid biosynthesis and blood pressure in the Dahl rat. *Nature Genet.* **3**: 346–353.

Cicila, G.T., Rapp, J.P., Bloch, K.D., *et al.* (1994) Cosegregation of endothelin-3 locus with blood pressure and relative heart weight in inbred Dahl rats. *J. Hypertension* **12**: 643–651.

Cicila, G.T., Dukhanina, O.I., Kurtz, T.W., Walder, R., Garrett, M.R., Dene, H. and Rapp, J.P. (1997) Blood pressure and survival of a chromosome 7 congenic strain bred from Dahl rats. *Mamm. Genome* **8**: 896–902.

Clark, J.S., Jeffs, B., Davidson, A.O., *et al.* (1996) Quantitative trait loci in genetically hypertensive rats. Possible sex specificity. *Hypertension* **28**: 898–906.

Dahl, L.K., Heine, M. and Tassinari, L. (1962) Role of genetic factors in susceptibility to experimental hypertension due to chronic excess salt ingestion. *Nature* **194**: 480–482.

Deng, Y. and Rapp, J.P. (1992) Cosegregation of blood pressure with angiotensin converting enzyme and atrial natriuretic peptide receptor genes using Dahl-sensitive rats. *Nature Genet.* **1**: 267–272.

Deng, Y. and Rapp, J.P. (1994) Evaluation of the angiotensin II receptor AT1B gene as a candidate gene for blood pressure. *J. Hypertension* **12**: 1001–1006.

Deng, Y., Dene, H. and Rapp, J.P (1994a) Mapping of a quantitative trait locus for blood pressure on rat chromosome 2. *J. Clin. Invest.* **94**: 431–436.

Deng, Y., Dene, H. and Pravenec, M., *et al.* (1994b) Genetic mapping of two new blood pressure quantitative trait loci in the rat by genotyping endothelin system genes. *J. Clin. Invest.* **93**: 2701–2709.

Deng, A.Y., Dene, H. and Rapp, J.P. (1997) Congenic strains for the blood pressure quantitative trait locus on chromosome 2. *Hypertension* **30**: 199–202.

Dubay, C., Vincent, M., Samani, N.J., *et al.* (1993) Genetic determinants of diastolic and pulse pressure map to different loci in lyon hypertensive rats. *Nature Genet.* **3**: 354–357.

Elston, R.C. (1997) Algorithms and Inferences: The challenge of multifactorial diseases. *Am. J. Hum. Genet.* **60**: 255–262.

Ely, D.L. and Turner, M.E. (1990) Hypertension in spontaneously hypertensive rats is linked to the Y-chromosome. *Hypertension* **16**: 282–289.

Ely, D.L., Daneshvar, H., Turner, M.E., *et al.* (1993) The hypertensive Y-chromosome elevates blood pressure in F11 normotensive rats. *Hypertension* **21**: 1071–1075.

Frantz, S.A., Kaiser, M., Gardiner, S.M., Gauguier, D., Vincent, M., Thompson, J.R., Bennet, T. and Samani, N.J. (1998) Successful isolation of a rat chromosome 1 blood pressure quantitative trait locus in reciprocal congenic strains. *Hypertension* **32**: 639–646.

Goldblatt, H., Lynch, J., Hanzal, R.F., *et al.* (1934) Studies on experimental hypertension: production of persistent evaluation of systolic blood pressure by means of renal ischemia. *J. Exp. Med.* **59**: 347–379.

Goldmuntz, E.A., Remmers, E.F., Zha, H., Cash, J.M., Mathern, P., Crofford, L.J. and Wilder, R.L. (1993) Genetic map of 12 polymorphic loci on rat chromosome 1. *Genomics* **16**: 761–764.

Group HDaF-UPC (1977) The hypertension detection and follow-up program: a progress report. *Circ. Res.* **40**: I-106–I-109.

Gu, L., Dene, H., Deng, A.Y., Hoebee, B., Bihoreau, M.T., James, M. and Rapp, J.P. (1996) Genetic mapping of two blood pressure quantitative trait loci on rat chromosome 1. *J. Clin. Invest.* **97**: 777–788.

Haldi, M.L., Lim, P., Kaphingst, K., *et al.* (1997) Construction of a large-insert yeast artificial chromosome library of the rat genome. *Mamm. Genome* **8**: 284.

Haley, C.S. and Knott, S.A. (1992) A simple regression method for mapping quantitative trait loci in line crosses using flanking markers. *Heredity* **69**: 315–324.

Hamet, P., Kong, D., Pravenec, M., *et al.* (1992) Restriction fragment length polymorphism of hsp70 gene, localized in the RT1 complex, is associated with hypertension in spontaneously hypertensive rats. *Hypertension* **19**: 611–614.

Harris, E.L., Dene, H. and Rapp, J.P. (1993) SA gene and blood pressure cosegregates using Dahl salt-sensitive rats. *Am. J. Hypertension* **6**: 330–334.

Harris, E.L., Phelan, E.L., Thompson, C.M., *et al.* (1995) Heart mass and blood pressure have separate genetic determinants in New Zealand genetically hypertensive (GH) rats. *J. Hypertension* **13**: 397–404.

Hilbert, P., Lindpaintner, K., Beckmann, J.S., *et al.* (1991) Chromosomal mapping of two genetic loci associated with blood pressure regulations in hereditary hypertensive rats. *Nature* **353**: 521.

Iwai, N., Tsujita, Y. and Kinoshita, M. (1998). Isolation of a chromosome 1 region that contributes to high blood pressure and salt sensitivity. *Hypertension* **32**: 636–638.

Jacob, H.J., Lindpaintner, K., Lincoln, S.E., *et al.* (1991) Genetic mapping of a gene causing hypertension in the stroke-prone spontaneously hypertensive rat. *Cell* **67**: 213–224.

Jacob, H.J., Brown, D.M., Bunker, R.K., *et al.* (1995) Genetic linkage map of the laboratory rat, *Rattus norvegicus*. *Nature Genet.* **9**: 63–69.

Jiang, J., Stec, D., Drummond, H., *et al.* (1997) Transfer of a salt-resistant renin allele raises blood pressure in Dahl salt-sensitive rats. *Hypertension* **2**: 619–627.

Kapusczinski, M., Charchar, F., Mitchell, G.A., *et al.* (1994) The nerve growth factor gene and blood pressure in the spontaneously hypertensive rat (abstract). *J. Hypertension* **12(suppl.3)**: S191.

Katsuya, T., Higaki, J. and Zhao, Y., *et al.* (1993) A neuropeptide Y locus on chromosome 4 cosegregates with blood pressure in the spontaneously hypertensive rat. *Biochem. Biophys. Res. Commun.* **192**: 261–267.

Koike, G., Jacob, H.J., Krieger, J.E., *et al.* (1995a) Investigation of the phenylethanolamine *N*-methyltransferase gene as a candidate gene for hypertension. *Hypertension* **26**: 595–601.

Koike, G., Winer, E.S., Horiuchi, L., *et al.* (1995b) Cloning, characterization and genetic mapping of the rat type 2 angiotensin receptor gene. *Hypertension* **26**: 998–1002.

Koike, G., Jacob, H.J., Cowley, A.W., *et al.* (1997) The comprehensive total genome scan to map genes responsible for high blood pressure in the rat. *FASEB J.* **11**: 2710.

Kovacs, P., Voigt, B. and Kloting, I. (1997a) Novel quantitative trait loci for blood pressure and related traits on rat chromosomes 1, 10, and 18. *Biochem. Biophys. Res. Commun.* **235**: 343–348.

Kovacs, P., Voigt, B. and Kloting, I. (1997b) Alleles of the spontaneously hypertensive rat decrease blood pressure at loci on chromosomes 4 and 13. *Biochem. Biophys. Res. Commun.* **238**: 586–589.

Kren, V., Pravenec, M., Lu, S., *et al.* (1997) Genetic isolation of a region on chromosome 8 that exerts major effects on blood pressure and cardiac mass in the spontaneously hypertensive rat. *J. Clin. Invest.* **99**: 577–581.

Kreutz, R., Huebner, N., James, M.R., *et al.* (1995a) Dissection of a quantitative trait locus for genetic hypertension on rat chromosome 10. *Proc. Natl Acad. Sci. USA* **92**: 8778–8782.

Kreutz, R., Hubner, N., Ganten, D. and Lindpaintner, K. (1995b) Genetic linkage of the ACE gene to plasma angiotensin-converting enzyme activity but not to blood pressure. *Circulation* **92**: 2381–2384.

Kreutz, R.. Kreutz, R., Struk, B., Rubattu, S., Hubner, N., Szpirer, J., Szpirer, C., Ganten, D. and Lindpaintner, K. (1997) Role of the alpha-, beta-, and gamma-subunits of epithelial sodium channel in a model of polygenic hypertension. *Hypertension* **29**: 131–136.

Krugliak, L. and Lander, E.S. (1995) Complex multipoint sib-pair analysis of qualitative and quantitative traits. *Am. J. Hum. Genet.* **57**: 439–454.

Kurtz, T., Simonet, L., Kabra, P.M., ***et al.*** (1990) Cosegregation of the renin allele of the spontaneously hypertensive rat with an increase in blood pressure. *J. Clin. Invest.* **85**: 1328–1332.

Lande, R. and Thompson, R. (1990) Efficiency of marker-assisted selection in the improvement of quantitative traits. *Genetics* **124**: 743–756.

Lander, E.S. and Botstein, D. (1986) Mapping complex genetic traits in humans: new methods using a complete RFLP linkage map. *Cold Spring Harb. Symp. Quant. Biol.* **1**: 49–62.

Lander, E. and Botstein, D. (1989) Mapping mendelian factors underlying quantitative traits using RFLP linkage maps. *Genetics* **121**: 185–199.

Lander, E.S. and Schork, N.J. (1994) Genetic dissection of complex diseases. *Science* **265**: 2037–2048.

Levan, G., Szpirer, J., Szpirer, C., Klinga, K., Hanson, C. and Islam, M.Q. (1991) The gene map of the Norway rat (*Rattus norvegicus*) and comparative mapping with mouse and man. *Genomics* **10**: 699–718.

Levan, G., Klinga-Levan, K., Spzirer, C., ***et al.*** (1992) Gene map of the rat (*Rattus norvegicus*). In: *Locus Maps of Complex Genomes* 6th ed (ed. S.J. O'Brien). Cold Spring Harbor Press.

Lifton, R.P., Dluhy, R.G., Powers, M., Rich, G.M., Cook, S., Ulick, S. and Lalouel, J.M. (1992) A chimaeric 11β-hydroxylase/aldosterone synthase gene causes glucocorticoid-remediable aldosteronism and human hypertension. *Nature* **355**: 262–265.

Lindpaintner, K., Hilbert, P., Ganten, D., ***et al.*** (1993) Molecular genetics of the SA-gene: cosegregation with hypertension and mapping to rat chromosome 1. *J. Hypertension* **11**: 19–23.

Matsumoto, C., Nara, Y., Ikeda, K., Tamada, T., Mashimo, T., Nabika, T., Sawamura, M. and Yamori, Y. (1996) Cosegregation of the new region on chromosome 3 with salt-induced hypertension in female F2 progeny from stroke-prone spontaneously hypertensive and Wistar–Kyoto rats. *Clin. Exp. Pharmacol. Physiol.* **23**:1028–1034.

Nakajima, S., Rioseco, N., Ma, L., ***et al.*** (1994) Candidate gene loci for hypertension in LaJolla colony SHR (lj) and WKY (lj) on chromosomes 2 and 4 (abstract). *J. Hypertension* **12(suppl. 3)**: S66.

Nara, Y., Nabika, T., Ikeda, K., ***et al.*** (1991) Blood pressure cosegregates with a microsatellite of angiotensin I converting enzyme (ACE) in F2 generation from across between original normotensive Wistar Kyoto (WKY) rat and stroke-prone spontaneously hypertensive rat (SHRSP). *Biochem. Biophys. Res. Commun.* **181**: 941–946.

Okamoto, K. and Aoki, K. (1963) Development of a strain of spontaneously hypertensive rats. *Jpn Circ. J.* **27**: 282–293.

Pravenec, M., Kren, V., Kunes, J., ***et al.*** (1991) Cosegregation of blood pressure with a kallikrein gene family polymorphism. *Hypertension* **17**: 242–246.

Pravenec, M., Gauguier, D., Schott, J.J., ***et al.*** (1995) Mapping of quantitative trait loci for blood pressure and cardiac mass in the rat by genome scanning of recombinant inbred strains. *J. Clin. Invest.* **96**: 1973–1978.

Rapp, J.P., Wang, S.M. and Dene, H. (1989) A genetic polymorphism in the renin gene of Dahl-sensitive rats cosegregates with blood pressure. *Science* **243**: 542–544.

Rapp, J.P., Wang, S. and Dene, H. (1990) Effect of genetic background on cosegregation of renin alleles and blood pressure in Dahl rats. *Am. J. Hypertension* **3**: 391–396.

Rapp, J.P., Garrett, M.R. and Deng, A.Y. (1998). Construction of a double congenic strain prove an epistatic interaction on blood pressure between rat chromosomes 2 and 10. *J. Clin. Invest.* **101**: 1591–1595.

Samani, N., Lodwick, D., Vincent, M., ***et al.*** (1993) A gene differentially expressed in the kidney of the spontanaously hypertensive rat cosegregates with increased blood pressure. *J. Clin. Invest.* **92**: 1099–2005.

Samani, N.J., Gauguier, D., Vincent, M., ***et al.*** (1996) Analysis of quantitative trait loci for blood pressure on rat chromosomes 2 and 13. *Hypertension* **28**: 1118–1122.

Schork, N.J., Krieger, J.K., Trolliet, M., *et al.* (1995) A biometrical genome search in rats reveals the multigenic basis of blood pressure variation. *Genome Res.* 5: 164–172.

Schork, N.J., Nath, S.P., Lindpaintner, K.L., *et al.* (1996) Extensions to QTL mapping in experimental organisms. *Hypertension* **28**: 1104–1111.

Serikawa, T., Montagutelli, X., Simon-Chazottes, D. and Guenet, J.L. (1992) Polymorphisms revealed by PCR with single, short-sized, arbitrary primers are reliable markers for mouse and rat gene mapping. *Mamm. Genome* **3**: 65–72.

Shimkets, R.A., Warnock, D.G., Bositis, C.M., *et al.* (1994) Liddle's syndrome: heritable human hypertension caused by mutations in the β subunit of the epithelial sodium channel. *Cell* **79**: 407–414.

Simon, J.S., Roman, R.J., Cowley, A.J., Greene, A. and Jacob, H.J. (1995) Rat alpha-2 adrenergic receptor class III is linked to systolic blood pressure after salt load in genetic hypertension. *Hypertension* **26**: 537 (abstract).

Simon, D.B., Karet, F.E., Hamdan, J.M., *et al.* (1996) Barrter's syndrome, hypokalaemic alkalosis with hypercalciuria, is caused by mutations in the Na–K–2Cl cotransporter NKCC2. *Nature Genet.* **13**: 183–188.

Simon, J.S., Deshmukh, G., Couch, F.J., *et al.* (1996) Chromosomal mapping of the rat Slc4a family of anion exchange genes, Ael, Ae2, and Ae3. *Mamm. Genome* **7**: 380–382.

Smirk, F.H. and Hall, W.H. (1958) Inherited hypertension in rats. *Nature* **182**: 727–728.

St. Lezin, E., Pravenec, M., Wong, A.L., *et al.* (1996) Effect of renin gene transfer on blood pressure and renin gene expression in a congenic strain of Dahl-resistant rats. *J. Clin. Invest.* **97**: 522–527.

St. Lezin, E., Liu, W., Wang, J.M., *et al.* (1997) Genetic isolation of a chromosome 1 region affecting blood pressure in the spontaneously hypertensive rat. *Hypertension* **30**: 854–859.

Sun, L., McArdle, S., Chun, M., *et al.* (1993) Cosegregation of the renin gene with an increase in mean arterial blood pressure in F2 rats of SHR–WKY cross. *Clin. Exp. Hypertension* **15**: 797–805.

Tanase, H., Suzuki, Y., Ooshima, *et al.* (1970) Genetic analysis of blood pressure in spontaneously hypertensive rats. *Jpn Circ. J.* **34**: 1197–1212.

Woon, P.Y., Osoegawa, K., Kaisaki, P.J., *et al.* (1998) Construction and characterization of a 10-fold genome equivalent rat P1-derived artificial chromosome library. *Genomics* **50**: 306–316.

Wright, S. (1968) The Genetics of Quantitative Variability. In: *The Genetics of Human Populations: A Treatise in Four Volumes.* University of Chicago Press, Chicago, pp. 373–420.

Yagil, C., Sapojnikov, M., Kreutz, R., Katni, G., Lindpaintner, K., Ganten, D. and Yagil, Y. (1998) Salt susceptibility maps to chromosomes 1 and 17 with sex specificity in the Sabra rat model of hypertension. *Hypertension* **31**: 119–124.

Yamada, J., Kuramoto, T. and Serikawa, T. (1994) A rat genetic linkage map and comparative maps for mouse and human homologous genes. *Mamm. Genome* **5**: 63–83.

Yamori, Y. (1984) Development of the spontaneously hypertensive rat (SHR) and of various spontaneous rat models, and their implications. In: *Handbook of Hypertension, Vol 4: Experimental and Genetic Models of Hypertension* (ed. W. Jong). Elsevier, Amsterdam, pp. 224–239.

Zha, H., Wilder, R.L., Goldmuntz, E.A., Cash, J.M. Crofford, L.J., Mathern, P. and Remmers, E.F. (1993) Linkage map of 10 polymorphic markers on rat chromosome 2. *Cytogenet. Cell Genet.* **63**: 117–123.

Zhang, Q., Dene, H., Jacob, H.J., *et al.* (1997) Interval mapping and congenic strains for a blood pressure QTL on rat chromosome 13. *Mamm Genome* **9**: 636–641.

3

Genetic and congenic mapping of loci for blood pressure and blood pressure-related phenotypes in the rat

Margit Knoblauch, Berthold Struk, Speranza Rubattu and Klaus Lindpaintner

1. Introduction

For the last 10 years, rat models of hypertension have become an important proving ground to test the feasibility of large-scale genetic screening approaches to detect quantitative trait loci that contribute to complex, multifactorial disease. The experience gained in studying these strains has significantly advanced our understanding of polygenic, complex disease, as well as of the difficulties of delineating the underlying, causative gene variants. This chapter describes major strategies currently used in rat genetics and discusses their application to genetic dissection of human hypertension. (See also Chapter 2 which addresses complementary issues).

2. The rat as a model organism for complex, polygenic disease

Historical differences in the orientation of scientific discovery that was pursued in mice and rats, respectively, had profound effects on the tools and resources that were developed for and from the two species over the last century. Scientific work in the mouse has traditionally had a strong focus on genetics, and genetic investigation in the mouse has been pursued with great sophistication. Because meaningful genetic experimentation was more or less restricted to monogenic

Molecular Genetics of Hypertension, edited by A.F. Dominiczak, J.M.C. Connell and F. Soubrier.

traits for most of this period, an extensive repository of single gene mutants, displaying categorical phenotypes, has been developed in the mouse, using specific programmed breeding schemes. While interest in models of complex disease in mice has increased lately (primarily diabetes, obesity, and cancer; Frankel, 1995), it was the availability of monogenic disease models through which mouse genetics made its most significant contributions to our understanding of mammalian genetics and development. In contrast, inherited disease models arising in the rat were primarily used in physiological research, and genetic aspects received little attention. The focus on physiology traditionally embraced by rat research has generated a wealth of experience and methodological sophistication for the accurate determination of quantitative phenotype measurements, such as blood pressure, that take advantage of the physical size of the rat compared to the mouse. Much of our current understanding of integrative physiology is based on these studies in rats. The comparatively little interest in, and relative lack of sophistication applied to genetics in the rat until recently, has, unwittingly and by serendipity, provided us with a number of disease models that bear particular relevance to human complex disease. These strains, which were created by selective inbreeding, but without further genetic manipulation, such as back-breeding onto the normotensive reference stain, represent examples of polygenic, multifactorial disease causation that resembles the human situation conceptually much more closely than the – for the detection of a particular gene more desirable – single gene disease models.

Thus, a substantial number of genetic models for a broad spectrum of complex pathological traits, including hypertension and a number of cardiovascular disorders, have been developed. These models have previously been used extensively – and until recently, almost exclusively – for comparative studies, by contrasting the disease strain with a strain that does not display the phenotype of interest, or that had been selection-bred for the opposite extreme of a quantitative phenotype. Most of this work is difficult or impossible to interpret: given both the confounding presence of many trait-unrelated differences among any two strains compared and the innate inability of association studies that use intermediate phenotypes (rather than direct genetic information) to determine causality, the major and lasting accomplishment of these studies may be the development and validation of a set of highly refined tools and methodologies for the precise determination of quantitative phenotypes such as blood pressure. As molecular genetic approaches have become more sophisticated and powerful, the ability to carry out highly accurate phenotype determination has long become one of the major limiting factors in quantitative trait mapping: while in monogenic disorders crude, categorical phenotype determinations usually suffice, the polygenic nature of complex diseases makes ultimate precision in characterizing the quantitative phenotype essential. Thus, these disease strains, along with the methods for high-precision phenotype measurements are uniquely suited for the investigation of polygenic disease genetics and the discovery of relevant genes; maybe even more importantly, they serve as complex (but, in comparison to humans, still reductionist) model systems for the testing of concepts and investigational approaches for similar diseases in man.

3. Genetic analysis of rat hypertension using intermediate phenotype markers

Although the futility of simple comparisons between inbred animal models and their 'control' strains has been pointed out for some time (Lindpaintner *et al.*, 1992; Rapp, 1983), only recently has there been an increased trend towards the application of genetically based strategies in experimental hypertension research, concomitant with the developments of more powerful molecular genetic tools. Early cosegregation studies in the rat consisted of experiments in which the co-inheritance of elevated blood pressure, and another phenotype trait that had been found to distinguish hypertensive from normotensive strains, was investigated. As pointed out above, this strategy, which infers a causal relationship between blood pressure and the 'intermediate' phenotype, presupposes that the latter is genetically controlled (i.e. that the phenotype represents an 'indirect' genotype as elaborated above; an assumption that could today be verified by performing a DNA-marker-based linkage study for the 'intermediate' phenotype). Among phenotypic traits which have been shown to cosegregate with blood pressure are a variety of physiological characteristics such as increased vascular smooth muscle responses to cations (Rapp, 1982) and to ouabain (Bruner *et al.*, 1986a), enhanced oscillatory activity of mesenteric resistance arteries (Mulvany, 1988) and sympathetic nerve activity (Judy *et al.*, 1979), decreased renal blood flow and glomerular filtration rate (Harrap and Doyle, 1988), and increased heart rate response to stress (Casto and Prinz, 1988) in the spontaneously hypertensive rat (SHR), augmented noradrenaline-induced oscillatory activity in tail arteries (Bruner *et al.*, 1986b) and increased lymphocyte potassium efflux (Furspan *et al.*, 1987) in the stroke-prone SHR (SHRSP), and increased red blood cell Na^+/K^+ co-transport (Ferrari *et al.*, 1987) in the Milan hypertensive strain. In addition, cosegregation of blood pressure and certain biochemical characteristics has been demonstrated, such as increased adrenal production of 18OH-DOC (18-hydroxy-11-deoxycorticosterone; Rapp and Dahl, 1976) in Dahl rats, renin–isoform patterns (Sessler *et al.*, 1986) in SHRSP, and esterase-4 isoforms (Yamori *et al.*, 1972) as well as nonspecific zymogram characteristics (Yamori and Okamoto, 1970) in the SHR. In contrast, independent segregation for blood pressure and salt appetite (Yongue and Myers, 1988), sodium balance (Harrap, 1986), and activity score (Casto and Prinz, 1988) has been documented in SHR (for a more extensive review and critique of this work see Rapp, 1991). Since, as pointed out above, it is not possible to prove that a cosegregating 'intermediate' phenotypic trait is inherited and thus causally related to hypertension, rather than representing simply a secondary phenomenon caused by elevations of blood pressure, the ultimate meaning of these studies, many of which were also conducted with inadequate statistical power, remains unclear. (It is intriguing, though, that a relatively large number of these studies found cosegregation with, in broad terms, cell membrane abnormalities, although exceptions exist (McLaren *et al.*, 1993); this might point either to a common, inherited etiopathological mechanism, or a rather homogeneous secondary effect of high blood pressure on biological membranes.)

A unique genetic experiment conducted prior to the dawn of the application of molecular genetics to hypertension research deserves special mention: to interrogate the potential role of (the) vasopressin (gene) in blood pressure elevations in the SHRSP, a congenic line was established where the diabetes insipidus phenotype (as carried by the Brattleboro rat) was introgressed onto an SHRSP background by selection for both morbid phenotypes over several generations. The resultant strain, termed SHRSP-DI, displayed both hypertension and diabetes insipidus (Ganten *et al.*, 1983), indicating that the vasopressin gene played no role in SHRSP hypertension; while, at the time, this could not be stated with absolute certainty (because the molecular defect responsible for diabetes insipidus in the rat was not known, and could, in theory, have been located elsewhere within the vasopressin system pathway), subsequent evidence for a point mutation in the vasopressin gene (Schmale and Richter, 1984) as a cause of the diabetes insipidus phenotype revealed that this was indeed the proper interpretation of this experiment.

4. Molecular genetic studies

With the advent of modern molecular genetics, new and powerful tools have become available that have obviously also found their application in the field of experimental animal research on hypertension in the rat. While the range of materials and resources available today for molecular genetic work in this species are still limited in comparison to mouse or human, a concerted effort currently undertaken on both sides of the Atlantic is expected to remedy this situation in the near future. The number of available microsatellite markers is now close to 1000, and the first large insert genomic library has recently gone on-line.

As in any other field of genetic investigation, two principal, different, yet complementary approaches have been used: one that tests known genes suspected to play a role in the pathogenesis of hypertension and vascular disorders, and another that makes no assumptions about involved genes but uses an *a priori* genetic linkage approach to screen the entire genome for, in first approximation, regions that are linked to the phenotype and that therefore must contain genes that are causally involved in its occurrence or modulation.

5. Candidate gene studies on hypertension

Among the candidate genes for which cosegregation studies have been reported using various systems of polymorphism detection in a variety of intercross and back-cross populations are the genes coding for renin (Dubay *et al.*, 1993; Kurtz *et al.*, 1990; Rapp *et al.*, 1989), atrial natriuretic peptide, angiotensin-converting enzyme (ACE) (Deng and Rapp, 1992), 11-beta-hydroxylase (Ciala *et al.*, 1993), kallikrein (Pravenec *et al.*, 1991), the heat shock protein *hsp70* (Hamet *et al.*, 1992), phospholipase C-delta 1 (Katsuya *et al.*, 1992), and the SA gene (Iwai and Inagami, 1992; Iwai *et al.*, 1992; Lindpaintner *et al.*, 1993; Samani *et al.*, 1993).

It is important to understand that while these findings represent positive linkage results, the inherent lack of precision of such studies in polygenic diseases allows, strictly speaking, only inferences about a relatively large chromosomal segment in which these candidate genes are localized. Thus, as long as sequence variants of these genes have not been found and functionally characterized – which, at this writing has not been accomplished for any of them – these genes can at best tentatively be regarded as the 'likely candidates': the broad array of pathways that could be envisioned as contributing to a complex disorder such as hypertension makes it likely that more than one potential 'candidate' gene will be present in any chromosomal region implicated. This holds true also for the two most intensely studied candidate genes, renin and the SA-gene, for which more extensive information from the study of several independent experimental crosses is available.

A restriction fragment length polymorphism (RFLP) from the renin gene (the marker used is a tandem repeat element in intron 1) was the first DNA-based polymorphism reported to show linkage to blood pressure in a salt-stressed intercross of salt-sensitive and salt-resistant Dahl-JR rats (Rapp *et al.*, 1989). A second investigation, performed in a hybrid cohort bred from SHR and Lewis rats, demonstrated elevated blood pressures in F_2 animals heterozygous for the renin allele, but similar low blood pressures in those homozygous for either the SHR or the Lewis allele, a finding interpreted as being, possibly, due to the effects of a recessive, hypotensive gene in linkage disequilibrium with the renin polymorphism (Kurtz *et al.*, 1990). A concomitant study in hybrids derived from Wistar–Kyoto (WKY) rats and SHRSP, in contrast, showed no cosegregation of the renin-locus and baseline or sodium-stimulated blood pressure (Lindpaintner *et al.*, 1990). Lastly, a cosegregation study in a cross generated from the Lyon hypertensive and normotensive strains again found evidence for linkage between a polymorphic marker at the renin locus and diastolic blood pressure (Dubay *et al.*, 1993). These data indicate that renin, or perhaps more likely a closely linked gene (no sequence differences have been found in the coding region and the 5′-untranslated region among Dahl SS/JR and Dahl SR/JR rats; Alam *et al.*, 1993), may indeed play a role in the pathogenesis of hypertension in selected rat strains, and emphasize the fact that depending on the particular genetic constellation, linkage to a gene may or may not be detectable. This variable etiopathological importance a given gene (mutation) may have, depending on the genetic reference it is compared with, is also illustrated by the finding of linkage between blood pressure and the ACE locus in a cross involving the Dahl SS/JR rat and the Milan normotensive rat, but not when the same hypertensive strain was crossed with the WKY rat, nor in several additional cross-bred cohorts involving a variety of different strains (Deng and Rapp, 1992).

The SA gene represents a somewhat unusual case of a candidate gene: while the nature and function of the encoded protein (the gene product) is still obscure, the gene (originally discovered based on differential hybridization techniques applied to cDNA libraries prepared from the kidneys of hypertensive SHR and normotensive Sprague–Dawley animals) shows significantly enhanced levels of expression in the kidneys of SHR as compared to the normotensive rats (Iwai and Inagami, 1991). Its potential contribution to rat hypertension has been implied based on the demonstration of cosegregation of a genomic polymorphism of the SA locus with blood pressure in independent F_2-cohorts derived from SHR/WKY (Iwai and Inagami, 1992; Iwai *et al.*, 1992; Samani *et al.*, 1993), SHRSP/WKY

(Lindpaintner *et al.*, 1993) and Dahl SS/Lewis (Harris *et al.*, 1993) crosses. Similar to previously cited examples, the makeup of the cross also impacts on the biological role of the SA gene: in a Dahl SS/WKY cross, no effect could be shown (Harris *et al.*, 1993). As much as these discrepancies may appear confusing, they reflect most likely genetic differences between strains that are nominally the same, but have been reared apart for years, often decades (Kurtz and Morris, 1987; Kurtz *et al.*, 1989; Louis and Howes, 1990; Samani *et al.*, 1990). Parenthetically, it also illustrates the need for precise definition of strains by laboratory or breeder, and for longitudinal quality control at each breeder to avoid gross breeding mishaps (fully realizing that spontaneously occurring mutations will be missed – unless they affect obvious phenotype parameters).

It should be emphasized that the limited power of candidate gene studies in polygenic experimental crosses is in contrast to the (potential for) considerable power of similar studies in humans: since meiotic recombination is limited to one generation in the rat experiments, linkage disequilibrium remains robust across relatively large stretches of chromosomal regions, thus, the power of these studies to discriminate among neighboring genes within a chromosomal region is comparatively poor. In outbred populations, for example most human samples, linkage disequilibrium is much more limited, yielding much higher localizing power; a positive study therefore implicates, in essence, only the gene from which the marker is derived.

6. Genome screening in rat hypertension

The genome-scanning or -screening approach for the dissection of polygenic traits is based on the principle of linkage analysis, using a large number of genetic markers that are polymorphic (that is, informative among the strains used to create a cross, and distributed as evenly as possible across all chromosomes). By testing each of these markers for cosegregation with the trait of interest the entire genome is evaluated, and chromosomal regions likely (within the confines of statistical error) to contain a gene contributing to trait variance can be identified. This approach has the advantage of being unbiased and nonparametric, but carries the burden of being based on a multiplicity of comparisons; thus, very small type 1 errors are required to declare linkage.

By now, a substantial number of such genome screens in different rat crosses have been completed. The approach was first applied to the SHRSP, a classical model strain for polygenic, multifactorial hypertension (Hilbert *et al.*, 1991; Jacob *et al.*, 1991). A large population of F_2 animals was bred by mating, brother-to-sister, F_1-progeny derived from cross-breeding SHRSP with a normotensive reference strain, the WKY rat. The F_2 animals underwent extensive characterization of a number of hemodynamic and morphometric phenotype parameters, such as blood pressure, heart rate and ventricular mass. Mini- and microsatellite markers were used for genetic characterization of the cross. Both are repetitive elements, the former between 10 and 100 base pairs, the latter 2–4 base pairs in length per repeat element, which show considerable less stability than nonrepetitive DNA, and therefore display a considerable degree of polymorphism among individual strains. They are visualized by different techniques, Southern blotting for the former (the initial, classic DNA fingerprinting method pioneered by Sir Alec Jeffreys *et al.*, 1985), size fractionation on sequencing gels and autoradiographic or fluorescent visualization for the latter (Weber and May, 1989). The first genome screens were performed with limited numbers of markers; since then, the repertoire – almost exclusively in microsatellites – has been greatly expanded (Jacob *et al.*, 1995; Serikawa *et al.*, 1992) and now comprises close to 1000 markers; still considerably short of what is available for other species, such as mouse or man. A concerted effort of developing rat genomic resources currently under way in both Europe and the USA is expected to remedy this situation within the next few years.

The initial genome screen in the SHRSP/WKY cross revealed three chromosomal loci which showed LOD scores in excess of 3.0 and thus fulfilled the generally accepted criteria for significant linkage. These regions were located on chromosome 10, 18 and X (Hilbert *et al.*, 1991; Jacob *et al.*, 1991). The locus on chromosome 10, termed *BP/SP–1*, identified by a microsatellite, gave the highest LOD score (5.3) and appears to account for about 20% of the blood pressure variance encountered in the F_2 population. Intriguingly, this locus is part of a linkage group which is homologous to a region on the long arm of human chromosome 17, which also contains the gene coding for ACE. Whereas subsequent analysis using a mouse microsatellite marker for ACE verified the location of this gene on rat chromosome 10, identifying ACE as a possible candidate gene, later, more sophisticated studies using congenic strains (see below) have recently provided strong evidence against ACE as the gene responsible for the *BP/SP–1* effect (Kreutz *et al.*, 1995), emphasizing the limited resolution of genetic mapping in complex disease. While the SHRSP allele at both the chromosome 10 and 18 loci confers higher blood pressures, the opposite was found for the locus on the X-chromosome. The hypertensive effect of this locus associated with the WKY allele may initially appear counterintuitive, but is certainly consistent with the nature of complex disease where individual gene effects are expected to differ not only in magnitude, but also in the direction with which they affect the trait of interest.

In fact, it is highly unlikely that a hypertensive strain would carry all possible blood pressure-raising gene variants (this would in all likelihood render such a strain biologically unfit and preclude its very existence). Rather, it is simply the net effect of all blood pressure-raising and all blood pressure-lowering alleles that ultimately determines phenotype; with an excess of blood pressure-raising alleles present in hypertensive animals, and a balance of pressure-raising and -lowering alleles (or a relative lack of both) in normotensive strains. This has important implications: depending on the strain examined, different hypertensive alleles (i.e. either blood pressure-raising molecular variants of different genes, or different molecular variants of the same gene) may be present; concomitantly, depending on the normotensive 'reference strain' that is used for the mapping cross, different results may be obtained since only those alleles that differ among the two strains can be mapped. Thus, if A, B, C, D are the blood pressure-raising alleles of genes a, b, c, d that are present in a given hypertensive strain, then a cross with a normotensive strain characterized by A, B, c, d, will only reveal genes c and d, while a cross with a phenotypically identical second normotensive strain characterized by a, B, c, D can of course only reveal a and c as genes with molecular variants that are causally involved in the pathogenesis of hypertension. The large number of available hypertensive strains, and the even larger number of normotensive strains therefore, represent an important resource that will allow us to capture at much greater breadth and depth the diversity of genes/alleles with a potential of raising blood pressure than would otherwise be possible.

Since the original study described above, a large number of studies using the genome screening approach in rat hypertension has appeared in the literature. A number of them, using the salt-sensitive substrain of the Dahl/JR rat as the hypertensive partner crossed with a series of different normotensive strains (Deng and Rapp, 1992, 1995; Deng *et al.*, 1994a,b), have indeed used the above-described approach to scan more thoroughly for the entirety of blood pressure raising alleles present in this strain. In such crosses, both consistent and inconsistent areas of genetic linkage have been found, just as expected. By now, taking the results of all available studies together, apparent blood pressure relevant loci have been identified on at least half of all chromosomes. While these results demonstrate the feasibility of a reverse genetic approach using random markers for the investigation of quantitative, polygenic traits in mammals, the results have also painfully demonstrated the limitations of this approach, particularly in a species where genomic resources have only recently begun to become available. Currently, much of the follow-up on the initial genome-screening work is to some extent on hold, awaiting the generation of informative congenic strains, and the availability of genomic resources for positional cloning efforts.

7. Congenic strains as a tool for high resolution mapping

Improving on the resolution offered by genetic mapping in polygenic disease models could, theoretically, be achieved with extremely large crosses but requires, for practical purposes, the conversion of polygenic into monogenic traits. This is achieved by creating congenic strains which represent a genetic composite of the disease strain and the reference strain such that the congenic strain is, ideally, identical to the reference strain except for a (single) chromosomal region of interest that is derived from the disease strain and has been grafted onto the reference strain background. In practice, this is achieved by back-crossing F_1 hybrids onto the wild-type or reference strain, followed by repeated rounds of similar back-crossing of the resultant progeny, always onto pure-bred reference strain animals. This results in a reduction by a factor of 2 of the genetic material derived from the disease strain in favor of the reference strain per round of back-crossing, thus diluting out the contribution of disease strain-derived genetic material in the hybrids. Meanwhile, as long as back-cross animals used for subsequent rounds of breeding continue to show the phenotype of interest, the relevant chromosomal region will still be disease strain ('donor' strain)-derived. After round 1, the donor strain-derived genetic material that amounted to 50% in the F_1 hybrids will have been reduced to 25%, by round 2 to 12.5%, and so on, until by round 10 – the number classically used for these experiments – only 0.05% of the genome will still be derived from the disease strain. For all practical purposes, then, these animals will differ from the wild-type only with regard to the region of interest, but will be otherwise genetically identical to the reference (or host) strain. As recombinations occur in the neighborhood of the disease gene during the back-crossing, the region of interest will become better and better defined, and restricted to an ever smaller chromosomal region: by comparing the position of crossover events which result in loss of the phenotype with those that continue to display it, one can determine the essential, minimal chromosomal region that is necessary to preserve the trait and that, therefore, must contain the relevant gene/mutation. Theoretically, the resolving power of this 'nested congenic' approach is absolute, that is, down to the point mutation that causes the phenotype to change.

While in monogenic disease models that are characterized by a categorical trait monitoring of the phenotype suffices to carry out such an experiment, it is important to realize that in polygenic traits, each of the loci involved contributes only a fraction of the overall phenotypic difference between the disease- and reference strains. Thus, the approach usually chosen incorporates the identification by genetic mapping of a chromosomal region (usually 10–30 cM in size) with high odds of containing a disease-relevant gene/mutation, and monitoring of the back-crossing process, initially for the preservation of this segment using strain-specific polymorphic markers (the back-crossing process can be somewhat accelerated by concomitant selection of hybrids that contain maximum of reference strain-specific markers elsewhere in the genome). Only after several rounds of back-crossing, when the majority of other, disease strain-related loci will have been lost, is it sensible to test the back-cross rats for the phenotype of interest (this is usually done after an additional round of sister–brother mating among the back-cross hybrids to create animals homozygous for the congenic segment). The congenic strain should display a 'partial' disease phenotype: depending on the relative contribution to the overall phenotype difference among disease and reference strain by the gene/mutation contained in the chromosomal region grafted onto the wild type background, a larger or smaller deviation of the quantitative phenotype from that typical for the reference strain towards that typical for the disease strain should be measurable. Because the effect may be rather small, highly accurate and sensitive methods are essential for phenotype determination. Since it is not possible to exclude donor-genome derived 'contamination' outside the congenic segment that could conceivably influence the trait, the phenotype comparisons are ideally made between the congenic line and a sub-congenic line bred from the former by a minimal number of additional rounds of back-crossing to produce loss of the congenic segment, other than between the congenic line and the wild type strain.

As indicated above, once a phenotype has been anchored to a congenic segment, further rounds of back-crossing are conducted with the aim of encountering recombinations within the congenic segment, thus producing a series of 'nested' congenic lines with progressively smaller congenic segments. Conservation and loss of the phenotype, respectively, are the critical events that define the localization of the gene in question between crossover points. The larger the number of such crossovers observed is, the higher the resolution of the approach will be (theoretically, if two crossover events that result in phenotype conservation and loss, respectively, were to occur one base-pair from each other, the causative point mutation itself would be localized!).

While this approach is extremely powerful, it is also very tedious, time-consuming, expensive, and often technically demanding (if the phenotypic effect attributable to the locus pursued is of modest magnitude), and not universally applicable (if an epigenetic interaction between two genes remote from each other is essential for phenotype expression the congenic route is, for all practical purposes, not feasible). For these reasons, and because the identification of quantitative trait loci (QTLs) in rat models of cardiovascular disease by molecular genetic mapping approaches only dates back a few years, only a few papers have been published on congenic experimentation in the applicable rat models of cardiovascular disease. One publication describes the transfer of a chromosomal region containing the renin gene from the Dahl/JR-salt-sensitive (hypertensive) strain to the normotensive reference strain. This experiment was based on the previously reported cosegregation of the renin locus with elevated blood pressure in an F_2 intercross between the two strains; surprisingly, the congenic strain was found to have lower, rather than, as expected, higher blood pressures than the reference (host) strain. Although it was concluded that the experiment demonstrated indeed the presence of blood pressure-relevant genes in the region and in the congenic segment, no more specific explanations for the unexpected outcome are presently available (essential epistatic interaction with a strain-specific allele of another gene might be one). The chance observation of a (normotensive) WKY strain that carries a small congenic segment derived from the hypertensive SHRSP allowed the dissection of what was previously perceived as a single hypertension-related QTL on chromosome 10 into two components (Kreutz *et al.*, 1995); the congenic strain was found to have moderately, but highly significantly elevated, borderline-hypertensive blood pressures when compared with the noncongenic strain. While identity of the noncongenic part of the genome between congenic and wild type strains was demonstrated by genotyping with more than 400 strain-specific markers, one cannot exclude with certainty, as in any other congenic experiment, the presence of small congenic grafts outside the region of interest that the markers fail to detect. Finally, a recent study exploited congenic experimentation to demonstrate cosegregation of loci influencing both biochemical and blood pressure phenotypes.

8. Blood pressure-related morbid phenotypes

The concept of primary hypertension as a disease causing various forms of 'end-organ' morbidity remains an assumption based strictly on association, but not on demonstrated causality (with the possible – but debated – exception of malignant hypertension). A different view, namely that elevated blood pressure is a common phenomenon, that accompanies a range of primary tissue disorders, possibly as a secondary event, needs to be considered, and is perhaps more likely. Such a possibility would greatly affect our approach to studying the disorder, as it would strongly favor a focus on actual, tissue-specific manifestations, such as stroke or heart failure, rather than on elevated blood pressure *per se*. One would predict that genetic loci should exist that cosegregate with these ailments, but not necessarily with blood pressure (although one cannot exclude a pleiotropic effect of disease-relevant genes on the tissue of interest and, concomitantly, on blood pressure: in such a case, the existence of independent effects will of course be difficult to prove).

Evidence supporting this view comes from a large body of clinical and experimental studies, as well as from molecular genetic studies in two disease models. The dissociation of so-called 'end-organ' disease and hypertension has long been noted in clinical practice: while individuals with hypertension are clearly at a heightened risk of these 'complications', only a small minority of them will actually develop them, even in the absence of treatment. Conversely, these morbid conditions, outwardly indistinguishably, are sometimes also seen in individuals without hypertension. These observations raise the possibility that elevated blood pressure *per se* is not sufficient to produce these morbid conditions, but that it may act, as a permissive or even essential cofactor in concert with organ- or disease-specific, equally essential/permissive gene variants, often termed 'susceptibility genes' or 'genetic predispositions' (the terms are commonly used to juxtapose complex disease genes and genes causing simple Mendelian conditions; they are somewhat misleading, and should be understood to denote not a qualitative, but rather a quantitative property with a differential degree of deterministic power, and a differential need for gene–environment interaction to cause expression of the phenotype).

To address the question whether genes other than the ones contributing to high blood pressure are operative in the pathogenesis of stroke, a study was conducted using the SHRSP as a model organism. The SHRSP provides an inbred animal model for a complex form of cerebrovascular disease resembling, in many aspects, the human disease. Stroke occurs only after high blood pressure has developed, and only if animals are exposed to a specific permissive dietary regimen (low in potassium and protein, high in sodium); under these conditions, there is an almost 100% incidence within 6–8 weeks (Nagaoka *et al.*, 1976; Okamoto *et al.*, 1974; Rubattu *et al.*, 1996; Slivka, 1991; Volpe *et al.*, 1990). The role of genetic factors in this model is highlighted by the resistance to stroke of a closely related strain, the SHR, which, despite a similar degree of hypertension after exposure to the same diet, remains stroke free. Previous cosegregation experiments, carried out in intercrosses of SHRSP and normotensive reference strains, were of limited utility because the concomitant segregation of blood pressure represented a major confounder. To avoid this complication, a cosegregation study was carried out in F_2 hybrids bred from SHRSP and SHR, thus removing blood pressure as a confounding variable (Rubattu *et al.*, 1996). It was hypothesized that an identical degree of hypertension in the two progenitor strains would remove the possible confounding effects of differential blood pressure, an important variable affecting stroke incidence, and allow to search for genetic loci with direct impact on the pathogenesis of stroke. Based on previous evidence indicating that the relative abundance of SHRSP-gene dosage correlates with the time to develop stroke in crossbred animals, time span (latency) after initiation of the permissive diet until occurrence of a cerebrovascular accident was considered the primary phenotype parameter to assess genetic susceptibility for stroke in the F_2 cohort. Latency until occurrence of a cerebrovascular accident among the hybrid animals showed indeed a large variance, ranging from 23 to 440 days, with a 50% cumulative incidence during the first 11 weeks and a 5% event free-rate at 8 months, consistent with the independent segregation of multiple genes.

The genetic proximity between the two strains used (the SHRSP was established by selective breeding from partially inbred SHR) presented a major challenge to identifying informative markers. A total of 1038 genetic markers had to be screened to yield a mapping panel of 112 polymorphic markers in the SHRSP/SHR cross, using previously described methods. While overall 77.3 and 94.9% of the genome was within 10 and 20 cM, respectively, of an informative marker, coverage was considerably sparser for certain chromosomes.

The cosegregation analysis revealed three loci in linkage with the stroke-latency phenotype. A locus on chromosome 1, centered at the anonymous marker, *D1Mit3*, was found to show highly significant linkage to the occurrence of stroke. This quantitative trait locus, termed *STR1*, strongly affected latency to stroke in a recessive mode with a LOD score of 7.4, accounting for 17.3% of overall phenotype variance. Additional consideration of age-adjusted blood pressure values as a covariate had no effects on the resultant statistic, indicating that this locus acts independently of blood pressure (presumably, given a certain permissive level). The 100:1 odds interval of placing the gene in question, which spans a 34 cM interval around the marker, *D1Mit3*, is not known to contain any candidate genes. In contrast to *STR1*, a locus on rat chromosome 5, termed *STR2*, was found to confer a significant protective effect against stroke in the presence of SHRSP alleles. The presence of one or two SHRSP alleles at this locus was associated with a proportional delay in the occurrence of stroke. *STR2* accounted for 9.6% of overall variance in stroke latency; while this quantitative trait locus occupies a broad confidence interval, the peak protective effect (LOD score 4.7) mapped close to the gene coding for atrial natriuretic factor (ANF), a hormone with important vasoactive properties and a potential candidate gene. *STR2* showed epistatic interaction with *STR1*, with the interaction term accounting for an additional 3% of overall phenotype variance. Thus, the protective effect of the SHRSP allele at *STR2* on stroke was maximal in animals homozygous for the SHRSP allele at *STR1*. An additional locus on chromosome 4, *STR3*, was also found to confer a similar, but less significant, recessive effect on preventing stroke in the presence of two SHRSP-derived alleles. The LOD score at *STR3* was 3.0, and the locus accounted for 4.7% of the phenotype variance.

Of note, none of the genetic markers tested, including those found to be linked to stroke latency, displayed either suggestive or significant linkage to baseline blood pressure, blood pressure after dietary exposure, or individual pre-event actual or age-adjusted blood pressure values. Histopathological diagnosis showed no association with blood pressure, stroke latency or zygosity at any of the stroke-linked markers.

The finding that several genetic loci, along with permissive environmental factors, affect predisposition to stroke validates the SHRSP as an appropriate model for the complex, polygenic and multifactorial disease that stroke represents in humans. The observations of two loci that confer a protective effect against stroke in the presence of SHRSP alleles, as well as of epistatic and ecogenetic interactions, are typical for complex diseases where the net cumulative effect of many genes that affect a quantitative trait in either direction, modulated by gene–gene and gene–environment interplay, determines the ultimate phenotype. While parallels drawn between the SHRSP and human stroke are, by necessity, speculative, a number of features, such as the dependence on dietary factors (in particular, sodium and potassium intake) and on elevated blood pressure for manifestation of stroke in this model bear striking resemblance to the characteristics of human disease (Alam *et al.*, 1993; Lindpaintner *et al.*, 1990).

These results did not only provide the first direct evidence for the existence of genes that specifically contribute to susceptibility to a complex polygenic form of stroke, identifying them as distinct of gene loci previously shown to contribute to hypertension in the same strain (Cicila *et al.*, 1993; Deng and Rapp, 1992; Pravenec *et al.*, 1991), but also lend clear support to the notion outlined above, namely that hypertension-associated morbid conditions do not simply represent mere mechanical sequelae (end-organ damage) of blood pressure elevation. The data also demonstrated the power of animal models in the study of complex traits. In a similar fashion, to test the possibility that hypertensive nephropathy may be caused by the coincident presence of genes causing hypertension and gene variants conferring susceptibility for renal damage, a cross involving the fawn-hooded rat, an animal model of hypertension that develops chronic renal failure was studied (Brown *et al.*, 1996). Using the genome screening approach in phenotypically well-characterized hybrid animals that showed a wide distribution of renal pathology, two QTLs linked to renal impairment, but not to blood pressure, and one locus linked to blood pressure, but not to indices of renal disease, were localized.

Additional morbid traits associated with hypertension that are currently pursued by a number of groups using similar experimental design strategies include congestive heart failure, left ventricular hypertrophy, and susceptibility to cardiac arrhythmias.

9. Utility of rat genetic investigation in hypertension

Clearly, almost all biomedical research conducted in animal models has, as its ultimate aim and purpose, the generation of information that will help address human health issues. On a rather simplistic and ambitious level, one might therefore hope that genes and their molecular variants that are found to contribute to disease in the rat will have human homologs that are analogously involved in the pathogenesis of hypertension/cardiovascular disease in man. We need to be cautious in considering this possibility, for several reasons.

First, even though hypertensive rat strains are models of polygenic and multifactorial disease origin, and are thus certainly more representative of complex disease than classically used monogenic disease, they differ in an important aspect from human hypertension, namely they are inbred, and thus genetic heterogeneity does not affect any of the experimental crosses done in rats. Thus, since the same limited number of loci are operative in any cross, their recognition is relatively easy. In contrast, the presence of genetic heterogeneity in humans, that is, the fact that different sets of 'polygenes' are operative in one individual or family as compared to another, dramatically lowers the statistical power and, thus, the likelihood of recognizing the effect of a gene identified in the animal model. Doing so requires the performance of very large-scale studies, with the size of the study population indirectly proportional to the frequency of the gene in question and its relative effect on phenotype. Of course, if we had a better means of characterizing genetically diverse hypertensive subgroups than by blood pressure, which represents a rather indirect and remote phenotype, a gene(-variant) could be tested with much greater sensitivity and power. Thus, attempts to measure so-called 'intermediate phenotypes' – biochemical or physiological characteristics that would allow such sub-stratification – are being undertaken by a number of investigators. While it is certainly reasonable to test in human populations any gene that has been identified as potentially blood pressure-relevant using the animal model, it is not very likely that this will yield positive results, unless very large and very well-characterized populations are studied.

This reservation refers to the second important limitation of applying rat data to human investigation: for none of the candidate genes so far suspected on account of their localization within a QTL to contribute to rat hypertension has proof been rendered as to their actual identity using the above-described congenic approaches (specific gene-replacement methodology would yield similar power and discrimination). Thus, the 'candidate gene' may in effect only represent a genetic marker for the locus, or chromosomal region, where an as yet unknown, actual disease-gene is localized. Transferring this data, then, to a human population faces the problem that only linkage studies, but not the more commonly used (since logistically much simpler to perform) case–control (association-, linkage-disequilibrium) studies would actually be applicable to testing the locus information derived from the rat; and even linkage approaches may prove *a priori* incapable of testing such markers if syntenic regions are not preserved among rat and man (in other words, if the chromosomal contiguity – or proximity – of the 'candidate gene' marker used for mapping and the actual disease-causing gene is not preserved across species, that is, from rat to man).

In addition, there is of course no guarantee that all human hypertension genes are in fact represented in (one of the available) rat strains; and unless extensive breeding experiments employing many different constellations of the various hypertensive and normotensive strains available are carried out, not even all rat-specific hypertensive loci will even be detectable.

The application of knowledge gained in the rat towards the discovery of homologous genes or pathways that contribute to human hypertension is an extremely attractive vista that would directly provide targets for the design and development of novel therapeutic and preventative strategies. However, it is important to realize that finding causal genes/pathways in rat hypertension may have major implications for human hypertension even if the homologous genes are not found to play a role in the pathogenesis of the disease in humans, as they would still provide potential novel targets for drug development: if pharmacological modulation of the respective pathway results in blood pressure lowering (or organ protection), these drugs would merely resemble currently used anti-hypertensive agents that are effective in the absence of targeting – as best as we can tell – pathways causally involved in the pathogenesis of hypertension (with the possible exception of the renin–angiotensin system). The development and use of such drugs would in effect replicate what we have done so far, but target novel pathways or mechanisms. Indeed, depending on our ultimate unravelling of the complexity of human hypertension, it may be difficult or impossible truly to administer what we envision today as the ideal therapy; namely, one that targets the true cause of hypertension, for a number of different reasons: first, although unlikely to be a generalized problem, certain individuals or social groupings may not allow the kind of genetic profiling necessary to determine the (predominant) cause of hypertension for individualized choice of the most appropriate agent; second, we may find that in the majority of hypertensives the disease is sufficiently polygenic to render the specific influence of any of the involved disease-genes so small as to not offer a clear choice of mechanism to be targeted; or, third, subgroups defined by disease causation may be so small as to make it economically impossible to develop specific agents for each and every one.

Lastly, an important caveat applies to most, if not all genetically hypertensive rat strains with regard to the extrapolation of data to humans: while they certainly provide at present the best model systems for elevated blood pressure, most of them show very little or none of the associated vascular and end-organ pathology that makes hypertension a clinically relevant and epidemiologically important issue in man. This is not altogether surprising, as selective breeding for – sometimes extremely – high blood pressure may have resulted in concomitant selection pressure for the elimination of other disease-genes the effects of which are augmented by hypertension. This leads us to the theoretical consideration whether hypertensive genes in man are independent of those causing actual vascular morbidity (although their effects may certainly either augment or even represent essential permissive factors for the manifestation of such morbidity), or whether certain genes cause hypertension concomitant with the clinically relevant vascular pathology (as primary pleiotropic manifestations of one gene, or with hypertension as a secondary phenomenon brought about by primary vascular changes; in both cases, hypertension might either further aggravate the pathological events, or, possibly, represent a mere, innocent epiphenomenon). If we subscribe to the latter notion, we are unlikely to find such genes using currently available rat models. If we assume that the former scenario is correct, then rat experimentation – within the constraints discussed above – has the potential to lead us to human genes; and although their role would be viewed primarily as a permissive or disease-enhancing one, this information would clearly be of major importance.

It comes, therefore, neither as a surprise, nor does it represent a robust finding that no convincing linkage or association could be demonstrated so far between polymorphisms of the ACE (Jeunemaitre *et al.*, 1992a), renin (Naftilan *et al.*, 1989), and SA (Nabika *et al.*, 1995) genes (all – variably – linked to hypertension in the rat) and blood pressure in panels of hypertensive sibpairs and case–control cohorts, respectively; and conversely, that no linkage exists between blood pressure and the angiotensinogen locus in the SHRSP/WKY cross-bred rats (Hubner *et al.*, 1994), as opposed to findings in human hypertension (Jeunemaitre *et al.*, 1992b).

Wherein, then, lies the value of studying the genetics of hypertensive rats? As progress is made in the field of molecular genetics in experimental hypertension, it is becoming increasingly evident that the true value of these animal systems lies at least in part in their paradigmatic nature as models of polygenic, multifactorial cardiovascular disease. While it is not *a priori* likely, as outlined above, that they will provide quick and direct access to genes causing hypertension in humans, they represent a probing ground for the investigation of complex human disease where methodologies can be evaluated and strategies developed. Thus, the lessons to be learned are probably mainly conceptual in nature: identification of rat hypertension genes will provide us with important new tools to study the mechanism by which blood pressure-regulating genes operate; this knowledge, in turn, may allow us to devise new and powerful algorithms for the stratification of intermediate phenotypes in the absence of knowing human genes, yet still enhancing the power attainable by studying such populations.

In addition, these experiments may help devise novel molecular and statistical strategies with impact on human studies; thus, their importance may extend well beyond the immediate topic, hypertension, into other aspects of cardiovascular disease, and of polygenic traits in mammalian systems in general. This has already been borne out by advances in the investigation of human disease made possible by original work in the rat model. To gain such insights requires not only the availability of molecular genetic tools, but – perhaps more importantly – of a model system in which methodologies for precise and sophisticated measurements of phenotype parameters are well established and have a proven track record. The detailed expertise accumulated in cardiovascular and hypertension research in the rat over past decades makes these animal models strong and extremely useful candidates in our endeavors to understand the workings of polygenic disease. Several examples from the limited number of experiments reported so far shall illustrate the notion that this research will broaden our perspectives and introduce new concepts that will enrich our understanding of human disease in much more profound ways than the discovery of a particular gene is likely to do. Thus, the demonstration that a hypertensive locus was contributed to by the normotensive progenitor in the SHRSP/WKY cross illustrates the fact that it is the net sum of influences derived from both blood pressure-raising and blood pressure-lowering (or, more general, of plus- and minus-) loci that determines ultimately an individual's phenotype. The finding also indicates that even very hypertensive animals like the SHRSP do not, by any means, express all potentially blood pressure-raising genes; in contrast, it is most likely that the balance of a number of hypotensive genes may be essential to prevent early lethality and permit the survival of the strain. As another example, the inconsistent findings regarding linkage of several candidate genes (i.e. renin) with blood pressure in different rat strains may be viewed as paradigmatic for the human situation; thus, what at first glance might appear a weakness of the experimental systems may in fact very well represent a strength by which these 'conflicting' models approximate – in certain aspects – quite closely the nature of human hypertension. The study revealing linkage between the carboxypeptidase B locus and pulse pressure demonstrates how important it will be to go beyond conventional concepts of hypertensive phenotype characterization and focus on a range of phenotype parameters, particularly less complex, intermediate ones, such as biochemical, metabolic, cell-biology, and so on, characteristics. Under the assumption that such intermediate phenotypes may be governed by a less complex integrative regulatory circuitry than blood pressure itself, they may provide more reliable measurements for phenotyping and enhanced statistical power for the detection of genetic linkage.

As indicated above, one of the most visible accomplishments greatly aided by rat experimentation has been the establishment of random-marker genome-screening techniques as feasible for the dissection of the genes involved in complex polygenic, multifactorial and quantitative traits. Thus, the accomplishments made so far in molecular genetic approaches to rat hypertension may certainly be viewed as providing important procedural and conceptual (although thus far not genetic) road maps towards the investigation of human hypertension.

10. Conclusion and outlook

Progress in molecular genetics has greatly enhanced our ability to find the genes causing polygenic disorders, of which hypertension is an important one. It has also opened our eyes about a number of misconceptions and assumptions about our categorization of human disease, as well as about the way experimental models of hypertension should be used and experimental results interpreted. Scientific discovery is dynamic, and we must expect that concepts which we embrace today, based on our present level of understanding, as valid and meaningful may, in turn, have to be revised, or abandoned, in the near or distant future as a consequence of newly emerging knowledge. This does not invalidate their importance, as they are the stepping stones to these future developments, just as past work represents an essential basis for our work today, as scientific pursuit not only (tries to) answer questions, but continually raises new ones, based on the more differentiated understanding it provides of the topic studied. Thus, it took many years of comparative studies in rat models of hypertension until most investigators realized that this approach would not provide causal evidence; but the data accumulated during this period are now extremely helpful in the design of the linkage studies we now conduct.

The final answer to the enigma that shrouds the etiology of hypertension may thus be still further away than our enthusiasm about newly found means of probing the genetics of the disease may lead us to believe. However, it is important to recognize that any advance in our understanding, no matter how trivial it may appear later on, does – directly or indirectly – translate into progress in the clinical arena. Thus, even with our present very limited understanding of the disease, tremendous progress has been accomplished in the management of hypertension and the prevention of some of its most crippling complications. It is reasonable to expect further, rapid progress in which the use of animal models will continue to play an important role.

References

Alam, K.Y., Wang, Y., Dene, H. and Rapp, J.P. (1993) Renin gene nucleotide sequence of coding and regulatory regions in Dahl rats. *Clin. Exp. Hypertension* **15**: 599–614.

Brown, D.M., Provoost, A.P., Daly, M.J., Lander, E.S. and Jacob, H.J. (1996) Renal disease susceptibility and hypertension are under independent genetic control in the fawn-hooded rat. *Nature Genet.* **12**: 44–51.

Bruner, C.A., Myers, J.H., Sing, C.F., Jokelainen, P.T. and Webb, R.C. (1986a) Genetic basis for altered vascular responses to ouabain and potassium-free solution in hypertension. *Am. J. Physiol.* **251**: H1276–H1282.

Bruner, C.A., Myers, J.H., Sing, C.F., Jokelainen, P.T. and Webb, R.C. (1986a) Genetic association of hypertension and vascular changes in stroke-prone spontaneously hypertensive rats. *Hypertension* **8**: 904–910.

Casto, R. and Printz, M.P. (1988) Genetic transmission of hyper-responsivity in crosses between spontaneously hypertensive and Wistar–Kyoto rats. *J. Hypertension* **6(suppl.)**: S52–S54.

Cicila, G.T., Rapp, J.P., Wang, J.M., St. Lezin, E., Ng, S.C. and Kurtz, T.W. (1993) Linkage of 11 beta-hydroxylase mutations with altered steroid biosynthesis and blood pressure in the Dahl rat. *Nature Genet.* **3**: 346–353.

Deng, Y. and Rapp, J.P. (1992) Cosegregation of blood pressure with angiotensin converting enzyme and atrial natriuretic peptide receptor genesusing Dahl salt-sensitive rats. *Nature Genet.* **1**: 267–272.

Deng, A.Y. and Rapp, JP. (1995) Locus for the inducible, but not a constitutive, nitric oxide synthase cosegregates with blood pressure in the Dahl salt-sensitive rat. *J. Clin. Invest.* **95**: 2170–2177.

Deng, A.Y., Dene, H. and Rapp, JP. (1994b) Mapping of a quantitative trait locus for blood pressure on rat chromosome 2. *J. Clin. Invest.* **94**: 431–436.

Deng, A.Y., Dene, H., Pravenec, M. and Rapp, JP. (1994) Genetic mapping of two new blood pressure quantitative trait loci in the rat by genotyping endothelin system genes. *J. Clin. Invest.* **93**: 2701–2709.

Dubay, C., Vincent, M., Samani, N.J., *et al.* (1993) Genetic determinants of diastolic and pulse pressure map to differnt loci in Lyon hypretensive rats. *Nature Genet.* **3**: 354–357.

Ferrari, P., Barber, B.R., Torielli, L., Ferrandi, M., Salardi, S. and Bianchi, G. (1987) The Milan hypertensive rat as a model for studying cation transport abnormality in genetic hypertension. (Review). *Hypertension* **10**: 132–136.

Frankel, W.N. (1995) Taking stock of complex trat genetics in mice. *Trends Genet.* **11**: 471–477.

Furspan, P.B., Jokelainen, P.T., Sing, C.F. and Bohr, D.F. (1987) Genetic relationship between a lymphocyte membrane abnormality and blood pressure in spontaneously hypertensive stroke prone and Wistar–Kyoto rats. *J. Hypertension* 5: 293–297.

Ganten, U., Rascher, W., Lang, R.E., *et al.* (1983) Development of a new strain of spontaneously hypertensive rats homozygous for hypothalamic diabetes insipidus. *Hypertension* 5: I119–I128.

Hamet, P., Kong, D. and Pravenec, M., *et al.* (1992) Restiction fragment length polymorphism of *hsp70* gene, localized in the RT1 complex, is associated with hypertensionin spontaneously hypertensive rats. *Hypertension* **19**: 611–614.

Harrap, S.B. (1986) Genetic analysis of blood pressure and sodium balance in spontaneously hypertensive rats. *Hypertension* **8**: 572–582.

Harrap, S.B. and Doyle, A.E. (1988) Genetic co-segregation of renal haemodynamics and blood pressure in the spontaneously hypertensive rat. *Clin. Sci.* **74:** 63–69.

Harris, E.L., Dene, H. and Rapp, J.P. (1993) SA gene and blood pressure cosegregation using Dahl salt-sensitive rats. *Am. J. Hypertension* **6**: 330–334.

Hilbert, P., Lindpaintner, K., Beckmann, J.S., *et al.* (1991) Chromosomal mapping of two genetic loci associated with blood-pressure regulation in hereditary hypertensive rats. *Nature* **353**: 521–529.

Hubner, N., Kreutz, R., Takahashi, S., Ganten, D. and Lindpaintner, K. (1994) Unlike human hypertension, blood pressure in a heredetary hypertensive rat strain shows no linkage to the angiotensinogen locus. *Hypertension* **23**: 797–801 (abstract).

Iwai, N. and Inagami, T. (1991) Isolation of preferentially expressed genes in the kidneys of hypertensive rats. *Hypertension* **17**: 161–169.

Iwai, N. and Inagami, T. (1992) Identification of a candidate gene responsible for the high blood pressure of spontaneously hypertensive rats. *J. Hypertension* **10**: 1155–1157.

Iwai, N., Kurtz, T.W. and Inagami, T. (1992) Further evidence of the SA gene as a candidate gene contributing to the hypertension in spontaneously hypertensive rat. *Biochem. Biophys. Res. Commun.* **188**: 64–69.

Jacob, H.E., Lindpaintner, K., Lincoln, S.E., *et al.* (1991) Genetic mapping of a gene causing hypertension in the stroke-prone spontaneously hypertensive rat. *Cell* **67**: 213–224.

Jacob, H., Brown, D.M., Bunker, R.K., *et al.* (1995) A genetic linkage map of the laboratory rat, *Rattus norvegicus*. *Nature Genet.* **9**: 63–69.

Jeffreys, A.J., Wilson, V. and Thein, S.L. (1985) Hypervariable 'minisatellite' regions in human DNA. *Nature* **314:** 67–73.

Jeunemaitre, X., Lifton, R.P., Hunt, S.C., Williams, R.R. and Lalouel, J. (1992a) Absence of linkage between the angiotensin converting enzyme locus and human essential hypertension. *Nature Genet.* **1**: 72–75.

Jeunemaitre, X., Soubrier, F., Kotelevtsev, Y.V., *et al.* (1992b) Molecular basis of human hypertension: role of angiotensinogen. *Cell* **71**: 169–180.

Judy, W.V., Watanabe, A.M., Murphy, W.R., Aprison, B.S. and Yu, P.L. (1979) Sympathetic nerve activity and blood pressure in normotensive backcross rats genetically related to the spontaneously hypertensive rat. *Hypertension* **1**: 598–604.

Katsuya, T., Higaki, J. and Miki, T., *et al.* (1992) Hypotensive effect associated with a phospholipase C-delta 1 gene mutation in the spontaneously hypertensive rat. *Biochem. Biophys. Res. Commun.* **187**: 1359–1366.

Kreutz, R., Hubner, N., James, M.R., *et al.* (1995) Dissection of a quantiative trait locus for genetic hypertension on rat chromosome 10. *Proc. Natl Acad. Sci. USA* **92**: 8778–8782.

Kurtz, T.W. and Morris, R.C., Jr. (1987) Biological variability in Wistar–Kyoto rats. Implications for research with the spontaneously hypertensive rat. *Hypertension* **10**: 127–131.

Kurtz, T.W., Montano, M., Chan, L. and Kabra, P. (1989) Molecular evidence of genetic heterogeneity in Wistar–Kyoto rats: Implications for research with the spontaneously hypertensive rat. *Hypertension* **13**: 188–192.

Kurtz, T.W., Simonet, L., Kabra, P.M., Wolfe, S. and Hjelle, B.L. (1990) Cosegregation of the renin allele in the spontaneously hypertensive rat with an increase in blood pressure. *J. Clin. Invest.* **85**: 1328–1332.

Lindpaintner, K., Takahashi, S. and Ganten, D. (1990) Structural alterations of the renin gene in stroke-prone spontaneously hypertensive rats: examination of genotype-phenotype correlations. *J. Hypertension* **8**: 763–773.

Lindpaintner, K., Kreutz, R. and Ganten, D. (1992) Genetic variation in hypertensive and 'control' strains. What are we controlling for anyway? *Hypertension* **19**: 428–430.

Lindpaintner, K., Hilbert, P., Ganten, D., Nadal-Ginard, B., Inagami, T. and Iwai, N. (1993) Molecular genetics of the SA-gene: cosegregation with hypertension and mapping to rat chromosome 1. *J. Hypertension* **11**: 19–23.

Louis, W.J. and Howes, L.G. (1990) Genealogy of the spontaneously hypertensive rat and Wistar–Kyoto rat strains: implications for studies of inherited hypertension. *J. Cardiovasc. Pharmacol.* **16(suppl. 7)**: S1–S5.

McLaren, Y., Kreutz, R., Lindpaintner, K., *et al.* (1993) Membrane microviscosity does not correlate with blood pressure: a cosegregation study. *Hypertension* **11**: 25–30.

Mulvany, M.J. (1988) Possible role of vascular oscillatory activity in the development of high blood pressure in spontaneously hypertensive rats. *J. Cardiovasc. Pharmacol.* **12(suppl. 6)**: S16–S20.

Nabika, T., Bonnardeaux, A., James, M., *et al.* (1995) Evaluation of the SA locus in human hypertension. *Hypertension* **25**: 6–13.

Naftilan, A.J., Williams, R., Burt, D., *et al.* (1989) A lack of genetic linkage of renin gene restriction fragment length polymorphisms with human hypertension. *Hypertension* **14**: 614–618.

Nagaoka, A., Iwatsuka, H., Suzuoki, Z. and Okamoto, K. (1976) Genetic predisposition to stroke in spontaneously hypertensive rats. *Am. J. Physiol.* **230**: 1354–1359.

Okamoto, K., Yamori, Y. and Nagaoka, A. (1974) Establishment of the stroke-prone spontaneously hypertensive rat (SHR). *Circ. Res.* **33/34(suppl. I)**: I-143–I-153.

Pravenec, M., Kren, V. and Kunes, J., *et al.* (1991) Cosegregation of blood pressure with a kallikrein gene family polymorphism. *Hypertension* **17**: 242–246.

Rapp, J.P. (1982) A gentic locus (*hyp-2*) controlling vascular smooth muscle response in spontaneously hypertensive rats. *Hypertension* **4**: 459–467.

Rapp, J.P. (1983) A paradigm for identification of primary genetic causes of hypertension in rats. *Hypertension* **5(suppl. I)**: I–198–I–203.

Rapp, J.P. (1991) Dissecting the primary causes of genetic hypertension in rats. *Hypertension* **18**: I18–I28.

Rapp, J.P. and Dahl, L.K. (1976) Mutant forms of cytochrome P–450 controlling both 18- and 11beta-steroid hydroxylation in the rat. *Biochemistry* **15**: 1235–1242.

Rapp, J.P., Wang, S.M. and Dene, H. (1989) A genetic polymorphism in the renin gene of Dahl rat cosegregates with blood pressure. *Science* **243**: 542–544.

Rubattu, S., Volpe, M., Kreutz, R., Ganten, U., Ganten, D. and Lindpaintner, K. (1996) Chromosomal mapping of quantitative trait loci contributing to stroke in a rat model of complex human disease [see comments]. *Nature Genet.* **3**: 429–434.

Samani, N.J., Swales, J.D., Jeffreys, A.J., *et al.* (1990) DNA fingerprinting of spontaneously hypertensive and Wistar–Kyoto rat: implications for hypertension research. *J. Hypertension* **7**: 809–816.

Samani, N.J., Lodwick, D., Vincent, M., *et al.* (1993) A gene differentially expressed in the kidney of spontaneously hypertensive rat cosegregates with increased blood pressure. *J. Clin. Invest.* **92**: 1099–1103.

Schmale, H. and Richter, D. (1984) Single base deletion in the vasopressin gene is the cause of diabetes insipidus in Brattleboro rats. *Nature* **308**: 705–709.

Serikawa, T., Kuramoto, T., Hilbert, P., *et al.* (1992) Rat gene mapping using PCR-analyzed microsatellites. *Genetics* **131**: 703–723.

Sessler, F.M., Jokelainen, P.T., Sing, C.F., Strack, A.M. and Malvin R.L. (1986) Renin heterogeneity in stroke-prone hypertensive and normotensive rats. *Am. J. Physiol.* **251**: E367–E372.

Slivka, A. (1991) Effect of antihypertensive therapy on focal stroke in spontaneously hypertensive rats. *Stroke* **22**: 884–888.

Volpe, M., Camargo, M.J. and Mueller, F.B., et al. (1990) Relation of plasma renin to end organ damage and to protection of K+ feeding in stroke-prone hypertensive rats. *Hypertension* **15**: 318–326.

Weber, J.L. and May, P.E (1989) Abundant class of human DNA polymorphisms which can be typed using the polymerase chain reaction. *Am. J. Hum. Genet.* **44**: 388–396.

Yamori, Y. and Okamoto, K. (1970) Zymogram analyses of various organs from spontaneously hypertensive rats. A gentico-biochemical study. *Lab. Invest.* **22**: 206–211.

Yamori, Y., Ooshima, A. and Okamoto, K. (1972) Genetic factors involved in spontaneous hypertension in rats an analysis of F 2 segregate generation. *Jpn Circ. J.–Engl. Ed.* **36**: 561–568.

Yongue, B.G. and Myers, M.M. (1988) Cosegregation analysis of salt appetite and blood pressure in genetically hypertensive and normotensive rats. *Clin. Exp. Hypertension – Part A: Theory and Practice* **10**: 323–343.

4

Molecular genetics of steroid biosynthesis in Dahl salt-sensitive and salt-resistant rats: linkage to the control of blood pressure

Theodore W. Kurtz, Michal Pravenec, Elizabeth St. Lezin and Synthia H. Mellon

1. Introduction

The Dahl salt-sensitive (S) rat is the most widely studied animal model of salt-sensitive hypertension (Rapp, 1982; Rapp and Dene, 1985). In the Dahl S strain, supplemental dietary NaCl increases blood pressure whereas in the Dahl salt-resistant (R) strain, increased dietary NaCl has little or no effect on blood pressure. In the Dahl model as in humans, variation in the blood pressure response to changes in dietary intake of NaCl involves the interaction of multiple environmental and genetic factors. Although the exact relevance of the Dahl model to blood pressure control in humans remains to be determined, it is hoped that identification of genes that influence the blood pressure response to dietary NaCl in S and R rats may shed light on the pathogenesis of human forms of salt-sensitive hypertension.

The Dahl S strain was originally derived by Dahl and colleagues from non-inbred Sprague–Dawley rats by recurrent selective breeding of animals that exhibited severe hypertension when fed a high NaCl diet (Dahl *et al.*, 1962; Rapp, 1982). The Dahl R strain was derived in parallel by recurrent selective breeding of Sprague–Dawley rats that exhibited unusually low blood pressures despite being fed a high NaCl diet (Dahl *et al.*, 1962; Rapp, 1982). Rapp subsequently derived inbred strains of Dahl salt-sensitive and salt-resistant rats (designated SS/JR and SR/JR, respectively) using S and R breeding stock obtained from Dahl. In this article, we

Molecular Genetics of Hypertension, edited by A.F. Dominiczak, J. M. C. Connell and F. Soubrier.

use the notation 'S' and 'R' in the generic sense to denote the phenotypes of salt-sensitivity and salt-resistance. In Dahl's original breeding studies, most of the Sprague–Dawley rats displayed at least some elevation in blood pressure when fed the high NaCl diet (Dahl *et al.*, 1962). In other strains of rats, high NaCl diets have also been reported to induce moderate increases in blood pressure (Meneely *et al.*, 1953; Preuss and Preuss, 1980; Smith-Barbaro *et al.*, 1980). Thus, both the robust salt-resistance of the Dahl R rat and the marked salt-sensitivity of the Dahl S rat constitute unusual cardiovascular phenotypes that have important genetic components. Accordingly, in the Dahl model, a search for genes that contribute to the strain differences in blood pressure may yield information on molecular variants that confer protection against salt-induced increases in blood pressure as well as on molecular variants that promote salt-sensitive hypertension.

In linkage studies in F_2 and backcross populations derived from Dahl rats, multiple genetic variants have been reported to cosegregate with effects on blood pressure (Deng and Rapp, 1992; Deng *et al.*, 1994a, 1994b; Gu *et al.*, 1996; Rapp *et al.*, 1989; Stec *et al.*, 1996). Most of these variants have involved noncoding sequences of unknown functional significance and may simply represent markers for linked genes involved in the pathogenesis of hypertension. Total genome scans have been very successful in identifying specific chromosome regions that regulate blood pressure in Dahl S rats and intense efforts are underway to dissect these regions genetically using congenic strains (Garret *et al.*, 1998). Recently, a Q276L variant in the α1 Na,K-ATPase gene that affects sodium–potassium pump activity by increasing the Na:K coupling ratio from 3:2 to 3:1 has been linked to increased blood pressure in the Dahl S rat (Herrera *et al.*, 1998). In addition, transgenic overexpression of the Dahl R variant of the α-1 Na,K-ATPase gene on the Dahl S background was reported to attenuate hypertension (Herrera *et al.*, 1998). These findings support a role for genetically determined alterations in sodium–potassium pump activity in the pathogenesis of hypertension in the Dahl S rat. Unfortunately, it is difficult to identify the Q276L variant by sequencing of PCR amplified DNA or by sequencing DNA from genomic libraries (Herrera *et al.*, 1998; Simonet and Kurtz, 1991). Moreover, the genetic contamination of commercially available Dahl S rats could complicate independent attempts to confirm the existence of this sequence variant (St. Lezin *et al.*, 1994). Thus, the precise status of the Q276L variant in the pathogenesis of hypertension in Dahl S rats requires further study. In addition to the putative mutation in the α1 Na,K-ATPase gene, sequence variants in genes involved in mineralocorticoid biosynthetic pathways have been linked to effects on blood pressure in the Dahl model. In the following sections, we discuss evidence suggesting that molecular variants in the coding sequences of key genes regulating mineralocorticoid biosynthesis may contribute to differences in mineralocorticoid levels and blood pressure between the Dahl S and R strains (Cicila *et al.*, 1993; Cover *et al.*, 1995).

2. Molecular basis of mineralocorticoid biosynthesis

The molecular basis for steroid hormone synthesis, including the steroidogenic pathways, the enzymes mediating steroidogenesis and the genes encoding these enzymes, is well known (Jamieson and Fraser, 1994; Miller, 1988). Most steroidogenic enzymes are cytochromes P450, a generic term for the group of oxidative enzymes which interact with specific steroids and which reduce oxygen with electrons from NADPH. The synthesis of 11-deoxycorticosterone (DOC) from cholesterol uses the same adrenal enzymes in both the adrenal glomerulosa and fasciculata/reticularis (Jamieson and Fraser, 1994; Miller, 1988; see also Chapter 7). DOC is then converted to mineralocorticoids in the glomerulosa and to glucocorticoids in the fasciculata/reticularis by the zone-specific expression of two P450c11 enzymes, P450c11β (11β hydroxylase) and P450c11AS (aldosterone synthase) (Lauber and Muller, 1989; Malee and Mellon, 1991; Ogishima *et al.*, 1989; Sander *et al.*, 1994). The CYP11Bl gene encoding P450c11β is regulated by ACTH, is expressed solely in the fasciculata/reticularis, and its encoded enzyme converts DOC to corticosterone or to 18-hydroxy–11-deoxycorticosterone (18-OH-DOC) (Ogishima *et al.*, 1992; Sander *et al.*, 1994). Thus, P450c11β has both 11-hydroxylase and 18-hydroxylase activity. The CYP11B2 gene is largely regulated by sodium and potassium via the renin–angiotensin system, is expressed solely in the zona glomerulosa, and encodes P450c11AS that converts DOC to aldosterone (Curnow *et al.*, 1991; Donnalik *et al.*, 1991; Malee and Mellon, 1991; Muller *et al.*, 1989; Sander *et al.*, 1994). Hence, this enzyme has 11 hydroxylase, 18 hydroxylase and 18 oxidase activities. Two other P450c11 genes, called CYP11B3 and CYP11B4, were recently cloned from a rat genomic library (Mellon *et al.*, 1995; Mukai *et al.*, 1993; Nomura *et al.*, 1993). CYP11B4 appears to be a pseudogene, as two exons are replaced by unrelated DNA. The protein encoded by CYP11B3, P450c11B3, closely resembles P450c11AS in mRNA and its predicted amino acid sequence of 498 amino acids. This protein has 11 hydroxylase and 18 hydroxylase activities but not 18 oxidase activity Mellon *et al.*, 1995. Relatively little is known about the functional significance of rat P450c11B3 and it is expressed for only a few weeks after birth. No gene corresponding to CYP11B3 or CYP11B4 has been found in humans (Zhang and Miller, 1996). Thus, it appears that P450c11β and P450c11AS are the principal enzymes responsible for mineralocorticoid biosynthesis in the rat and human.

3. Molecular variation in CYP11B1 as a determinant of blood pressure in the Dahl model

Based on a classic series of biochemical and genetic studies of steroid biosynthesis in the Dahl S and R strains, Rapp and Dahl proposed that strain differences in the adrenal synthesis of 18-OH-DOC stemming from genetically determined differences in 11β hydroxylase contribute to the strain differences in blood pressure observed with administration of a high NaCl diet (Rapp and Dahl, 1971, 1972, 1976). Rapp and Dahl not only observed lower circulating levels of 18-OH-DOC and reduced adrenal synthesis of 18-OH-DOC in Dahl R rats versus S rats, they also found that in segregating populations derived from R and S rats, the differences in adrenal synthesis of 18-OH-DOC cosegregated with differences in blood pressure; rats that exhibited the low rates of adrenal synthesis of 18-OH-DOC characteristic of the Dahl R strain exhibited lower levels of blood pressure than rats that exhibited greater rates of hormone synthesis (Rapp and Dahl, 1971, 1972, 1976; Rapp *et al.*, 1978). Although 18-OH-DOC has much lower affinity for the mineralocorticoid receptor than aldosterone (Feldman and Funder, 1973), circulating levels of 18-OH-DOC far exceed those of aldosterone (Nicholls *et al.*, 1979; Rapp *et al.*, 1978). Moreover, administration of a high NaCl diet suppresses circulating levels of aldosterone but does not affect the production of 18-OH-DOC. Thus, in rats fed large amounts of NaCl, it has been proposed that 18-OH-DOC may act as the principal steroid that promotes sodium retention even though its mineralocorticoid effects are much weaker than those of aldosterone (Feldman and Funder, 1973). These observations, together with studies in uni-nephrectomized NaCl-fed rats in which parenteral administration of physiological amounts of 18-OH-DOC has been found to increase blood pressure (Carroll *et al.*, 1981), are consistent with the proposal that in Dahl S versus Dahl R rats, genetically determined differences in adrenal production of 18-OH-DOC contribute to the strain differences in blood pressure.

To investigate the molecular genetic basis for the differences in 18-OH-DOC production between Dahl S and R rats, Cicila *et al.* cloned and sequenced near full length cDNAs for P450c11β from the adrenals of inbred Dahl S and R rats (Cicila *et al.*, 1993). The predicted protein sequence of P450c11β in the Dahl S rat was found to be identical to that reported in a noninbred Sprague–Dawley rat. However, five amino acid substitutions were found in the deduced P450c11β sequence of the Dahl R rat (*Table 1*) with two of the substitutions being located at positions 381 and 384 near the putative steroid binding site of the enzyme. In a backcross population derived from Dahl S and R rats, Cicila *et al.* further demonstrated that an intragenic RFLP marking the CYP11B1 variant of the Dahl R rat

Table 1. Molecular variants causing amino acid substitutions in 11β hydroxylase and aldosterone synthase in the Dahl model

Gene:	CYP11B1					CYP11B2	
Protein:	P450c11β (11β hydroxylase)					P450c11AS (aldosterone synthase)	
Location of mutation							
Nucleotide#	379	1052	1141	1150	1327	408	752
Residue#	127	351	381	384	443	136	251
Rat strain							
Sprague–Dawley							
Codon	CGT	GTT	GTA	ATC	GTG	GAA	CAG
Amino acid	Arg	Val	Val	Ile	Val	Glu	Gln
Dahl SS/JR							
Codon	CGT	GTT	GTA	ATC	GTG	GAA	CAG
Amino acid	Arg	Val	Val	Ile	Val	Glu	Gln
Dahl SR/JR							
Codon	TGT	GCT	TTA	CTC	ATG	GAC	CGG
Amino acid	Cys	Ala	Leu	Leu	Met	Asp	Arg

The nucleotide variants giving rise to each amino acid substitution in the Dahl salt-resistant (SR/JR) rat vs. the Dahl salt-sensitive (SS/JR) rat and a Sprague–Dawley rat are underlined. Modified from Cicila G.T. *et al.* (1993) Linkage of 11β-hydroxylase mutations with altered steroid biosynthesis and blood pressure in the Dahl rat. *Nature Genetics*, vol. 3, pp. 346–353. Reprinted by permission of Nature America Inc.

on chromosome 7 cosegregated with significant decreases in the adrenal synthesis of 18-OH-DOC and in blood pressure (Cicila *et al.*, 1993). Matsukawa and colleagues detected the same amino acid substitutions described by Cicila *et al.* in the Dahl R strain and found significantly lower conversion of DOC to 18-OH-DOC in heterologous COS–7 cells expressing P450c11β from the R rat than in those expressing P450c11β from the S rat (Matsukawa *et al.*, 1993). Transfection experiments with chimeric enzymes have suggested that the amino acid substitutions at positions 381, 384, and possibly 443 are likely to be responsible for the altered P450c11β activity of the Dahl R rat (Matsukawa *et al.*, 1983). Thus, consistent with the original proposal of Rapp and Dahl, molecular genetic studies have confirmed that coding sequence variants in the CYP11B1 gene give rise to altered activity of P450c11β and reduced adrenal synthesis of 18-OH-DOC in Dahl R rats versus Dahl S rats.

Recently, Cicila *et al.* derived a congenic strain of Dahl S rats in which a segment of chromosome 7 that includes the CYP11B1 gene was replaced by the corresponding chromosome region from the Dahl R rat (Cicila *et al.*, 1997). In experiments using diets that contained 0.2–4% NaCl, the blood pressures and heart weights of the Dahl S congenic rats carrying the CYP11B1 gene of the Dahl R strain were significantly lower than those of the progenitor Dahl S rats (Cicila *et al.*, 1997). The Dahl S congenic rats fed a 4% NaCl diet also survived much longer than the Dahl S progenitor rats fed a 4% NaCl diet. The differences in survival between the two strains could be accounted for by the differences in blood pressure. These findings, together with the results in backcross populations demonstrating that RFLPs in CYP11B1 cosegregate with effects on blood pressure and adrenal synthesis of 18-OH-DOC, provide further support for the hypothesis that differences in blood pressure between Dahl S and R rats may be due in part to genetically determined differences in 18-OH-DOC synthesis mediated by structural differences in P450c11β.

The CYP11B1 gene of the Dahl R rat differs from those of 12 other commonly used strains of laboratory rats and is associated with uniquely reduced capacity to synthesize 18-OH-DOC (Cicila *et al.*, 1993). In contrast, the CYP11B1 gene of the Dahl S rat appears similar to that in other strains of rats and is associated with normal activity of P450c11β (Cicila *et al.*, 1993). Thus, it would appear that the difference in blood pressure between Dahl S and R rats is related at least in part to abnormally reduced synthesis of 18-OH-DOC by P450c11β in the Dahl R rat rather than abnormally increased production of 18-OH-DOC by the Dahl S rat.

P450c11β also catalyzes the formation of 19-OH-DOC, the precursor for the potent mineralocorticoid 19-nor-DOC (Nonaka and Okamoto, 1991). The urinary excretion of 19-nor-DOC is significantly lower in Dahl R rats than in sex-matched Dahl S rats (Dale *et al.*, 1985). Accordingly, it is possible that the difference in blood pressure between Dahl S and R rats involves differences in P450c11β mediated synthesis of 19-nor-DOC as well as 18-OH-DOC. Gomez-Sanchez and Gomez-Sanchez have suggested that because urinary excretion of 19-nor-DOC in female Dahl R rats is lower than in female Dahl S rats, but greater than in male Dahl S rats, the strain differences in blood pressure cannot be clearly related to strain differences in production of 19-nor-DOC (Gomez-Sanchez and

Gomez-Sanchez, 1988). However, this interpretation is based on the assumption that the hypertensinogenic effect of 19-nor-DOC in female rats is identical to that in male rats. Because it is possible that 19-nor-DOC might have different effects on blood pressure in male versus female rats, one cannot use comparisons of male versus female rats to discount the possibility that 19-nor-DOC contributes to the differences in blood pressure between the S and R strains.

4. Molecular variation in CYP11B2 in the Dahl model

Because the aldosterone synthase gene (CYP11B2) encoding P450c11AS is tightly linked to the CYP11B1 gene on rat chromosome 7 (Cicila *et al.*, 1993), it is possible that the blood pressure results observed in linkage and chromosome transfer studies in Dahl rats are reflecting hemodynamic effects of molecular variants in CYP11B2 (or another linked gene such as CYP11B3) rather than CYP11B1. In the Dahl S rat, adrenal and plasma levels of aldosterone are lower than in the Dahl R rat and, therefore, it has been assumed that genetically determined alterations in aldosterone biosynthesis do not contribute to the strain differences in blood pressure (Rapp *et al.*, 1978). The lower aldosterone levels in Dahl S versus Dahl R rats have been proposed to be a physiological consequence of the lower renin concentrations in Dahl S rats. However, in Dahl S rats, the ratio of aldosterone:renin in the adrenals and in the plasma is greater than in Dahl R rats (Kusano *et al.*, 1986; Rapp *et al.*, 1978). This observation suggests that in Dahl S and Dahl R rats, differences in aldosterone biosynthesis may involve something more than just differences in plasma renin activity. To investigate this possibility, we cloned, sequenced and expressed P450c11AS cDNAs from Dahl S and R rats and tested for strain differences in the structure and regulation of P450c11AS that might contribute to strain differences in aldosterone levels and aldosterone:renin ratios (Cover *et al.*, 1995).

The amount of adrenal P450c11AS mRNA in Dahl S rats was found to be similar to that in Dahl R rats regardless of the amount of NaCl in the diet (Cover *et al.*, 1995). In both strains, NaCl depletion increased P450c11AS mRNA and NaCl loading decreased P450c11AS mRNA in a similar fashion (Cover *et al.*, 1995). Thus, the reduced adrenal and plasma levels of aldosterone in S rats versus R rats cannot be readily attributed to physiological reductions in aldosterone synthase gene expression secondary to reduced levels of adrenal and plasma renin. Indeed, the finding that P450c11AS mRNA levels are similar in Dahl S and R rats despite the fact that adrenal and plasma renin levels are known to be reduced in the S strain suggests that Dahl S and R rats may differ with respect to the transcriptional regulation and or message stability of aldosterone synthase. That is, enhanced transcription or message stability of P450c11AS in Dahl S rats versus R rats could account for similar levels of P450c11AS message despite reduced renin levels in the S strain. This could serve to explain the greater aldosterone:renin ratios that have been reported in Dahl S versus R rats.

Given that P450c11AS mRNA levels are similar in Dahl S versus Dahl R rats, it appears that the lower aldosterone levels in S rats are not simply a consequence of reduced gene expression. However, sequence analysis has revealed structural alterations in CYP11B2 encoding P450c11AS that could definitely contribute to differences in aldosterone synthase activity between Dahl S and R rats (Cover *et al.*, 1995). We found that in the Dahl S rat, the mRNA for P450c11AS is identical to that of a normotensive Sprague–Dawley rat whereas the P450c11AS mRNA of the Dahl R rat contains seven mutations that result in two amino acid substitutions in aldosterone synthase (Cover *et al.*, 1995). Both of the substitutions in the Dahl R rat (Glu136–Asp and Gln251–Arg) generate the same amino acids found at the corresponding residues in the gene encoding P450c11β. Transfection studies demonstrated that these two amino acid substitutions produce a form of P450c11AS with an increased apparent V_{max} and decreased apparent K_m, resulting in an enzyme that catalyses the conversion of DOC to aldosterone at a greater rate in Dahl R rats than the P450c11AS in Dahl S rats or Sprague–Dawley rats (Cover *et al.*, 1995). In transfection studies with mutant forms of human P450c11AS, the same substitutions resulted in 50–80% more aldosterone production than was observed with wild-type P450c11AS (Fardella *et al.*, 1995).

Given that the CYP11B2 allele of the Dahl R rat promotes greater aldosterone levels than the CYP11B2 allele of the Dahl S rat, one might expect increased aldosterone levels and increased blood pressures in the congenic Dahl S rats that carry the corresponding segment of chromosome 7 from the Dahl R strain (Cicila *et al.*, 1997). However, the differential chromosome segment in the congenic Dahl S strain includes multiple genes (e.g. CYP11B1 as well as CYP11B2) and therefore, the blood pressure effect conferred by this segment of chromosome 7 is not necessarily the result of a single gene (Cicila *et al.*, 1997). If the CYP11B2 allele of the Dahl R strain promotes greater aldosterone synthesis and hence greater blood pressure than the CYP11B2 allele of the Dahl S strain, then the Dahl R strain must carry other alleles in this region of chromosome 7 that promote lower blood pressure than their S counterparts, because transfer of this region of chromosome 7 from the R strain onto the S background resulted in a decrease in blood pressure. A likely example is the R CYP11B1 allele that results in reduced production of 11-OH-DOC, explaining the importance of this steroid to the syndrome in the Dahl strain. Blood pressure is under the control of multiple genes and it is well recognized that normotensive strains can harbor variants that promote increased blood pressure as well as variants that promote decreased blood pressure. Alternatively, it is conceivable that the CYP11B2 allele of the Dahl S rat promotes greater blood pressure than the CYP11B2 allele of the R rat even though the R allele encodes a more active form of aldosterone synthase. As previously noted, aldosterone: renin ratios are higher in Dahl S rats than in Dahl R rats, perhaps as a consequence of differences in the transcription of aldosterone synthase. This raises the possibility that in Dahl S rats versus R rats, increased blood pressure may be due in part to overproduction of aldosterone relative to the activity of the renin–angiotensin system (i.e. despite lower absolute levels of aldosterone, the increased ratio of aldosterone:renin in the Dahl S rat may be promoting hypertension). As emphasized by Gordon, patients with hypertension due to primary aldosteronism often present with increased aldosterone:renin ratios and normal plasma levels of aldosterone (Gordon *et al.*, 1993a, 1993b).

5. Implications for future research

Are the genes that regulate blood pressure in animal models of hypertension relevant to those involved in the heritable control of blood pressure in humans? Lifton and colleagues have demonstrated that fusion of the CYP11B1 (11β hydroxylase) and CYP11B2 (aldosterone synthase) genes causes glucocorticoid-remediable aldosteronism (GRA), a rare form of salt-sensitive hypertension in humans (Lifton *et al.*, 1992). Recent studies by Brand *et al.* have also suggested that a variant in the promoter region of aldosterone synthase is associated with essential hypertension (Brand *et al.*, 1998). Although the precise molecular lesions present in humans with GRA or with essential hypertension are different from those in the Dahl model, it appears that in both humans and in rats, molecular variants affecting 11β hydroxylase and aldosterone synthase are involved in the inherited control of blood pressure and mineralocorticoid biosynthesis. Of course, one does not necessarily expect to find the same molecular variants in animal models of hypertension that are present in humans with increased blood pressure. However, the findings in the Dahl model and those in patients with GRA or with essential hypertension confirm that loci contributing to the genetic control of blood pressure in animals can also be involved in the pathogenesis of at least some forms of human hypertension. Based on these observations, it indeed appears that genes regulating blood pressure in animal models can be relevant to the inherited control of blood pressure in humans.

References

Brand, E., Chatelain, N., Mulatero, P., Fery, I., Curnow, K., Jeunemaitre, X., Corvol, P., Pascoe, L. and Soubrier, F. (1998) Structural analysis and evaluation of the aldosterone synthase gene in hypertension. *Hypertension* **32**: 198–204.

Carroll, J., Komanicky, P. and Melby, J.C. (1981) The relationship between plasma 18-hydroxy-11-deoxycorticosterone levels and production of hypertension in the rat. *J. Steroid. Biochem.* **14**: 989–995.

Cicila, G.T., Rapp, J.P., Wang, J.-M., St.Lezin, E., Ng, S.C. and Kurtz, T.W. (1993) Linkage of 11β-hydroxylase mutations with altered steroid biosynthesis and blood pressure in the Dahl rat. *Nature Genet.* **3**: 346–353.

Cicila, G.T., Dukhanina, O.I., Kurtz, T.W., Walder, R., Garrett, M.R., Dene, H. and Rapp, J.P. (1997) Blood pressure and survival of a chromosome 7 congenic strain bred from Dahl rats. *Mamm. Genome*. **8**: 896–902.

Cover, C.M., Wang, J.M., St.Lezin, E., Kurtz, T.W. and Mellon, S.H. (1995) Molecular variants in the P450c11AS gene as determinants of aldosterone synthase activity in the Dahl rat model of hypertension. *J. Biol. Chem.* **270**: 16555–16560.

Curnow, K.M., Luna, M.T.T., Pascoe, L., Natarajan, R., Gu, J.L., Nadler, J.L. and White, P.C. (1991) The product of the CYP11B2 gene is required for aldosterone biosynthesis in the human adrenal cortex. *Molec. Cell. Endocrinol.* **5**: 1513–1522.

Dahl, L.K., Heine, M. and Tassinari, L. (1962) Effects of chronic excess salt ingestion: evidence that genetic factors play an important role in susceptibility to experimental hypertension. *J. Exp. Med.* **115**: 1173–1190.

Dale, S.L., Holbrook, M.M. and Melby, J.C. (1985) 19-Nor-deoxycorticosterone excretion in rats bred for susceptibility and resistance to the hypertensive effects of salt. *Endocrinology* **117**: 2424–2427.

Deng, Y. and Rapp, J.P. (1992) Cosegregation of blood pressure with angiotensin converting enzyme and atrial natriuretic peptide receptor genes using Dahl salt-sensitive rats. *Nature Genet.* **1**: 267–272.

Deng, A.Y., Dene, H., Pravenec, M. and Rapp, J.P. (1994a) Genetic mapping of two new blood pressure quantitative trait loci in the rat by genotyping endothelin system genes. *J. Clin. Invest.* **93**: 2701–2709.

Deng, A.Y., Dene, H. and Rapp, J.P. (1994b) Mapping of a quantitative trait locus for blood pressure on rat chromosome 2. *J. Clin. Invest.* **94**: 431–436.

Domalik, L.J., Chaplin, D.D., Kirkman, M.S., Wu, R.C., Liu, W., Howard, T.A., Seldin, M.F. and Parker, K.L. (1991) Different isozymes of mouse 11-hydroxylase produce mineralocorticoids and glucocorticoids. *Molec. Endocrinol.* 5: 1853–1861.

Fardella, C.E., Rodriguez, H., Hum, D.W., Mellon, S.H. and Miller, W.L. (1995) Artificial mutations in P450c11AS (aldosterone synthase) can increase enzymatic activity: a model for low-renin hypertension? *J. Clin. Endocrinol. Metab.* **80**: 1040–1043.

Feldman, D. and Funder, J.W. (1973) The binding of 18-hydroxydeoxycoticosterone and 18-hydroxycorticosterone to minerlocortoid and glucocortoid recepters in the rat kidney. *Endocrinology* **92**: 1389–1396.

Garrett, M.R., Dene, H., Walder, R., Zhang, Q.Y., Cicila, G.T., Assadnia, S., Deng, A.Y. and Rapp, J.P. (1998) Genome scan and congenic strains for blood pressure QTL using Dahl salt-sensitive rats. *Genome Res.* **8**: 711–723.

Gomez-Sanchez, E.P. and Gomez-Sanchez, C.E. (1988) 19-nordeoxycorticosterone excretion in male and female inbred salt-sensitive (S/JR) and salt-resistant (R/JR) Dahl rats. *Endocrinology* **122**: 1110–1113.

Gordon, R.D., Ziesak, M.D., Tunny, T.J., Stowasser, M. and Klemm, S.A. (1993a) Evidence that primary aldosteronism may not be uncommon: 12% incidence among antihypertensive drug trial volunteers. *Clin. Exp. Pharmacol. Physiol.* **20**: 296–298.

Gordon, R.D., Klemm, S.A., Stowasser, M., Tunny, T.J., Storie, W.J. and Rutherford, J.C. (1993b) How common is primary aldosteronism? Is it the most frequent cause of curable hypertension. *J. Hypertension* **11(suppl. 5)**: S310–S311.

Gu, L., Dene, H., Deng, A.Y., Hoebee, B., Bihoreau, M.T., James, M. and Rapp, J.P. (1996) Genetic mapping of two blood pressure quantitative trait loci on rat chromosome 1. *J. Clin. Invest.* **97**: 777–788.

Herrera, V.L., Xie, H.X., Lopez, L.V., Schork, N.J. and Ruiz-Opazo, N. (1998) The 1 Na,K-ATPase gene is a susceptibility hypertension gene in the Dahl salt-sensitive HSD rat. *J. Clin. Invest.* **102**: 1102–1111.

Jamieson, A. and Fraser, R. (1994) Developments in the molecular biology of corticosteroid synthesis and action: implications for an understanding of essential hypertension. *J. Hypertension* **12**: 503–509.

Kusano, E., Baba, K., Rapp, J.P., Franco-Saenz, R. and Mulrow, P.J. (1986) Adrenal renin in Dahl salt-sensitive rats: a genetic study. *J. Hypertension* **4(supp. 5)**: s20–s22.

Lauber, M. and Muller, J. (1989) Purification and characterization of two distinct forms of rat adrenal cytochrome P–450–11B. Functional and structural aspects. *Arch. Biochem. Biophys.* **274**: 109–119.

Lifton, R.P., Dluhy, R.G., Powers, M., Rich, G.M., Cook, S., Ulick, S. and Lalouel, J.M. (1992) A chimeric 11B-hydroxylase/aldosterone synthase gene causes glucocorticoid-remediable aldosteronism and human hypertension. *Nature* **355**: 262–265.

Malee, M.P. and Mellon, S.H. (1991) Zone-specific regulation of two messenger RNA's for P450c11 in the adrenals of pregnant and nonpregnant rats. *Proc. Natl Acad. Sci. USA* **88**: 4731–4735.

Matsukawa, N., Nonaka, Y., Higaki, J., Nagano, M., Mikami, H., Ogihara, T. and Okamoto, M. (1993) Dahl's salt-resistant normotensive rat has mutations in cytochrome P450(11B), but the salt-sensitive hypertensive rat does not. *J. Biol. Chem.* **266**: 9117–9121.

Mellon, S.H., Bair, S.R. and Monis, H. (1995) P450c11B3 mRNA, transcribed from a third P450c11 gene, is expressed in a tissue-specific, developmentally, and hormonally regulated fashion in the rodent adrenal and encodes a protein with both 11-hydroxylase and 18-hydroxylase activities. *J. Biol. Chem.* **270**: 1643–1649.

Meneely, G.R., Tucker, R.G., Darby, W.J. and Auerbach, S.H. (1953) Chronic sodium chloride toxicity in the rat. *J. Exp. Med.* **98**: 71–79.

Miller, W.L. (1988) Molecular biology of steroid hormone synthesis. *Endocr. Rev.* **9**: 295–318.

Mukai, K., Imai, M., Shimada, H. and Ishimura, Y. (1993) Isolation and characterization of rat CYP11B genes involved in the late steps of mineralo- and glucocorticoid syntheses. *J. Biol. Chem.* **268**: 9130–9137.

Muller, J., Meuli, C., Schmid, C. and Lauber, M. (1989) Adaptation of aldosterone biosynthesis to sodium and potassium intake in the rat. *J. Steroid Biochem.* **34**: 271–277.

Nicholls, M.G., Brown, W.C.B., Hay, G.D., Mason, P.A. and Fraser, R. (1979) Arterial levels and mineralocorticoid activity of 18-hydroxy-11-deoxycorticosterone in the rat. *J. Steroid Biochem.* **10**: 67–70.

Nomura, M., Morohashi, K., Kirita, S., Nonaka, Y., Okamoto, M., Nawata, H. and Omura, T. (1993) Three forms of rat CYP11B genes: 11 beta hydroxylase, aldosterone synthase, and a novel gene. *J. Biochem.* **113**: 144–152.

Nonaka, Y. and Okamoto, M. (1991) Functional expression of the cDNAs encoding rat 11B-hydroxylase [cytochrome P450(11B)] and aldosterone synthase [cytochrome P450(11B, aldo)]. *Eur. J. Biochem.* **202**: 897–202.

Ogishima, T., Mitani, F. and Ishimura, Y. (1989) Isolation of aldosterone synthase cytochrome P-450 from zona glomerulosa mitochondria of rat adrenal cortex. *J. Biol. Chem.* **264**: 10935–10938.

Ogishima, T., Suzuki, H., Hata, J., Mitani, F. and Ishimura, Y. (1992) Zone-specific expression of aldosterone synthase cytochrome P-450 and P45011B in rat adrenal cortex: histochemical basis for the functional zonation. *Endocrino*logy **130**: 2971–2977.

Preuss, M.B. and Preuss, H.G. (1980) The effects of sucrose and sodium on blood pressures in various substrains of wistar rats. *Lab. Invest.* **43**: 101–107.

Rapp, J.P. (1982) Dahl salt-susceptible and salt-resistant rats. A review. *Hypertension* **4**: 753–763.

Rapp, J.P. and Dahl, L.K. (1971) Adrenal steroidogenesis in rats bred for susceptibility and resistance to the hypertensive effect of salt. *Endocrinology* **88**: 52–65.

Rapp, J.P. and Dahl, L.K. (1972) Mendelian inheritence of 18-and 11-steroid hydroxylase activities in the adrenals of rats gentically susceptible or resistant to hypertension. *Endocrinology* **90**: 1435–1446.

Rapp, J.P. and Dahl, L.K. (1976) Mutant forms of cytochrome P-450 controlling both 18-and 11-steroid hydroxylation in the rat. *Biochemistry* **15**: 1235–1242.

Rapp, J.P. and Dene, H. (1985) Development and characteristics of inbred strains of Dahl salt-sensitive and salt-resistant rats. *Hypertension* **7**: 340–349.

Rapp, J.P., Tan, S.Y. and Margolius, H.S. (1978) Plasma mineralocorticoids, plasma renin, and urinary kallilrein in salt-sensitive and salt-resistant rats. *Endocrine Res. Commun.* 5: 35–41.

Rapp, J.P., Wang, S.-M. and Dene, H. (1989) A genetic polymorphism in the renin gene of Dahl rats cosegregates with blood pressure. *Science* **243**: 542–544.

Sander, M., Ganten, D. and Mellon, S.H. (1994) Role of adrenal renin in the regulation of adrenal steroidogenesis by corticotropin. *Proc. Natl Acad. Sci. USA* **91**: 148–152.

Simonet, L., St and Kurtz, T.W. (1991) Sequence analysis of the alpha 1 Na^+,$K(^+)$-ATPase gene in the Dahl salt-sensitive rat. *Hypertension* **18**: 689–693.

Smith-Barbaro, P.A., Quinn, M.R., Fisher, H. and Hegsted, D.M. (1980) Pressor effects of fat and salt in rats. *Proc. Soc. Exp. Biol. Med.* **165**: 283–290.

St. Lezin, E.M., Pravenec, M., Wong, A., *et al.* (1994) Genetic contamination of Dahl SS/JR rats: Impact on studies of salt-sensitive hypertension. *Hypertension* **23**: 786–790.

Stec, D.E., Deng, A.Y., Rapp, J.P. and Roman, R.J. (1996) Cytochrome P4504A genotype cosegregates with hypertension in Dahl S rats. *Hypertension* **27**: 564–568.

Zhang, G. and Miller, W.L. (1996) The human genome contains only two CYP11B (P450c11) genes. *J. Clin. Endocrinol. Metab.* **81**: 3254–3256.

5

Transgenic rats and hypertension

Cathy Payne, Linda J. Mullins, Donald Ogg and John J. Mullins

1. Introduction

Essential hypertension is a multifactorial or complex genetic trait. An individual's susceptibility to high blood pressure is influenced not only by the many genetic factors which effect control through biochemical and physiological pathways, but also by environmental determinants. By this token, the multitude of contributory factors make identification of the underlying etiology very difficult. Transgenesis may be defined as the introduction of exogenous DNA into the genome (by microinjection into the pronucleus of a fertilized oocyte; see *Figure 1*) such that it is stably maintained in a heritable manner. The genetic modification of animals through transgenic technology is playing an increasingly important role in the field of hypertension research, allowing analysis of gene function and regulation *in vivo*. In this chapter, we will briefly discuss general considerations for designing a transgenic experiment. We will then outline the usefulness of transgenic rat models in the context of hypertension research, reviewing the most important transgenic lines that have been generated to date. Finally, we will allude to future possibilities for transgenic research in this field.

2. General considerations in experimental design

2.1 Choice of species

The mouse has traditionally been the species of choice for transgenic research, and with the emergence of embryonic stem (ES) cell technology, the versatility of the mouse for generating disease models is unsurpassed. However, for certain physiological and biochemical studies, the rat may be preferable, because of the inherent size constraints of the mouse. This is true for techniques such as radiotelemetry and echocardiography, where miniaturization and resolution must be improved before the methods can be readily and reproducibly applied to the mouse. Though such improvements are inevitable, present limitations prevent many researchers from taking full advantage of the molecular genetics available in the mouse.

Molecular Genetics of Hypertension, edited by A.F. Dominiczak, J.M.C. Connell and F. Soubrier.

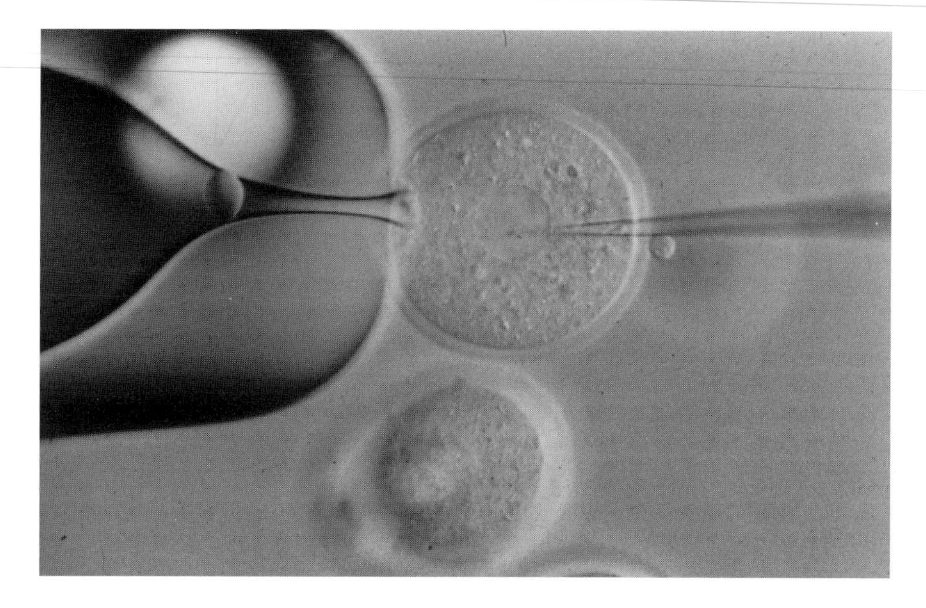

Figure 1. Microinjection of the fertilized oocyte. The fertilized egg is held by gentle suction using a holding pipette (left). The microinjection needle is gently pushed through the zona pellucida and the nuclear membrane of one pronucleus (right) and DNA is introduced.

The rat has historically been the animal of choice in hypertension research by virtue of the number of available models including the spontaneously hypertensive rat (SHR), the spontaneously hypertensive stroke-prone rat (SHRSP), the Dahl salt-sensitive, the Milan hypertensive and the Lyon hypertensive strains. Extensive study has generated a large body of data regarding the progression of hypertension in each model, and it is clear that the observed phenotypic manifestation reflects pleiotropic changes in interconnected homeostatic pathways, in an attempt to redress the balance affected by the underlying (and often unknown) etiology. Transgenesis allows targeting of key components in order to determine their role in basic physiological processes leading to the overall control of blood pressure.

Apart from general considerations of space and cost when undertaking a transgenic rat program, one must also bear in mind the relative values of using inbred versus outbred strains. The benefits of fecundity of an outbred strain such as Sprague–Dawley must be balanced against the poorer breeding efficiency, but perhaps greater suitability, of a particular inbred strain for a given experiment. Additionally, the genetic background can play a highly significant part in the phenotypic manifestation of the expression of a given transgene, as will be highlighted in specific examples later in the chapter.

2.2 Candidate transgenes

Genes which are potentially involved in the development of hypertension have been identified by a number of different strategies, including genetic, biochemical and linkage analyses, apparent involvement in classical hypertensive rat models or association with pathological findings. Mutations in aldosterone synthase, 11β-hydroxysteroid dehydrogenase, and subunits of the amiloride-sensitive epithelial sodium channel have been identified as the genetic defects underlying the Mendelian forms of hypertension – glucocorticoid-remediable aldosteronism (Lifton *et al.*, 1992), apparent mineralocorticoid excess syndrome (Mune *et al.*, 1995) and Liddle's syndrome (Hansson *et al.*, 1995; Shimkets *et al.*, 1994), respectively.

Linkage analysis in both animal models and man has been used to identify factors contributing to hypertension based on their chromosomal location. Both partial and whole-genome searches using microsatellite markers have been used to correlate blood pressure and hypertension with genetic regions. Mapping DNA polymorphisms in a cross between SHRSP and Wistar–Kyoto (WKY) revealed a locus, *Bp1*, which is closely linked to the rat gene encoding ACE (Jacob *et al.*, 1991). Crosses involving Dahl salt-sensitive rats have suggested that loci near the calmodulin-dependent protein kinase II-d locus (Deng *et al.*, 1994) and the 11β-hydroxylase locus (Cicila *et al.*, 1993) may influence rat blood pressure. Furthermore, variants of the *SA* gene have been found to cosegregate with blood pressure in a cross between SHR and WKY rats (Samani *et al.*, 1993). Though linkage analysis identifies the chromosomal location of factors contributing to hypertension, an obvious candidate gene falling within this region may not be the causitive agent. It is therefore necessary to substantiate the linkage analysis with gene expression studies. Once a suitable candidate has been identified, transgenic studies can aid in determining its role in the development of hypertension.

2.3 Anomolies due to strain and species specificity

One must be cautious in attributing the influence of a particular candidate gene to blood pressure. When the contribution of several different renin alleles to blood pressure was analyzed in appropriate F_2 populations, only one allele, the r allele of Dahl salt-resistant rats, was found to cosegregate with blood pressure (Rapp *et al.*, 1994). The net result of multiple alleles, differences in genetic background, linkage relationships and genotype–environment interactions is to make cosegregation results strain- and cross-specific. The need for caution when searching for disease-relevant genetic loci is even greater due to anomolies between species. For example, the association between angiotensinogen (AGT) and hypertension demonstrated in humans with different *Agt* alleles (Jeunemaitre *et al.*, 1992) was not confirmed by a study on F_2 hybrids between SHRSP and WKY (Hubner *et al.*, 1994). Such examples suggest caution in the extrapolation of transgenic experiments across strain and species boundaries.

2.4 Transgenic experiments

Transgenic animals derived by microinjection share a common problem – namely that the researcher has no control over the site of insertion of the transgene. Positional effects may adversely affect transgene expression – integration near a strong enhancer may lead to overexpression, possibly in a novel range of tissues, while integration into a silent region of the genome may silence an otherwise active transgene. Clearly, expression of the transgene is also dependent on the presence of appropriate control elements within it. Since control elements can lie at some distance from the gene, it may be necessary to include large amounts of flanking DNA within the transgene, through the use of PACs, BACs or YACs (Linton *et al.*, 1993; Mullins *et al.*, 1997; Schedl *et al.*, 1993).

Additionally, a phenotype may only be observed if transgene expression achieves sufficient levels – hypertension in transgenic mice carrying the rat AGT gene was dependent on expression level (Kimura *et al.*, 1992; Ohkubo *et al.*, 1990). Finally, the introduction of a foreign gene may produce no phenotype, or alternatively an unexpected phenotype due to differences in substrate specificity or action.

All the above must be borne in mind when designing or interpreting transgenic experiments. However, despite these caveats, an immense amount can be learnt from specific transgenic models, as reviewed below.

3. Transgenic rat models relevant to cardiovascular research

The following review of transgenic rat lines is not exhaustive, but has been chosen to exemplify points raised in the previous section. The first three lines demonstrate how animal models can be modified in order to answer specific questions raised by the phenotypic analysis of the parental transgenic strain.

3.1 TGR(mRen2)27

The TGR(mRen2)27 transgenic rat harbors a 24 kb transgene spanning the mouse *Ren2* gene (*Figure 2*) and is a monogenic model of fulminant hypertension (Mullins *et al.*, 1990). The exact mechanisms involved in the onset of hypertension in this transgenic rat are still unclear, although the high circulating mouse prorenin levels, the enhanced expression of renin in the adrenal gland and the corresponding increase in the levels of circulating mineralocorticoids (Sander *et al.*, 1992), have all been implicated. Fulminant hypertension develops within 10 weeks of age (Mullins *et al.*, 1990), with a systolic blood pressure of 230–265 mmHg in heterozygotes and up to 300 mmHg in homozygotes. Animals exhibit secondary complications associated with hypertension, such as stroke, heart failure and renal sclerosis, but can be maintained when treated with ACE inhibitor (Mullins *et al.*, 1990; Tokita *et al.*, 1995). The Ang II receptor (AT_1) antagonist DuP753 has a similar protective effect (Bader *et al.*, 1992; Sander *et al.*, 1992), demonstrating that the renin–angiotensin system (RAS) plays a key role in the hypertensive phenotype. Further, the AT_1 receptor antagonist, Telmisartan, is able to reduce cardiac hypertrophy and renal glomerulosclerosis in the TGR(mRen2)27 rats (Bohm *et al.*, 1995), indicating a specific involvement of Ang II.

Renin is barely detectable in the kidney of TGR(mRen2) rats (Mullins *et al.*, 1990; Zhao *et al.*, 1993), indicating that both endogenous- and transgene-derived renin gene expression are down-regulated. *In situ* hybridization and

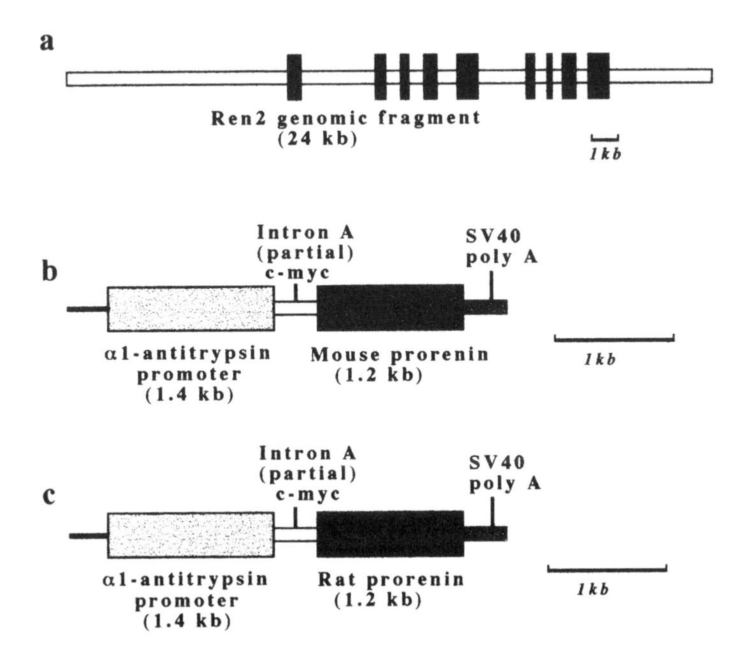

Figure 2. Schematic diagram comparing three transgenes. (a) A 24 kb genomic fragment spanning *Ren2* from the DBA/2 mouse strain, used to generate TGR(mRen2)27. (b) Ren2 cDNA fused to the human α1 antitrypsin promoter, used to generate TGR(α1AT mRen2). (c) Rat renin cDNA fused to the human α 1 antitrypsin promoter, used to generate TGR(hAT rpR).

immunohistochemical studies show that the majority of juxtaglomerular afferent arterioles are almost devoid of renin expression, and the JG cells are relatively dedifferentiated (Bachmann *et al.*, 1992). Additionally, *Ren*-2 is highly expressed in the adrenal cortex (Véniant *et al.*, 1995; Zhao *et al.*, 1993), and is moderately expressed in other extrarenal tissues including the gastrointestinal tract, lung, brain and vasculature, but not in the submaxillary gland (SMG), despite high expression of *Ren2* in the SMG of the mouse (Mullins *et al.*, 1989). This is probably due to the lack of essential transacting factors necessary for expression in this tissue.

Components of the RAS, were analyzed by radioimmunoassay (Hermann *et al.*, 1988; Schelling *et al.*, 1980). There is some controversy about plasma renin concentrations, though recent evidence suggests that the level of circulating renin increases with age, up to 50 ng Ang I ml^{-1} h^{-1} at 7 weeks compared to 20 ng Ang I ml^{-1} h^{-1} in the controls (Véniant *et al.*, 1995), and is approximately 70% transgene-derived (Peters *et al.*, 1993). The observation that the kinetics of the reaction between mouse renin and rat AGT is 10 times that of rat renin on the rat substrate (Tokita *et al.*, 1994), suggests that the enhanced kinetics associated with mouse renin might be involved in the development of hypertension. Plasma angiotensin I (Ang I) concentrations are significantly lower than in control animals, while plasma AGT and Ang II levels are not significantly changed. However, the plasma prorenin concentration is approximately 20-fold higher than in age-matched

Table 1. Comparison of the three transgenic rat lines expressing renin or prorenin

Transgenic line	Transgene	BP (mmHg)	Plasma prorenin (ng Ang I ml^{-1} h^{-1})	Pathology	References
TGR(mRen2)27	Mouse *Ren-2* genomic sequence	230–265 (at 10 weeks)	850*	Stroke Heart failure Left ventricular hypertrophy Hypertrophic vascular myocytes Severe glomerulosclerosis Thickening of walls of arcuate interlobular arteries	Mullins *et al.*, 1990 Bohm *et al.*, 1995 Bachmann *et al.*, 1992 Lee *et al.*, 1991
TGR(α1ATmRen2)	α1AT -promoter fused to mouse Ren-2 cDNA	185 (at 7 weeks)	17000***	Concentric left ventricular hypertrophy LV:BW at 3 weeks: 6.8+/–0.26: 1* Hypertrophic cardiomyocytes Kidney lesions	Ogg, 1997
TGR(hAT rpR)	α1AT- promoter fused to rat cDNA	Normal (at 10 weeks)	9557***	Left ventricular hypertrophy HW/BW: 0.38+/–0.02** Hypertrophic cardiomyocytes Subendocardial fibrosis Kidney sclerosis and thickening of arterial walls	Véniant *et al.*, 1996
Sprague–Dawley control	–	110, 130 (at 7 and 10 weeks respectively)	20–30	LV:BW (3 weeks): 4.4+/–0.21: 1 HW/BW: 0.29+/–0.02	

BP measured by tail plethysmography under light anesthesia; plasma prorenin analysis for males only; LV: BW is the left ventricular mass to body weight ratio × 10^{-3}); HW/BW is the heart weight to body weight ratio × 100); * $P < 0.05$, ** $P < 0.01$, ***$P < 0.001$ *vs.* nontransgenic rats.

Sprague–Dawley controls (see *Table 1*; Peters *et al.*, 1993). The plasma prorenin and renin are partly derived from the adrenal gland, as shown by bilateral adrenalectomy which decreased the concentrations by 20% and 40%, respectively (Bachmann *et al.*, 1992; Tokita *et al.*, 1995). It is important to note that the mouse zymogen can be activated in rats despite amino acid differences at the pre-proenzyme cleavage site (Morris, 1992).

The pathological alterations arising from prolonged hypertension in TGR(mRen2)27 rats are shown in *Figure 3* and summarized in *Table 1*. The mechanisms behind the cardiovascular hypertrophy are unknown, although a direct

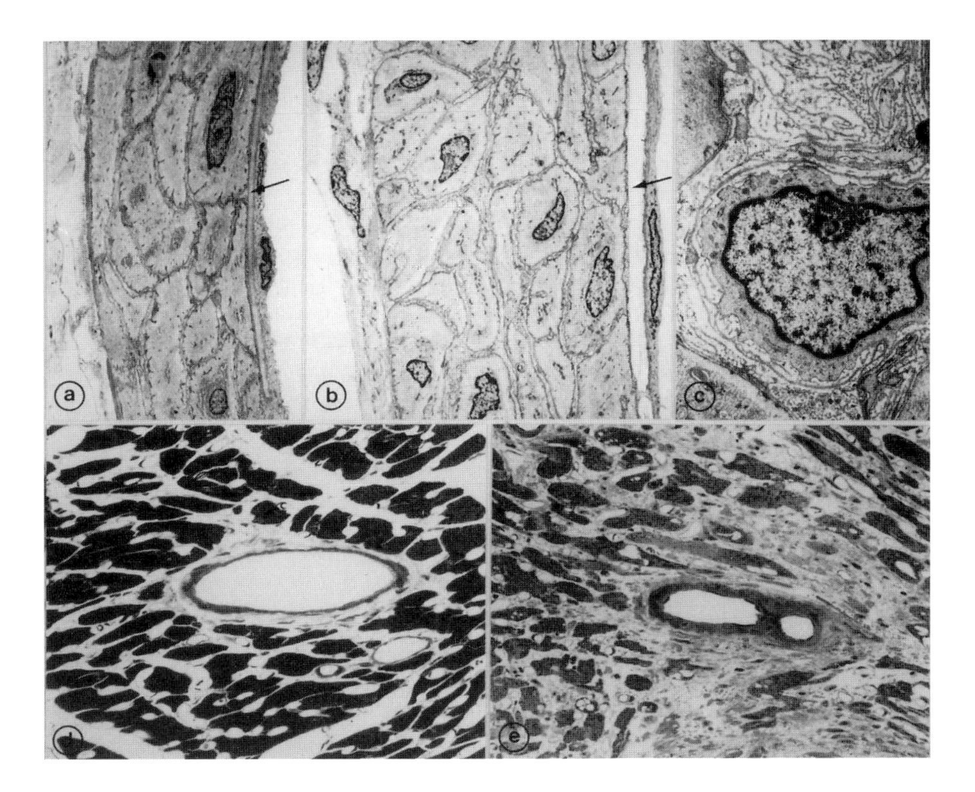

Figure 3. Hypertension-induced morphological alterations in TGR(mRen2)27 rats. Compared with a control (a), the tunica media of the renal arcuate artery in TGR (b) is significantly thickened and there is expansion of the interstitial matrix. Note the difference in thickness of the internal elastic lamina (arrows). There is no obvious size difference of media myocytes between the hypertensive and the normotensive. (a, b: ultrathin sections, × 2166. (c) Higher magnification of a TGR renal arcuate artery revealing increase in basement membrane-like material and collagen fibers between myocytes of the tunica media. (Ultrathin section, × 10 355.) Coronary arteries of a similar external diameter in a control (d) and a TGR heart (e); note the increase in tunica media thickness as well as the obvious myocyte hypertrophy in the TGR arteriole. The TGR arteriole reveals perivascular fibrosis. (d, e: semithin sections, toluidine blue, × 243.) Reprinted with permission from Bachmann S., Peters J., Engler E. *et al.* Transgenic rats carrying the mouse renin gene – morphological characteristics of a low renin hypertensive model. *Kidney International* 1992; 41: 24–36. © 1992 Blackwell Scientific Inc.

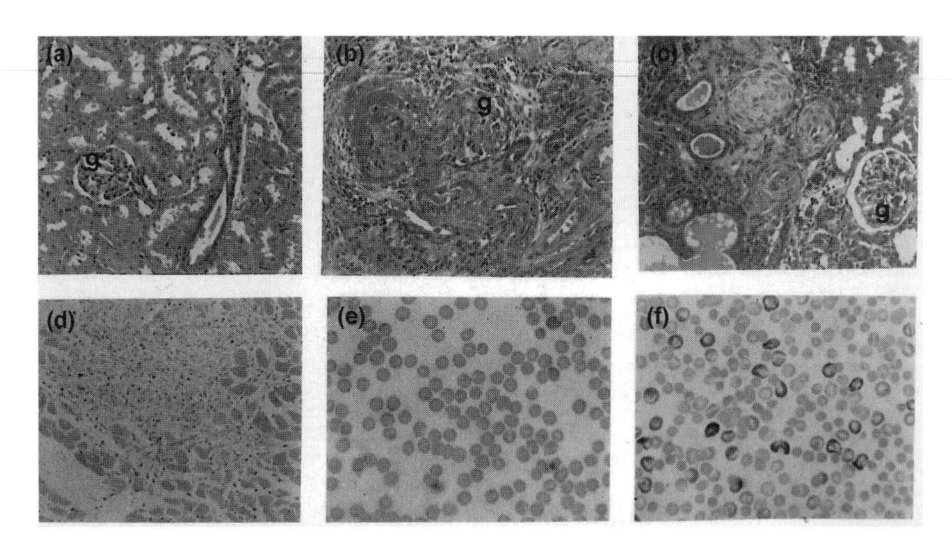

Figure 4. Pathological consequences of malignant hypertension in TGR(mRen2)27 rats. Kidneys from control rats (a) and rats exhibiting malignant hypertension (MH) (b, c) were paraffin-embedded, sectioned (3 μm) and stained with hematoxylin and eosin (g = glomeruli). Fibrinoid necrosis (b) and myointimal proliferation (c) can be observed in the MH samples. Two further features of the MH group were the occurrence of microscopic infarcts in heart (d), and microangiopathic hemolytic anemia (f) which was absent from the controls (e). Reprinted with permission from Whitworth C., Fleming S., Kotelevtsev Y. *et al.* A genetic model of malignant phase hypertension in rats. *Kidney International* 1995; 47: 529–535. © 1995 Blackwell Scientific Inc.

growth-promoting effect of Ang II on cardiomyocytes and vascular smooth muscle cells has been demonstrated (Schelling *et al.*, 1991), suggesting that the local increase in Ang II concentration may contribute to the cardiac hypertrophy in TGR(mRen2)27.

Malignant phase hypertension. Malignant hypertension (MH) is a rare clinical complication of human essential hypertension, and is characterized by a rapid elevation of blood pressure, accompanied by cellular myointimal proliferation, endothelial swelling and fibrinoid necrosis (Kincaid–Smith, 1982; Wilson *et al.*, 1939). The renal RAS is presumed to be activated by the renal afferent vascular pathology, and sodium and water excretion further increases blood pressure, leading to progressive renal damage.

A small percentage of the original TGR(mRen2)27 animals maintained on the 'Hannover' Sprague–Dawley (HanSD) genetic background developed MH. The penetrance of MH increased when HanSD rats homozygous for the *Ren2* transgene were crossed with 'Edinburgh' Sprague–Dawley rats (EdinSD). Seventy-six per cent of the offspring showed a spontaneous appearance of MH (see *Figure 4*), with similar clinical features to human MH, providing a potential animal model of the human condition. Further studies have identified genetic factors associated with the increased incidence of MH in this model (Kantachuvesiri *et al.*, 1999; Whitworth *et al.*, 1994; Whitworth *et al.*, 1995).

3.2 TGR(α1AT mRen2)

The TGR(α1AT mRen2) rats were generated in order to study the role of elevated circulating mouse prorenin levels on the hypertensive phenotype. Expression of mouse Ren2 was directed to the liver since an abundance of nonactivated prorenin would be secreted into the circulation. Liver-specific expression (assessed by RNase protection) was achieved by fusing the Ren2 cDNA to a previously characterized promoter fragment from human α1 antitrypsin (*Figure 2*) (Jallet *et al.*, 1990; Ogg, 1997; Sifers *et al.*, 1987). One founder died at 5 weeks of age, exhibiting severe left ventricular hypertrophy, and a second founder developed malignant phase hypertension, as assessed by fibrinoid necrosis of the kidney. Blood pressure measurements, determined by tail-cuff plethysmography, demonstrated a steep rise in systolic blood pressure from 4 to 7 weeks of age (see *Table 1*). The maximum blood pressure (approximately 200 mmHg systolic), was similar to that reached by the TGR(mRen2)27 rats, albeit at an accelerated rate. The plasma RAS levels were measured at 4 weeks of age by radioimmunoassay. Prorenin levels were significantly elevated, whereas plasma active renin, Ang I and Ang II levels were not significantly different from the negative controls. Immunoprecipitation with a rabbit antimouse-renin antibody revealed that 75% of active renin was of Ren2 origin, although the site of mouse prorenin activation in this model is as yet unknown. All pathological alterations observed (*Table 1*) were consistent with either chronic or acute blood pressure effects.

To estimate the timing of onset of left ventricular hypertrophy, left ventricular mass:body weight (LV:BW) ratios were determined (Jones *et al.*, 1992). By 3 weeks of age the rats demonstrated marked cardiac hypertrophy even though no statistical difference in systolic blood pressures was observed between transgenic and nontransgenic animals at this age. The LV:BW ratios clearly show that left ventricular hypertrophy occurs prior to the development of hypertension in this line.

3.3 TGR(hAT rpR)

A third line of transgenic rat, TGR(hAT rpR), which expresses rat prorenin exclusively in the liver, was established to facilitate comparison between species-specific kinetic differences in the RAS (Véniant *et al.*, 1996). The rat prorenin cDNA was fused to the same promoter as that described for line TGR(α1AT mRen2) (*Figure 2*) and RNase protection analysis, using a rat renin riboprobe, showed liver-specific expression. *In situ* hybridization studies demonstrated reduced endogenous renal renin expression in male animals, as observed in TGR(mRen2)27, although normal levels of renin staining were seen in female transgenics and nontransgenic controls. Plasma prorenin, renin, and angiotensinogen concentrations, measured by radioimmunoassay, showed that plasma prorenin levels were significantly higher for both male (400-fold increase) and female (2–3-fold increase) animals (*Table 1*). Plasma active renin and plasma AGT levels were not elevated for either sex compared to nontransgenic controls. Importantly, despite having normal systolic blood pressure as young adults, the males exhibited cardiac and renal histopathological alterations (*Table 1*) (Véniant *et al.*, 1996).

In conclusion, since TGR(hAT rpR) rats have similar prorenin levels to TGR(α1AT mRen2) but show no increase in blood pressure until the cardiac hypertrophy and renal pathology are well advanced, this suggests that increased plasma prorenin is sufficient to cause cardiac and renal histopathology independent of hypertension. The increased kinetics of mouse renin with rat substrate contributes to the advanced hypertensive phenotype seen in TGR(mRen2)27 and TGR(α1AT mRen2) animals.

4. Other cardiovascular-related transgenic rat strains

4.1 Human renin and angiotensinogen transgenics

When the human renin gene was introduced into the rat genome, the rats remained normotensive, confirming that human renin did not recognize rat AGT as a substrate (Ganten *et al.*, 1992). Likewise, rats expressing the human AGT gene failed to develop hypertension, again indicating species specificity of the RAS system. This was further borne out by the observation that infusion of human renin into the TGR(hAOGEN) animals caused elevation of blood pressure. When the two lines were crossed together, the progeny remained normotensive, but this was probably a function of low transgene expression levels (Ganten *et al.*, 1992). An equivalent series of transgenic mouse lines, expressing rat renin and AGT (Ohkubo *et al.*, 1990), *did* result in elevated blood pressure in the double transgenic animals, and exemplifies the possibilities for using nonprimate animal models expressing human transgenes, for primate-specific *in vivo* analyses.

4.2 Human endothelin-2

A transgenic rat line overexpressing the human endothelin-2 gene in the kidney has been described (Hocher *et al.*, 1996). Since long-term activated paracrine endothelin has been implicated in chronic renal failure, due to progressive glomerular injury, the kidneys were examined in detail. Using *in situ* hybridization, the human endothelin-2 gene was found to be almost exclusively expressed within the glomeruli. This resulted in significantly increased glomerular injury and protein excretion, though glomerular filtration rate was unaltered, and blood pressure was normal. This model should be useful for studying renal endothelin in relation to kidney pathophysiology.

4.3 Human α_2Na,K-ATPase

A transgenic rat line expressing the chloramphenicol acetyltransferase reporter gene, driven by the human α_2Na,K-ATPase regulatory region was generated (Ruiz-Opazo *et al.*, 1997), in order to investigate transcriptional pressure response mechanisms. Parallel tissue-specific regulation of the transgene and the endogenous rat gene was demonstrated, as was coordinate load-induced deinduction of both cardiac and vascular expression of the two genes. This suggests that both the rat and the human have similar systolic pressure gradient-dependent responses.

4.4 ACE

Recently, a transgenic line over-expressing human ACE under the control of the rat MLC2 promotor, was described (Pinto *et al.*, 1997). The 50-fold increase in cardiac specific ACE activity had no effect on left ventricular weight under normal conditions. However, left ventricular hypertrophy was seen in response to suprarenal aortic stenosis. This suggests that increased myocardial ACE could be an independent risk factor in the pathogenesis of cardiac hypertrophy.

5. Future directions

Transgenic technology has greatly enhanced the development of animal models in cardiovascular research, providing the pharmaceutical industry with better tools for the testing and development of new drugs. The relative speed with which transgenic models can be characterized, given that the underlying genetic alteration is known, is pivotal to the design of more refined models in the future. Despite the subtleties of species specificity, or in some cases because of it, much has been learnt about many aspects of hypertension and cardiovascular biology, which may ultimately lead to the identification of those at risk of developing hypertension.

To improve the design of transgenic models, there are a number of strategies available. The inclusion of extensive flanking sequences reduces position effects, and increases the possibility that important locus control elements will be retained within the transgene, making its expression site independent. Alternatively, one can include regulatory elements within the construct, which allow the researcher to turn transgene expression on or off at will. This is exemplified by the tetracycline-regulated transcription system (Gossen *et al.*, 1995; Kistner *et al.*, 1996) and applications of the cre-*loxP* recombination system (Sauer, 1998).

Ever more ingenious and elegant strategies are being published for the refinement of gene control, but many of these depend upon gene targeting in order to replace the endogenous copy of a given gene with a mutated or mutatable transgene (Brocard *et al.*, 1997; Li *et al.*, 1996; Meyers *et al.*, 1998). The importance of ES cell technology, and homologous recombination as tools for generating disease models in the mouse has already been alluded to. Since the development of ES cell technology in the rat would enormously increase the possibilities for manipulating the rat genome, many groups are striving towards this goal (Iannaconne *et al.*, 1994). Ultimately, one would envisage being able to derive ES cells from rat strains such as SHR, so that 'correction' of candidate genes, through homologous recombination, might modify or cure the hypertensive phenotype, and positively identify the underlying etiology, with obvious repercussions for the classification and treatment of essential hypertension in man.

References

Bachmann, S., Peters, J., Engler, E., Ganten, D. and Mullins, J. (1992) Transgenic rats carrying the mouse renin gene – morphological characteristics of a low renin hypertensive model. *Kidney Int.* **41**: 24–36.

Bader, M., Zhao, Y., Sander, M., *et al.* (1992) Role of renin in the pathophysiology of hypertension in TGR(mRen2)27 rats. *Hypertension* **19**: 681–686.

Böhm, M., Lee, M., Kreutz, R. *et al.* (1995) Angiotensin II receptor blockade in TGR(mRen2)27: effects of renin-angiotensin-system gene expression and cardiovascular function. *J. Hypertens.* **13**: 891–899.

Brocard, J., Warot, X., Wendling, O., Messaddeq, N., Vonesch, J.-L., Chambon, P. and Metzger, D. (1997) Spatio-temporally controlled site-specific somatic mutagenesis in the mouse. *Proc. Natl Acad. Sci. USA* **94**: 14559–14563.

Cicila, G.T., Rapp, J.P., Wang, J.M., St. Lizin, E., Chung Ng, S. and Kurtz, T.W. (1993) Linkage of 11β-hydroxylase mutations with altered steroid biosynthesis and blood pressure in the Dahl rat. *Nature Genet* **3**: 346–353.

Deng, A.Y., Dene, H. and Rapp, J.P. (1994) Mapping of a quantitative trait locus for blood pressure on rat chromosome 2. *J. Clin. Invest.* **94**: 431–436.

Ganten, D., Wagner, J., Zeh, K. *et al.* (1992) Species specificity of renin kinetics in transgenic rats harbouring the human renin and angiotensinogen genes. *Proc. Natl Acad. Sci. USA* **89**: 7806–7810.

Gossen, M., Freundlieb, S., Bender, G., Muller, G., Hillen, W. and Bujard, H. (1995) Transcriptional activation by tetracyclines in mammalian cells. *Science* **268**: 1766–1769.

Hansson, J.H., Nelson-Williams, C., Suzuki, H., Schild, L., Shimkets, R., Lu,Y., Canessa,C., Iwasaki, T. and Rossier, B. (1995) Hypertension caused by a truncated epithelial sodium channel γ subunit: genetic heterogeneity of Liddle's syndrome. *Nature Genet.* **11**: 76–82.

Hermann, K., Ganten, D., Unger, T., Bayer, C. and Lang, R.E. (1988) Measurement and characterisation of angiotensin peptides in plasma. *Clin. Chem.* **34**: 1046–1051.

Hocher, B., Liefeldt, L., Thone-Reineke, C., Orzechowski, H.-D., Distler, A., Bauer, C. and Paul, M. (1996) Characterisation of the renal phenotype of transgenic rats expressing the human endothelin-2 gene. *Hypertension* **28**: 196–201.

Hubner, N., Kreutz, R., Takahashi, S., Ganten, D. and Lindpaintner, K. (1994) Unlike human hypertension, blood pressure in a hereditary hypertensive rat strain shows no linkage to the angiotensinogen locus. *Hypertension* **23**: 797–801.

Iannaccone, P.M., Taborn, G.U., Garton, R.L., Caplice, M.D. and Brenin, D.R. (1994) Pluripotent embryonic stem cells from the rat are capable of producing chimeras. *Dev. Biol.* **163**: 288–292.

Jacob, H.J., Lindpaintner, K., Lincoln, S.E., Kusumi, K., Bunker, R.K., Mao, Y.P., Ganten, D., Dzau, V.J. and Lander, E.S. (1991) Genetic mapping of a gene causing hypertension in the stroke-prone spontaneously hypertensive rat. *Cell* **67**: 213–224.

Jallet, S., Perraud, F., Dalemans, W., Balland, A., Dieterle, A., Faure, T., Meulien, P. and Pavirani, A. (1990) Characterisation of recombinant human factor IX expressed in transgenic mice and in derived trans-immortalised hepatic cell lines. *EMBO J.* **9**: 3295–3301.

Jeunemaitre, X., Soubrier, F., Kotelevtsev, Y. *et al.* (1992) Molecular basis of human hypertension: role of angiotensinogen. *Cell* **71**: 169–180.

Jones, E.F., Harrap, S.B., Calafiore, P. and Tonkin, A.M. (1992) Development and validation of echocardiographic methods for estimating left ventricular mass in rats. *Clin. Exp. Pharm. Physiol.* **19**: 361–364.

Kantachuvesiri, S., Haley, C., Fleming, S., Kurian, K., Whitworth, C.E., Wenham, P., Kotelevtsev, Y. and Mullins, J. (1999) Genetic mapping of modifier loci affecting malignant hypertension in TGRmRen2 rats. *Kidney Int.* **56**: (in press).

Kimura, S., Mullins, J., Bunnemann, B. *et al.* (1992) High blood pressure in transgenic mice carrying the rat angiotensinogen gene. *EMBO J.* **11**: 821–827.

Kincaid-Smith, P. (1982) Renal pathology in hypertension and the effects of treatment. *Br. J. Clin. Pharmacol.* **13**: 107–115.

Kistner, A., Gossen, M., Zimmermann, F., Jerecic, J., Ullmer, C., Lubbert, H. and Bujard, H. (1996) Doxycycline-mediated quantitative and tissue-specific control of gene expression in transgenic mice. *Proc. Natl Acad. Sci. USA* **93**: 10933–10938.

Lee, M., Zhao, Y. and Peters, J. (1991) Preparation and analysis of transgenic rats expressing the mouse *Ren2* gene. *J. Vasc. Med. Biol.* **3**: 50–54.

Li, Z.-W., Stark,G., Gotz, J., Rulicke, T. and Muller, U. (1996) Generation of mice with a 200-kb amyloid precursor protein gene deletion by Cre recombinase-mediated site-specific recombination in embryonic stem cells. *Proc. Natl Acad. Sci. USA* **93**: 6158–6162.

Lifton, R.P., Dluhy, R.G., Powers, M., Rich, G.M., Cook, S., Ulick, S. and Lalouel, J.-M. (1992) A chimaeric 11 beta-hydroxylase/aldosterone synthase causes glucocorticoid-remediable aldosteronism and human hypertension. *Nature* **355**: 262–265.

Linton, M.F., Farese, R.V. Jr, Chiesa, G., Grass, D.S., Chin, P., Hammer, R.E., Hobbs, H.H. and Young, S.G. (1993) Transgenic mice expressing high plasma concentrations of human apolipoprotein B100 and lipoprotein (a). *J. Clin. Invest.* **92**: 3029–3037.

Meyers, E.N., Lewandoski, M. and Martin, G.R. (1998) An *Fgf8* mutant allelic series generated by Cre- and Flp-mediated recombination. *Nature Genet.* **18**: 136–141.

Morris, B.J. (1992) Molecular biology of renin 1: Gene and protein structure, synthesis and processing. *J. Hypertens.* **10**: 209–214.

Mullins, J.J., Sigmund, C.D., Kane-Haas, C. and Gross, K.W. (1989) Expression of the DBA/2J *Ren2* gene in the adrenal gland of transgenic mice. *EMBO J.* **8**: 4065–4072.

Mullins, J.J., Peters, J. and Ganten, D. (1990) Fulminant hypertension in transgenic rats harbouring the mouse *Ren-2* gene. *Nature* **344**: 541–544.

Mullins, L.J., Kotelevtseva, N., Boyd, C. and Mullins, J. (1997) Efficient Cre-lox linearisation of BACs: applications to physical mapping and generation of transgenic animals. *Nucleic Acids Res.* **25**: 2539–2540.

Mune, T., Rogerson, F.M., Nikkila, H., Agarwal, A.K. and White, P.C. (1995) Human hypertension caused by mutations in the kidney isozyme of 11β-hydroxysteroid dehydrogenase. *Nature Genet.* **10**: 394–399.

Ogg, D. (1997) Characterisation of rat lines transgenic for the mouse Ren-2^d cDNA. Ph.D. thesis, Edinburgh University.

Ohkubo, H., Kawakami, H., Kakehi, Y. *et al.* (1990) Generation of transgenic mice with elevated blood pressure by introduction of the rat renin and angiotensinogen genes. *Proc. Natl Acad. Sci. USA* **87**: 5153–5157.

Peters, J., Munter, K., Bader, M., Hackenthal, E., Mullins, J. and Ganten, D. (1993) Increased adrenal renin in transgenic hypertensive rats TGR(mRen2)27 and its regulation by cAMP, angiotensin II and calcium. *J. Clin. Invest.* **91**: 742–747.

Pinto, Y.M., Tian, X.L., Costerousse, O. et al. (1997) Cardiac overexpression of angiotensinogen-converting enzyme in transgenic rats augments cardiac hypertrophy. *Circulation* **96**: 3521.

Rapp, J.P., Dene, H. and Deng, A.Y. (1994) Seven renin alleles in rats and their effects on blood pressure. *J. Hypertens.* **12**: 349–355.

Ruiz-Opazo, N., Xiang, X.H. and Herrera, L.M. (1997) Pressure-overload deinduction of human α_2Na,K-ATPase gene expression in transgenic rats. *Hypertension* **29**: 606–612.

Samani, N.J., Lodwick, D., Vincent, M. *et al.* (1993) A gene differentially expressed in the kidney of the spontaneously hypertensive rat cosegregates with increased blood pressure. *J. Clin. Invest.* **92**: 1099–1103.

Sander, M., Bader, M., Djavidani, B., Maser-Gluth, C., Vecsai, P., Mullins, J., Ganten, D. and Peters, J. (1992) The role of the adrenal gland in hypertensive transgenic rat TGR(mRen2)27. *Endocrinology* **131**: 807–814.

Sauer, B. (1998) Inducible gene targeting in mice using the cre/lox system. *Methods Companion Methods Enzymol* **14:** 381–392.

Schedl, A., Montoliu, L., Kelsey, G. and Schutz, G. (1993) A yeast artificial chromosome covering the tyrosinase gene confers copy number-dependent expression in transgenic mice. *Nature* **362**: 258–260.

Schelling, P., Ganten, U., Sponer, G., Unger, T. and Ganten, D. (1980) Components of the renin-angiotensin system in the cerebrospinal fluid of rats and dogs with special consideration of the origin and the fate of angiotensin II. *Neuroendocrinology* **31**: 297–308.

Schelling, P., Fischer, H. and Ganten, D. (1991) Angiotensin and cell growth: a link to cardiovascular hypertrophy? *J. Hypertens.* **9**: 3–15.

Shimkets, R.A., Warnock, D.G., Bositis, C.M. *et al.* (1994) Liddle's syndrome: heritable human hypertension caused by mutations in the β subunit of the epithelial sodium channel. *Cell* **79**: 407–414.

Sifers, R., Carlson, J., Clift, S., DeMayo, F., Bullock, D. and Woo, S. (1987) Tissue specific expression of the human alpha-1-antitrypsin gene in transgenic mice. *Nucleic Acids Res.* **15**: 1459–1475.

Tokita, Y., Franco-Saenz, R., Reimann, E.M. and Mulrow, P.J. (1994) Hypertension in the transgenic rat TGR(mRen2)27 may be due to enhanced kinetics of the reaction between mouse and rat angiotensinogen. *Hypertension* **23**: 422–427.

Tokita, Y., Franco-Saenz, R. and Mulrow, P.J. (1995) Reversal of the suppressed kidney renin level in the hypertensive transgenic rat TGR(mRen2)27 by angiotensin converting enzyme inhibition. *Am. J. Hypertens.* **8**: 1031–1039.

Véniant, M., Whitworth, C.E., Ménard, J., Sharp, M., Gonzales, M., Bruneval, P. and Mullins, J. (1995) Developmental studies demonstrate age dependent elevation of renin activity in TGR(mRen2)27 rats. *Am. J. Physiol.* **8**: 1167–1176.

Véniant, M., Ménard, J., Bruneval, P., Morley, S., Gonzales, M.F. and Mullins, J. (1996) Vascular damage without hypertension in transgenic rats expressing prorenin exclusively in the liver. *J. Clin. Invest.* **98**: 1966–1970.

Whitworth, C., Fleming S., Cumming, A., Morton, J.J., Burns, N.J.T., Williams, B.C. and Mullins, J.J. (1994) Spontaneous development of malignant phase hypertension in transgenic *Ren2* rats. *Kidney Int.* **46**: 1528–1532.

Whitworth, C., Fleming, S., Kotelevtsev, Y., Manson, L., Brooker, G., Cumming, A. and Mullins, J. (1995) A genetic model of malignant phase hypertension in rats. *Kidney Int.* **47**: 529–535.

Wilson, C. and Byrom, F. (1939) Renal changes in malignant hypertension. *Lancet* **1**: 136–139.

Zhao, Y., Bader, M., Kreutz, R. *et al.* (1993) Ontogenic regulation of mouse *Ren2*^{d} renin gene in transgenic hypertensive rats, TGR(mRen2)27. *Am. J. Physiol.* **265**: E699–E707.

6

Mutating genes to study hypertension

John H. Krege

1. Introduction

Normal blood pressures are maintained by homeostatic regulatory systems in the face of widely varying environmental conditions. However, in some individuals, a pathological interaction of environmental and/or genetic factors results in hypertension. Several complementary approaches are being taken to determine the causes of this disease. One important approach is to analyze the blood pressure effects of environmental factors, such as dietary sodium intake (Dahl *et al.*, 1962). Other approaches include searching in hypertensive and control humans or animal models for disturbances in biochemical blood pressure-controlling systems or in blood pressure-controlling genes (Rapp, 1983).

Our approach to studying the genetics of hypertension is to study the blood pressure effects of experimentally mutating candidate blood pressure-controlling genes in mice (Krege *et al.*, 1995d; Smithies and Maeda, 1995). With careful attention to experimental design, it is possible to study the effects of these mutations as single variables free of interference from environmental influences, unlinked genes and linked genes. The combination of this and the above described approaches should help to clarify the complex genetic determination of hypertension and related diseases. We will describe some of the theoretical considerations involved in this approach, summarize some of illustrative results from some gene targeting studies, and present a classification of candidate genes based on experimental results.

2. Theoretical considerations related to gene targeting

This section will discuss some of the theoretical considerations relevant to using gene targeting to study blood pressure regulation. These considerations include the selection of genes for gene targeting, the choice of mutation to be generated, genetic background considerations, and environmental considerations.

Molecular Genetics of Hypertension, edited by A.F. Dominiczak, J.M.C. Connell and F. Soubrier.

2.1 Selection of genes for gene targeting

The selection of target genes may be directed at testing specific hypotheses arising from the results of genetic analysis experiments in humans or in animal models. For example, our mouse studies of the angiotensinogen gene were motivated by the association of human angiotensinogen gene variants with increased circulating levels of angiotensinogen and increased blood pressure (Jeunemaitre *et al.*, 1992b). Alternatively, the selection of target genes may be based on the hypothesized functions of the protein encoded by the gene.

2.2 Choice of mutation to be generated

Almost any mutation can be generated in mice to address relevant hypotheses (Bronson and Smithies, 1994). For example, mutations can be induced to eliminate gene function, alter gene function quantitatively, or to alter gene function qualitatively.

Often, the most initially relevant mutation is inactivation ('knockout') of the target gene to determine what critical functions can be ascribed to the encoded protein (Capecchi, 1994; Koller and Smithies, 1992). This is in general accomplished by insertion of disruptive sequences (usually a selectable marker gene) into the coding sequences of the target, with or without deletion of some or all of the target gene. Two general comments bear mentioning. First, the strategy employed can affect the completeness of inactivation of a target gene. For example, several groups have inactivated the mouse *Cftr* gene which encodes the murine cystic fibrosis transporter protein; groups employing strategies that completely inactivated the gene observed a severe cystic fibrosis phenotype (Colledge *et al.*, 1992; Snouwaert *et al.*, 1992), while a group employing a 'leaky' mutation which may have allowed alternate splicing and the generation of some intact protein (Smithies, 1993) observed a mild cystic fibrosis phenotype (Dorin *et al.*, 1992). Second, the common practice of inserting a selectable marker gene, such as the neomycin-resistance gene, into the target locus may have unexpected effects on the transcription of neighboring genes. An example of such neighborhood effects emerged from work involving the myogenic basic-helix–loop-helix gene *Mrf4*; the phenotypes of homozygous mutant mice for three different insertional disruption mutations of this gene ranged from complete viability to complete lethality, with these differences being most likely due to differing effects of the selectable markers used on the level of expression of the adjacent *Myf5* gene (Olson *et al.*, 1996). Thus, investigators should consider gene inactivation strategies that include removal of the selectable marker gene and its promoter; this can be accomplished for example using the Cre-loxP system (see, e.g., Jung *et al.*, 1993). If this precaution is not taken, it is important to consider the identities and activities of neighboring genes in interpreting experimental results (Olson *et al.*, 1996).

However, complete inactivation of important blood pressure genes is probably uncommon in the population. Much more commonly, genetic variants affect the level of function of blood pressure genes quantitatively. To study whether quantitative changes in the level of gene function impact on phenotypes of interest, Smithies and Kim have published a strategy using gene targeting to induce graded quantitative change in gene function (Smithies and Kim, 1994). This 'gene titration' strategy involves the generation of a strain carrying an inactivation of the target gene as well as the generation of a second strain carrying a duplication of the target gene at its normal chromosomal location (*Figure 1*). This strategy does not require any knowledge of the relevant target gene regulatory sequences although successful increase of gene expression in the duplication mice will only occur if sufficient regulatory sequences are present in each of the two tandem copies of the duplicated gene. Using appropriate breeding strategies, the investigator can study the blood pressures of mice having one, two (normal mice), three, and four copies of a target gene and directly determine whether quantitative change (in both a positive and negative direction) in the level of candidate gene expression influences blood pressure. Sucessful demonstration of a statistically significant slope in the relationship between gene copy number and blood pressure establishes that quantitative change in the level of target gene expression directly influences blood pressure.

Genetic variants may also exist in humans that exert more qualitative changes in gene function, resulting in an altered protein or a normal gene product expressed in differential locations. For example, the Dahl salt-resistant (R) rat has coding sequence mutations of the gene encoding 11-β-hydroxylase that reduce the capacity of the R rat to synthesize 18-hydroxy–11-deoxycorticosterone (18OH-DOC). It has been proposed that this mutation may explain the robust salt resistance of the Dahl R rat (Cicilia *et al.*, 1993). To test the physiological relevance of mutations of this type, it is possible to either mutate a mouse gene in a precise fashion, such as by altering specific nucleotides, or to replace the homologous mouse gene with that from another species (Bronson and Smithies, 1994). However, species differences will sometimes complicate this approach; for example, neither human renin nor human angiotensinogen enzymatically

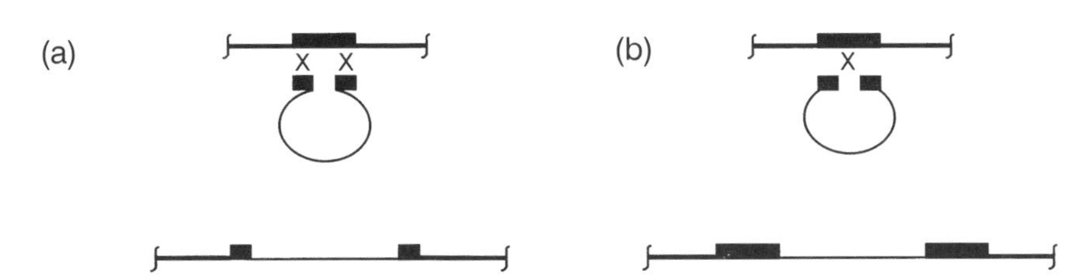

Figure 1. Schematic diagram of the homologous recombination events used to generate two strains of mice having disruption (a) or duplication (b) of the target gene at its normal chromosomal location for a gene titration experiment. The target gene is shown as a black box (top), the targeting constructs are shown with areas of homology to the target gene indicated by smaller black boxes and plasmid sequences indicated by thin lines (middle). Sites of homologous recombination are indicated by Xs. (a) With an Ω-type replacement targeting construct, the gap between the two arms of homology is deleted and replaced by the plasmid sequences resulting in the targeted chromosome shown at the bottom. (b) With an O-type insertional recombination event, the gap is repaired from chromosomal information and the entire targeting construct is inserted resulting in the targeted locus shown at the bottom having two copies of the target gene separated by plasmid sequences.

interacts with the endogenous murine renin–angiotensin system (Fukamizu *et al.*, 1993).

2.3 Genetic background considerations

If possible, the breeding strategy employed should result in the experimentally induced mutation being the only difference between experimental groups. If this ideal cannot be obtained, various controls can be employed to test for confounding influences of linked or unlinked genes on the experimental results (Krege *et al.*, 1995d, Krege, 1996; Smithies and Maeda, 1995).

Theoretically, experiments should be performed in a single inbred strain of mice so that the experimental mice differ only in genotype at the target locus. For example, mutations are typically generated *in vitro* in embryonic stem cells from strain 129. If chimeric mice generated from these cells transmit a strain 129 gamete to a strain 129 mate, completely inbred and genetically uniform offspring are obtained with the exception of being either wild-type (+/+) or heterozygous (+/–), depending on whether the sperm or egg from the 129 carries the wild-type or mutant target gene. Heterozygous mice can be mated together to generate mice of all genotypes (+/+, +/–, and –/–). In practice, however, strain 129 mice breed poorly, are susceptible to disease, and in our experience are difficult to study in physiological experiments. These practical problems suggest that it will be necessary to use embryonic stem cells from other inbred strains in experiments requiring genetic rigor.

Because of the above limitations of strain 129 animals, chimeras (or a strain 129 +/– male) are frequently mated to a second inbred strain such as C57BL/6 (B6). Offspring from this particular mating receive one copy of each chromosome from strain 129 and one copy of each chromosome from strain B6. F_1 mice of this type are large and healthy due to 'hybrid vigor' and are genetically identical except at the target locus (Smithies and Kim, 1994). Thus, +/+ and +/– F_1 mice can be experimentally compared with rigorously interpretable results. To investigate the impact of a mutation in a variety of genetic backgrounds, mating of the target strain to several inbred strains can be used to generate a panel of F_1 mice having different genetic backgrounds.

To obtain homozygous mutant (–/–) mice, investigators frequently intercross +/– F_1 mice to obtain F_2 mice of all three genotypes (+/+, +/–, –/–). Unfortunately, F_2 mice of this type may suffer from the problems of linked or unlinked genes. With regard to the chromosomal region near the target gene, all mutant targeted genes and their linked sequences will originate from strain 129 while all non-targeted genes and their linked sequences will originate from strain B6 (or whatever second inbred strain is used). Thus, +/+ mice will have both wild-type genes and linked sequences originating from B6, +/– mice will have one wild-type gene and linked sequences from B6 and one targeted gene and linked sequences from 129, and –/– mice will have two targeted genes and all linked sequences originating from 129. In an experiment of this type, it is important to note that observed phenotypic differences may arise either from the genotype at the target locus or from inheritance of differing sequences linked to the target gene (or unlinked genes, as discussed below) that impact the phenotype under investigation. To underscore the importance of this point, genes with similar function are in fact frequently linked. For example, the genes encoding atrial natriuretic peptide and brain natriuretic peptide are separated by only 15 kb of DNA (Huang *et al.*, 1996), and the genes for the β and γ subunits of the epithelial sodium channel as well as the potentially important *SA* locus exist on the same segment of human chromosome 16 (Hansson *et al.*, 1995).

A genetic strategy to evaluate the effects of linked genes in an F_2 set in which a mutation is segregating involves the generation of a parallel F_2 set of mice which is wild-type at the target gene. Mice of this second F_2 set can be evaluated for the phenotype of interest, such as blood pressure, and then genotyped at the target locus for presence of 129 or B6 wild-type target gene and linked sequences using restriction fragment or simple sequence length polymorphisms tightly linked to the target gene. The presence or absence of a phenotypic impact of these 129 or B6 wild-type target and linked sequences will then allow a rigorous interpretation of the results in the F_2 set involving mutant mice (*Figure 2*). For example, we have used this approach to demonstrate that the increased blood pressure observed in mice lacking the endothelial nitric oxide synthase (eNOS) gene is not due to other strain 129 linked genes that cosegregate with the mutant eNOS locus (Shesely *et al.*, 1996).

Another way to eliminate the systematic effects of linked genes in experiments involving F_2 animals is to generate two mouse strains having different mutations in the same target gene and compare mice heterozygous for one mutation with mice heterozygous for the other mutation. For example, mice heterozygous for a disruption mutation and mice heterozygous for a duplication mutation of the same gene differ systematically only at the targeted gene locus and are directly comparable (*Figure 3*).

In experiments involving F_2 animals, unlinked sequences will usually segregate randomly and independently, and will not systematically influence the results. By chance, however, unlinked sequences may sometimes segregate in unexpected fashion and influence the observed results. The maximal effect of unlinked genes on the blood pressures of a group is proportional to the number of blood pressure-affecting loci that differ between the two strains and the magnitude of the effect imparted by each locus. Although these numbers are currently unknown, the probability of unlinked genes affecting the mean of a group by chance is reduced by increasing the size of the group.

Investigations of the impact of a mutation in varying genetic backgrounds can be undertaken by backcross breeding of the mutation into other inbred genetic backgrounds. For example, a mutation generated in strain 129 can be transferred into the B6 genetic background. The transfer of a chromosomal segment of strain 129 DNA including the induced mutation is accomplished by backcross breedings in which offspring of successive generations are genotyped and those carrying the mutation are selected for further backcrossing to B6. Studies in backcross 'congenic' mice do not suffer from problems of unlinked genes. However, even after many generations of backcrossing, the congenic strain will have a relatively large piece of strain 129 DNA, including many genes linked to the mutant gene of interest. Thus, observed differences between congenic and wild-type mice are due to the transferred chromosomal segment, but the effects of the mutation cannot be distinguished from those of linked sequences which may differ in the two strains. A theoretical solution to this problem would be to transfer the chromosomal segment containing the strain 129 mutated gene into the B6 background by backcross breeding, and in parallel to transfer the wild-type strain 129 chromosomal segment into the B6 background. This creates two mutant strains that are effectively B6, except for having either the targeted or nontargeted strain 129 chromosomal segment. However, the two congenic strains will also differ at the location of chromosomal crossover on each side of the target gene, so that the transferred strain 129 sequences linked to the target gene will differ in length in the two strains.

The rapid generation of mice having mutations in cardiovascular genes raises the exciting possibility of combining mutations through breeding. In the even more complex situation in which two or more loci are involved, rigorous interpretation of the impact of each mutation will only be possible if very careful attention to breeding strategies is observed. As an example of a rigorous experiment, we investigated the impact of genetic variation of the *Ace* (angiotension converting enzyme) gene on atherosclerosis by introducing an insertional mutation of the *Ace* gene (Krege *et al.*, 1995c) into the atherogenic background (Zhang *et al.*, 1994) of heterozygosity for a disrupted apolipoprotein e gene (*Apoe*) and high fat diet. Strain 129 males +/– for the *Ace* gene mutation (Krege *et al.*, 1995c) were bred to female mice homozygous for an insertional disruption of the *Apoe* gene backcrossed six generations to strain C57BL/6J. Resulting F_1 offspring were uniformly heterozygous for the *Apoe* mutation and systematically differed only in their

Figure 2. A rigorous test for assessing the blood pressure (or other phenotype) effect of a mutation in two parallel sets of F_2 mice. The target gene is shown as a *box* with *delta* indicating a mutation; *absence of hatching* indicates a 129 gene and *hatching* indicates a B6 gene. The 129 chromosome is shown as a *straight line*, the B6 chromosome as a *wavy line*. Centromeres are not shown because of recombination. (a) The target gene and linked sequences of the three possible categories of mice for the two F_2 sets are shown. The first set (Cross I) involves mice from a +/– × +/– F_1 mating in which a mutation and linked strain 129 sequences are segregating; the second set (Cross II) involves mice from a +/+ × +/+ F1 mating in which strain 129 sequences linked to a nontargeted target gene are segregating. (b) By comparing the blood pressure results from F_2 set I with the results from F_2 set II, the investigator can interpret the effect of the mutation on blood pressure. The *arrows* indicate the observed effects of the mutation or linked sequences from comparison of F_2 set I animals A, B, and C or comparison of F_2 set II animals D, E, and F. The interpreted effect of the mutation is deduced from the cross I and cross II results. Up, down, and side arrows indicate increase, decrease, and no effect respectively. For example, consider permutation 2. If analysis of set I animals A, B, and C indicate that the mutation, strain 129 target gene, and linked sequences increase blood pressure and analysis of set II animals D, E, and F show no effect of the strain 129 target gene and linked sequences on blood pressure, the mutation can rigorously be interpreted to increase blood pressure. Additionally, note that animal pairs B,E and C,F differ only in the presence or absence of the mutation and are rigorously comparable if the F_2 sets were analyzed concurrently. Reprinted from *Trends in Cardiovascular Medicine,* vol. 6, Krege, J.H., Mouse systems for studying the genetics of hypertension and related disorders, pp. 232–238, Copyright 1996, with permission from Elsevier Science.

genotypes (+/+ or +/–) at the *Ace* locus (*Figure 4*). This experimental design therefore allowed a direct assessment of the impact of *Ace* genotypes on atherosclerosis (Krege *et al.*, 1996).

2.4 Environmental considerations in evaluating mouse blood pressures

In otherwise unstressed and intact animals, the effects of quantitative or subtle mutations are likely to be small. To demonstrate small differences in cardiovascular parameters induced by genetic changes, the investigator should strive to

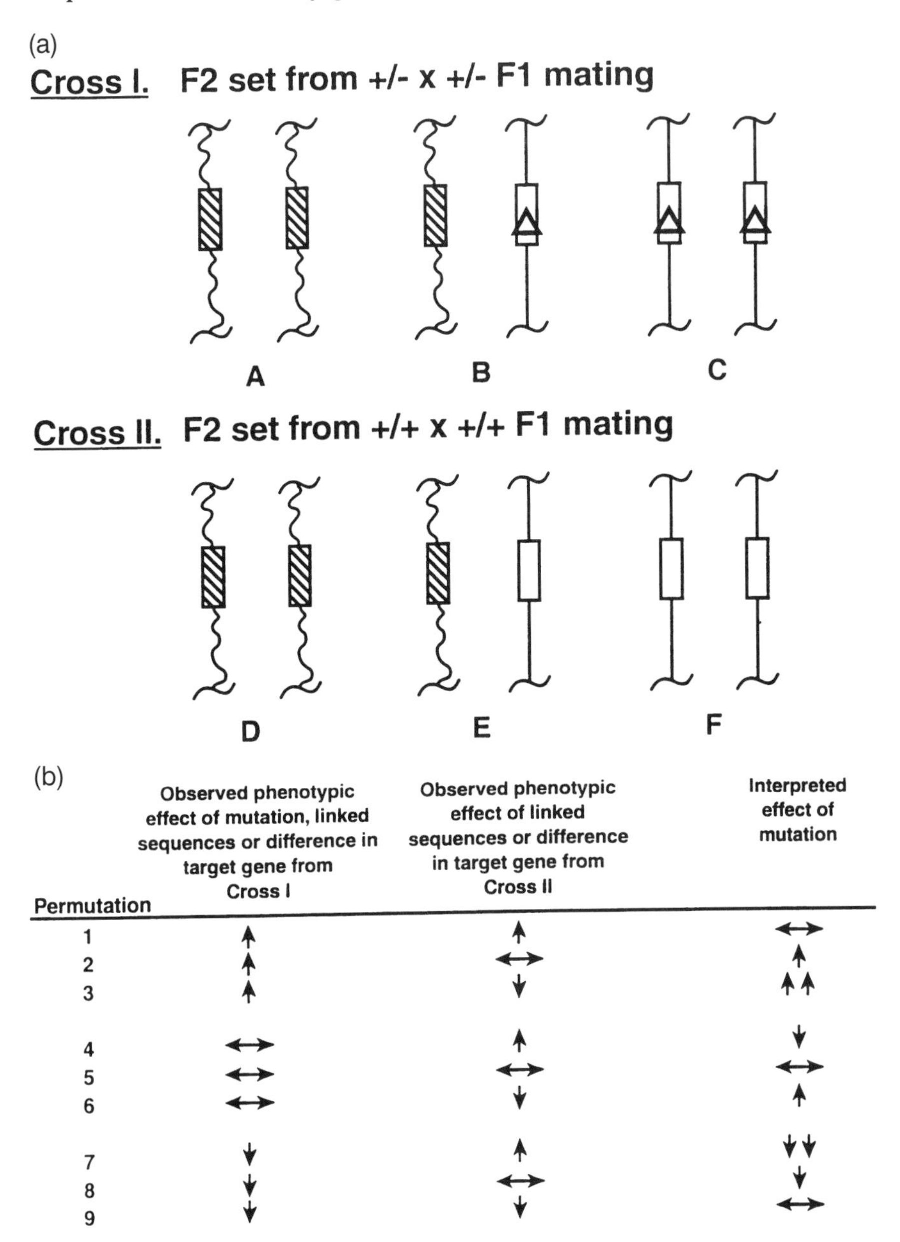

Permutation	Observed phenotypic effect of mutation, linked sequences or difference in target gene from Cross I	Observed phenotypic effect of linked sequences or difference in target gene from Cross II	Interpreted effect of mutation
1	↑	↑	↔
2	↑	↔	↑
3	↑	↓	↑↑
4	↔	↑	↓
5	↔	↔	↔
6	↔	↓	↑
7	↓	↑	↓↓
8	↓	↔	↓
9	↓	↓	↔

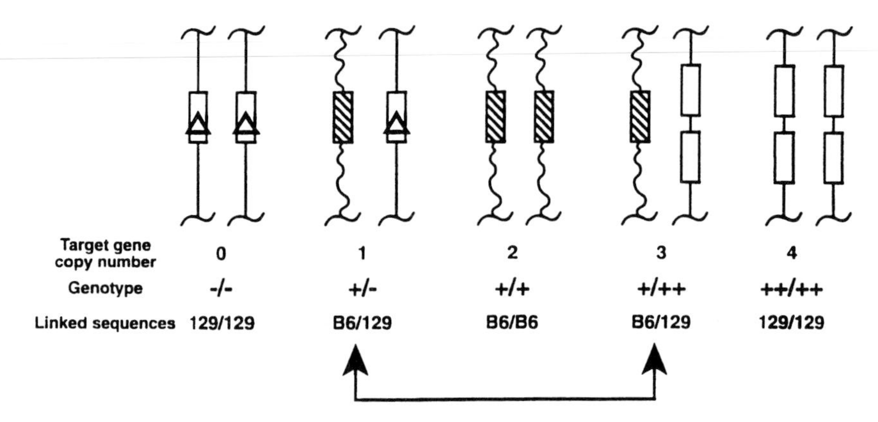

Figure 3. The gene titration approach results in F_2 animals having zero, one, two, three and four copies of the target gene. The target gene is shown as a *box* with *delta* indicating a mutation; *absence of hatching* indicates a 129 gene and *hatching* indicates a B6 gene. The 129 chromosome is shown as a *straight line*, the B6 chromosome as a *wavy line*. Centromeres are not shown because of recombination. Note that animals having one and three copies of the target gene differ systematically only in inheriting a disrupted or a duplicated strain 129 target gene. Additionally, animals with zero and four copies are rigorously comparable. Reprinted from *Trends in Cardiovascular Medicine*, vol. 6, Krege, J.H., Mouse systems for studying the genetics of hypertension and related disorders, pp. 232–238, Copyright 1996, with permission from Elsevier Science.

eliminate all other variables between experimental groups. Attention should be given to age, diet, time of day, smells and sounds, and uniformity of handling procedures.

We have described a noninvasive computerized tail-cuff system developed for the mouse as well as an invasive intra-arterial blood pressure protocol (Krege *et al.*, 1995a). These two complementary methods allow blood pressure evaluations under two different environmental conditions. The tail-cuff system requires restraint, heating, and training of the mice; the intra-arterial approach is in awake, unanesthetized mice at least 4 h after surgery. We have established a correlation ($R=0.86$) of blood pressures in mice given enalapril, L-NAME or nothing in their drinking water first evaluated by tail-cuff and subsequently by intra-arterial methods while awake and unrestrained (Krege *et al.*, 1995a). Nevertheless, it is not anticipated that these differing environmental conditions will always give concordant results. Data Sciences (St Paul, Minnesota, USA) has developed a radiotelemetry system for the mouse, a smaller version of the system available for the rat (Brockway *et al.*, 1991), that should provide new experimental options and precision in using the laboratory mouse for blood pressure investigation.

3. Gene targeting studies of the renin–angiotensin system

This section will describe results from gene targeting studies involving components of the renin–angiotensin system.

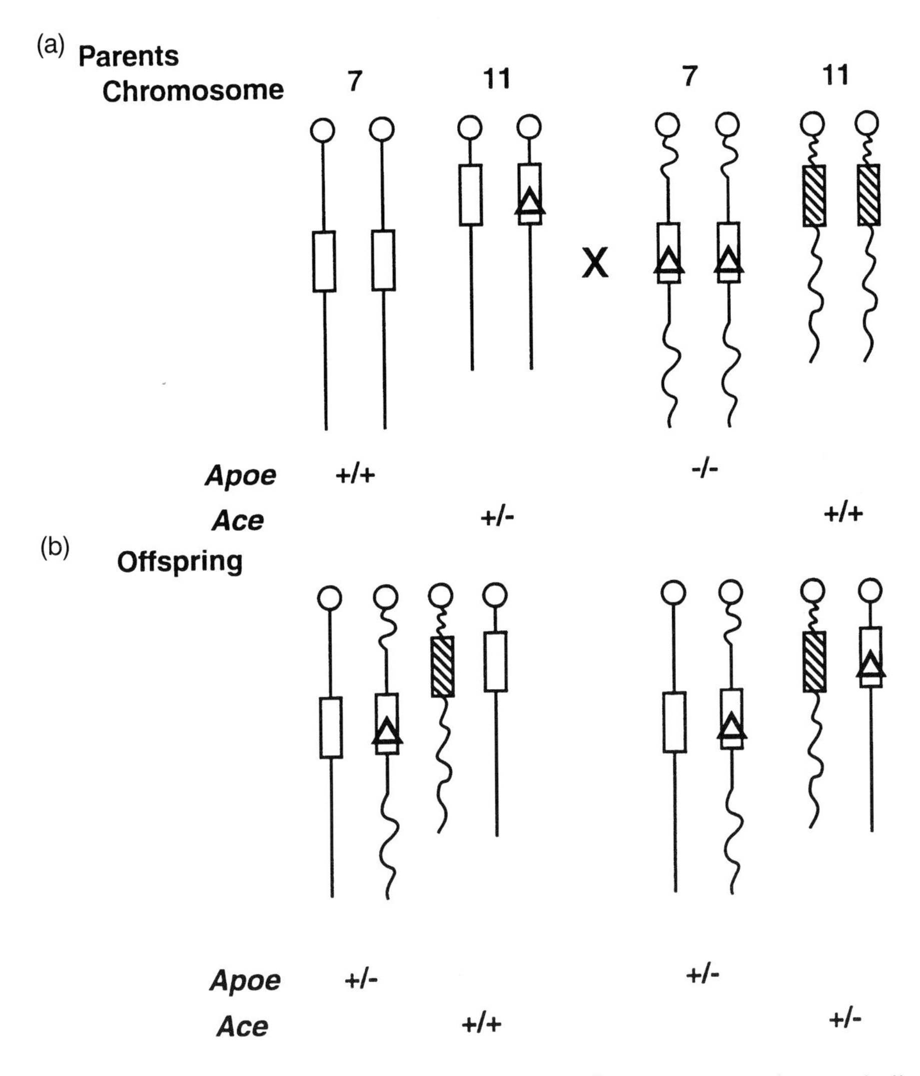

Figure 4. An example strategy for combining two mutations to generate mice genetically uniform except at a target locus. The target gene is shown as a *box* with *delta* indicating a mutation; *absence of hatching* indicates a 129 gene and *hatching* indicates a B6 gene. The 129 chromosome is shown as a *straight line*, the B6 chromosome as a *wavy line*. Centromeres are shown as *circles* in some mice and are not shown in other mice because of recombination. (a) Parental mice are a strain 129 male mouse +/– for a disruption of the chromosome 11 *Ace* gene (and +/+ for *Apoe*) and a strain B6 (backcrossed) female –/– for a disruption of the chromosome 7 *Apoe* gene (and +/+ for *Ace*). (To be precise, it should be noted that the strain 129 sequences flanking the mutation may differ in size in the chromosome 7 pair of the strain B6 backcrossed female.) (b) The two types of offspring are shown and differ only in the presence or absence of the chromosome 11 mutation of the *Ace* gene. Note that all offspring receive a B6 X chromosome; females will also inherit a 129 X chromosome while males will also inherit a 129 Y chromosome. Reprinted from *Trends in Cardiovascular Medicine*, vol. 6, Krege, J.H., Mouse systems for studying the genetics of hypertension and related disorders, pp. 232–238, copyright 1996, with permission from Elsevier Science.

3.1 Angiotensinogen gene

After reviewing six lines of evidence suggesting the angiotensinogen (AGT) gene (*AGT* in human, *Agt* in mouse) as a hypertension candidate gene, Jeunemaitre *et al.* (1992b) found that siblings sharing hypertension also shared genotypes at the *AGT* locus more frequently than would occur by chance. Further, the M235T variant, which causes a threonine rather than a methionine to be present at position 235 of the protein, was significantly more prevalent in hypertensive patients than in controls. Finally, inheritance of the M235T variant compared to M235 was found to be associated with approximately 20% higher steady state plasma AGT concentrations. These studies suggested that genetic variation of the *AGT* gene associated with increased circulating levels of angiotensinogen may confer increased risk for hypertension.

The observed quantitative change in level of AGT being associated with increased blood pressure motivated the development of the gene titration approach to directly test experimentally whether changes in the level of *Agt* gene expression were blood pressure determining (Smithies and Kim, 1994). Mice engineered to have one and three copies of the murine *Agt* gene demonstrated steady state circulating AGT levels of 35% and 124% of that observed in wild-type mice. These results demonstrated that altering gene copy number quantitatively influences the level of gene product. Smithies and Kim concluded that being able to genetically increase or decrease the expression of a gene over physiological ranges without altering the chromosomal location or sequence of the gene or its regulatory elements offered a promising approach to analyzing complex quantitative genetic traits. (The less than 150% levels of AGT in mice having three copies of the *Agt* gene suggests that the two tandem genes introduced by the gap repair gene targeting function at a less than normal level, due either to the presence of the neomycin-resistance gene between the two tandem copies or else to inadequate regulatory sequences for one or both duplicated copies.)

In independently generated strains, two groups (Kim *et al.*, 1995; Tanimoto *et al.*, 1994) found that blood pressures in mice completely lacking the coding sequences for angiotensin were about 25 mm Hg lower than those in wild-type mice, indicating that AGT is essential for the maintenance of normal blood pressure. Additionally, Kim *et al.* reported a severe kidney pathology in mice lacking AGT characterized by thickening of the medial layer of blood vessels of the kidney and cortical atrophy, indicating that angiotensinogen is necessary for normal kidney vasculature and maintenance of renal cortex in the mouse. The presence of this kidney pathology was studied in more detail in a third independently generated strain having a functionally similar *Agt* mutation (Niimura *et al.*, 1995).

Because complete absence of AGT is probably exceedingly uncommon, the blood pressure results of mice having quantitative variation in the number of *Agt* genes is probably more relevant to understanding blood pressure variation in the human population. In mice having from one to four copies of the *Agt* gene, we observed that the slope of the relationship between intra-arterial blood pressure and gene copy number was 8.3 mm Hg per *Agt* copy (Kim *et al.*, 1995). These results in F_2 mice were not due to effects of sequences linked to the *Agt* locus because mice heterozygous for the duplication (three-copy mice) had significantly higher blood pressures than mice heterozygous for the disruption (one-copy mice). Because these mice differ systematically only in *Agt* genotype, it was established that *Agt* copy number directly and causatively influences blood pressure.

Mice having mutations in important blood pressure regulatory genes would be expected to demonstrate homeostatic adjustments directed at normalizing blood pressure. The above-mentioned observation of 35% rather than the expected 50% circulating angiotensinogen levels in mice heterozygous for the *Agt* disruption suggest the possibility of relatively increased consumption (conversion to angiotensin I) of angiotensinogen generated from the remaining functional *Agt* copy. This hypothesis was supported by the finding that mice having only one functional *Agt* gene had elevated plasma renin levels; the levels in F_1 heterozygous mice were 250% those of otherwise identical F_1 wild-type mice (Kim *et al.*, 1995). This finding suggests that *Agt* genotype might have different effects in other genetic backgrounds. For example, it is reasonable to hypothesize that mutations in the gene for angiotensinogen might have an increased (or decreased) effect within genetic backgrounds differing at the renin locus.

3.2 Renin gene

The genes encoding renin have been of considerable interest in the investigations of hypertension both in humans and in animal models (St. Lezin and Kurtz, 1993). The mouse renin locus has the interesting feature that some inbred strains, such as strain B6, have a singleton renin gene (*Ren-1c*), while other strains such as 129 have two tandemly arranged renin genes (*Ren-2* and *Ren-1d*; Abel and Gross, 1990). These genes differ in conferring different patterns of tissue renin expression; for example, the two-renin-gene locus confers high levels of renin expression in the submandibular gland and adrenal gland, while the one renin gene locus confers very little renin gene expression at these sites (Sigmund and Gross, 1991). The mouse *Ren-2* gene is at least partially regulated by androgens (Wagner *et al.*, 1990; Wilson *et al.*, 1981) with expression of *Ren-2* being higher in some tissues in males than in females.

Interestingly, rats transgenic for the mouse *Ren-2* gene expressed predominantly in the adrenal gland develop fulminant hypertension (Mullins *et al.*, 1990). Thus, because rats transgenic for the *Ren-2* gene are hypertensive and because the *Ren-2* gene is expressed in some tissues at higher levels in male than in female mice, a testable hypothesis is that inheritance of the two-renin-gene locus in mice might confer increased blood pressure, especially in males. Experimentally, this was indeed the case in two F_2 sets of strain 129 × B6 mice, in which sequences at or linked to the mouse two-renin-gene locus on chromosome 1 were linked to increased blood pressures in male but not female mice (Krege *et al.*, 1995b).

Targeted inactivations of each renin gene has been accomplished (Clark *et al.*, 1997; Sharp *et al.*, 1996). Sharp *et al.* observed that mice lacking the *Ren-2* gene are healthy and viable, and have tissues and resting blood pressures that were indistinguishable from wild-type. Interestingly, compared to wild-type mice having both renin genes, mice lacking *Ren-2* had increased plasma renin concentrations and reduced plasma prorenin concentrations. This data suggests that in the absence of tissue expression of *Ren-2*, increased circulating renin is generated from *Ren-1* resulting in maintenance of normal blood pressure homeostasis.

Clark *et al.* (1997) inactivated the *Ren-1d* gene by replacing part of exon 3 and intron 3 with the neomycin-resistance gene. Although blood pressures in male mice lacking *Ren-1d* were indistinguishable from wild-type, blood pressures in female mice lacking *Ren-1d* were reduced by 12.7 mmHg compared to wild-type. Sexually dimorphic changes were also observed in plasma renin concentrations: males lacking the *Ren-1d* gene had plasma renin concentration indistinguishable from wild-type, while females lacking the *Ren-1d* gene had reduced plasma renin concentrations. In both male and female mice lacking *Ren-1d*, plasma prorenin concentration was increased compared to wild-type. These overall data indicate that in mice having only *Ren-2* genes, males but not females are able to maintain their plasma renin concentrations and blood pressures. A major lesson from these complex results is clear – the functional effects of mutations in genes that have sexually dimorphic expression patterns may frequently also be sexually dimorphic.

An additional finding of importance was that absence of *Ren-1d* resulted in altered morphology of macula densa cells and complete absence of juxtaglomerular cell granulation. Compared to the granularity of juxtaglomerular cells from wild-type mice, mice lacking *Ren-1d* exhibited diffuse, uniform, low-level cytoplasmic renin staining consistent with constitutive secretion rather than high-level storage in granules. The authors concluded that *Ren-1d* gene expression is essential for secretory granule formation and maturation.

In these gene targeting studies of the renin genes (Clark *et al.*, 1997; Sharp *et al.*, 1996), the investigators maintained an entirely inbred genetic background by breeding their male chimeras, which were derived from typical strain 129/Ola embryonic stem cells, with stain 129/Ola females. By examination of coat color, the investigators selected offspring that had received a strain 129/Ola gamete from the chimera and were therefore inbred strain 129/Ola mice. With this approach, the investigators achieved genetic rigor in that the experimental mice differed genetically only in genotype at the target locus. In our hands, the difficulty of this approach is that inbred strain 129/Ola mice are small and breed poorly.

Because the *Ren-2* and *Ren-1d* genes are tightly linked, it is unlikely that the above described mutations in the renin genes can be combined through breeding. However, it may be possible to generate a targeting construct designed to delete the whole of the *Ren-2*, *Ren-1d* locus and hence inactivate both genes.

3.3 Angiotensin-converting enzyme gene

The angiotensin-converting enzyme (ACE) gene (*ACE* in humans, *Ace* in mice) is composed of two homologous regions and codes for both a somatic isozyme found throughout the body, including bound to endothelial cells and in the circulation, and a testis isoenzyme found only in post-meiotic developing and mature sperm (Bernstein *et al.*, 1989; Sibony *et al.*, 1993; Soubrier *et al.*, 1988). Pharmacological inhibitors of ACE are commonly used for the treatment of hypertension in human patients. Experiments with hypertensive rats (Hilbert *et al.*, 1991; Jacob *et al.*, 1991) and some (Duru *et al.*, 1994; Morise *et al.*, 1994; Zee *et al.*, 1992), but not other (Harrap *et al.*, 1993; Higashimori *et al.*, 1993; Jeunemaitre *et al.*, 1992a; Schmidt *et al.*, 1993), studies of humans suggest that sequences at or linked to the gene influence blood pressure.

We reported the generation and physiological evaluation of wild-type, heterozygous, and homozygous mutant mice for an insertional disruption of both ACE isozymes (Krege *et al.*, 1995c). Serum ACE activities were reduced in the heterozygous mutant mice and were not significantly different from zero in the homozygous mutant mice. The homozygous mutant animals showed three major phenotypes: first, their blood pressures were reduced about 33 mm Hg compared to wild-type, demonstrating that ACE is essential for the maintenance of normal blood pressures. Second, homozygous mutant males and females demonstrated a severe kidney pathology indistinguishable from that observed in mice lacking the gene for angiotensinogen. Third, male but not female mice lacking ACE demonstrated significantly reduced fertility.

Another group generated and studied mice having a functionally similar disruption of the *Ace* gene (Esther *et al.*, 1996). The three findings above were confirmed with the additional finding that homozygous mutant kidneys have a thin medulla with dilated renal calyces. Additional experiments revealed that mice lacking ACE demonstrated an approximately 2-fold increased creatinine. Also, urine output of mice lacking ACE under both normal and fluid restricted conditions was approximately 2-fold increased; this finding was accompanied by an approximately 2-fold decreased urine osmolality and urine aldosterone concentration.

Because a common polymorphism of the human *ACE* gene exists which is associated with about a 35% change in circulating ACE levels (Tiret *et al.*, 1992), the results of quantitative change in *Ace* gene expression are probably more relevant to understanding the physiological role of ACE in regulating blood pressure. Towards this end, both groups studied the blood pressures of mice heterozygous for *Ace* gene inactivation. Krege *et al.* (1995c) found that the blood pressures of heterozygous mutant female mice were not significantly different from wild-type, but the blood pressures of heterozygous mutant male mice were lower than wild-type controls. Esther *et al.* (1996) found no difference in the blood pressures of heterozygous and wild-type males or females. In our study of the role of the *Ace* gene in atherosclerosis, we also did not observe blood pressure differences between +/+ and +/– males or females (Krege *et al.*, 1996).

We have subsequently used the gene titration approach to more definitively assess the cardiovascular effects of quantitative variation in *Ace* gene copy number (Krege *et al.*, 1997). Study of more than 150 mice having one, two (wild-type), or three copies of the murine *Ace* gene showed that quantitative increase in *Ace* gene copy number resulted in a roughly linear increase in circulating ACE levels, but did not directly influence blood pressure. However, we have observed three phenotypes demonstrating that quantitative change in *Ace* gene copy number does influence the cardiovascular system. *Ace* gene copy number was inversely related to modest but statistically significant changes in heart rate and heart weight. Also, mice heterozygous for the *Ace* gene inactivation had significantly higher kidney renin mRNA levels than otherwise genetically identical wild-type controls. Clearly, additional work is needed to determine whether in different environmental or genetic backgrounds, full homeostatic compensations of this type become inadequate and demonstrable blood pressure changes become observable.

3.4 Angiotensin type 1A receptor gene

The physiological effects of angiotensin II are mediated by two pharmacologically distinct receptor types designated AT1 and AT2. In rodents, two subtypes of AT1 exist, with AT1A being present in higher amounts than AT1B in most tissues except the adrenal gland and the anterior pituitary. In a study of humans with essential hypertension, an association between several AT1 receptor gene polymorphisms and hypertension but not linkage of a highly polymorphic AT1 marker in 267 hypertensive sib pairs was observed (Bonnardeaux *et al.*, 1994). The authors concluded that their results were compatible with a common variant of the AT1 receptor gene imparting a small effect on human blood pressure.

Two groups used gene targeting to inactivate the murine gene encoding the AT1A receptor (*Agtr1A*) (Ito *et al.*, 1995; Sugaya *et al.*, 1995). Studies by Ito *et al.* showed that complete absence of the AT1A receptor resulted in virtual absence of AT1 binding in the kidney. In contrast to mice lacking AGT or ACE, mice lacking the AT1A receptor had normal kidney histology. In mice lacking this receptor, acutely infused angiotensin II had virtually no effect on blood pressure, and resting blood pressures were 24–43 mm Hg lower than wildtype controls. Sugaya *et al.* also observed this blood pressure result, and also dramatically increased kidney renin mRNA and plasma renin concentration in mice lacking the AT1A receptor.

To study quantitative change in *Agtr1a* gene function, both groups also compared the blood pressures of wild-type and heterozygous mutant mice. Both groups observed a significant reduction ranging from 10 to 17 mm Hg in the blood pressures of heterozygous mice, indicating that quantitative change in the level of function of the *Agtr1a* gene is indeed blood pressure determining.

3.5 Angiotensin type 1B receptor gene

The targeted replacement (and hence elimination) of the mouse AT1B gene (*Agtr1b*) with the lacZ reporter gene has been reported (Chen *et al.*, 1997). As an assay for identifying cells having *Agtr1b* gene transcriptional activities, tissues of homozygous mutant mice were stained with β-galactosidase as an assay for lacZ activity. β-Galactosidase staining was observed in adrenal zona glomerulosa cells and in the testis, with lower levels in anterior pituitary cells and choroidal plexus vessel walls; activity was not observed in the kidney, liver, lung, or heart. Homozygous mutant mice were indistinguishable from wild-type mice in assays of histology, blood pressures, and plasma aldosterone levels. These results indicate that the AT1B receptor does not appear to be essential for normal cardiovascular homeostasis.

In the initial characterization of mice lacking *Agtr1b*, no important roles for the gene were identified. In cases of this type, additional genetic or environmental perturbations of the mutant mice may be necessary to identify functional significance for the gene. For example, as noted above, mice lacking *Agt* or *Ace* but not mice lacking *Agtr1a* have reduced survival and a severe kidney defect. However, mice generated through breeding to lack both *Agtr1a* and *Agtr1b* have reduced survival and a severe kidney phenotype similar to that present in mice lacking *Agt* or *Ace* (Tsuchida *et al.*, 1998). Thus, the presence of either *Agtr1a* or *Agtr1b* is essential for normal survival and kidney morphology.

3.6 Angiotensin type 2 receptor gene

The high levels of expression of the AT2 receptor in fetal tissues (Grady *et al.*, 1991) and certain brain nuclei (Tsutsumi and Saavedra, 1991) suggested possible roles for the receptor in growth, development, and neuronal functions. The cardiovascular significance of this receptor was previously unclear.

In situations involving genes of unclear function, the most informative mutation is the gene inactivation. To determine the physiological role of the AT2 receptor, two groups (Hein *et al.*, 1995; Ichiki *et al.*, 1995) disrupted the X chromosome *Agtr2* gene in embryonic stem cells. Mice lacking this gene grew and developed normally and were histologically indistinguishable from wild-type. However, mice lacking the receptor had significantly reduced exploratory behavior (Ichiki *et al.*, 1995) or spontaneous movements (Hein *et al.*, 1995). Because the AT2 receptor has been implicated in the dipsogenic response, Hein *et al.* studied the water intake of hemizygous and wild-type males after cessation of a 40-h period of water deprivation, and observed that mice lacking the receptor drank significantly less water than normal mice.

Regarding the role of the receptor in cardiovascular regulation, both teams of investigators studied the baseline blood pressures and responses to captopril or infused angiotensin II in male hemizygous (lacking their only X-chromosome located *Agtr2* gene) and wild-type mice. Hein *et al.* reported that absence of the receptor did not influence baseline blood pressures or the response to acutely infused angiotensin II, while Ichiki *et al.* found that the baseline blood pressures of mice lacking the receptor were significantly increased compared to wild-type. Both groups also evaluated the blood pressure responses of hemizygous and wild-type males to infusions of angiotensin II given under conditions in which endogenous production of angiotensin II was blocked by concomitantly administered ACE inhibitors. In these experiments, mice lacking the AT2 receptor showed significantly increased blood pressure responses to infused angiotensin II compared to normal mice. The results of these experiments suggest that the blood pressure effect of the AT2 receptor is to antagonize the AT1-mediated pressor action of angiotensin II.

For illustrative purposes, we will discuss two possible genetic explanations for the different conclusions regarding the impact of the *Agtr2* mutation on baseline blood pressures. First, the results in one or both experiments may have been systematically influenced by linked genes: the targeted *Agtr2* gene and linked sequences were always from strain 129, while the nontargeted *Agtr2* gene and linked sequences were from B6 in the Hein *et al.* (1995) study and from strain FVB/N in the Ichiki *et al.* (1995) study. To evaluate the potential effects of linked genes, the investigators could compare the baseline blood pressures of hemizygous males with those of otherwise genetically similar wild-type males whose nontargeted *Agtr2* gene and linked X-chromosome sequences were also from strain 129. As a second explanation, the breeding strategy used by Hein *et al.* gave a genetic background of approximately 50% strain 129 and 50% strain B6, while the breeding strategy of Ichiki *et al.* gave a genetic background of approximately 25% strain 129 and 75% strain FVB/N. Thus, the impact of the functionally similar *Agtr2* mutations may have differed within these genetic backgrounds.

4. A classification of candidate blood pressure genes

This section will describe a way to think about the outcomes of cardiovascular gene targeting experiments. In general, the first investigation of a cardiovascular gene is gene inactivation; studies in mice lacking a cardiovascular gene are useful in determining whether the gene is essential or not essential for normal blood pressure regulation. As described in the preceding section, overt phenotypes were observed for almost all null mutations of renin–angiotensin system genes, supporting the central importance of this system in blood pressure homeostasis.

Because complete inactivation of cardiovascular genes is probably unusual, the cardiovascular effects of quantitative changes in gene function are particularly relevant to predicting the blood pressure effects of genetic variation within the human population. We propose that candidate genes may be classified by the observed cardiovascular effects of quantitative variation in their function in mice (*Table 1*).

As discussed above, gene titration experiments in mice having one, two, three or four copies of the murine gene for angiotensinogen have demonstrated that quantitative variation in this gene directly influences blood pressures (Kim *et al.*, 1995), indicating that the angiotensinogen gene is of the type we will call Class I. Sequence variation that quantitatively affects the level of function of homologous human genes classified in mice as class I would therefore be expected to directly affect blood pressures.

However, studies in mice having quantitative variation in the level of ACE gene function revealed that compensatory homeostatic adjustments were present in the heart rate and kidney renin mRNA concentrations of mutant mice; it is likely that these and other compensatory adaptations normalize the blood pressures (Krege *et al.*, 1997). The ACE gene is therefore of the type we will call class II. The significance of class II genes is that quantitative changes in the expression of these genes, when accompanied by additional genetic or environmental factors, may measurably affect blood pressures by stressing the homeostatic machinery beyond its limits. For example, mice wild-type and heterozygous for an inactivation mutation of the atrial natriuretic peptide gene had indistinguishable blood pressures in experiments

Table 1. A classification of candidate genes based on the cardiovascular effects of quantiatively varying their function

Class I:	Measurably affects blood pressures
Class II:	Compensatory adaptations normalize the blood pressures
Class III:	No measurable compensatory adaptations or changes in blood pressures

involving a 0.8% sodium chloride diet; however, the blood pressures of heterozygous mice significantly exceeded those of wild-type mice in experiments involving a 8% sodium chloride diet (John *et al.*, 1995). Thus, the gene encoding atrial natriuretic peptide is a class II gene because quantitative changes in its expression affect blood pressures when accompanied by the additional factor of a high salt diet. The existence of class II genes emphasizes the need to explore both genetic variants and the context in which they exist.

In some cases, it may be informative to find that quantitative change in the level of function or even complete inactivation of a gene unexpectedly does *not* influence cardiovascular phenotypes. For example, we have studied the cardiovascular effects of genetic reduction and complete inactivation of the gene encoding estrogen receptor α. Mice lacking this receptor have quite severe reproductive abnormalities (Lubahn *et al.*, 1993). However, estradiol inhibits the normal proliferative response to carotid artery injury in mice lacking estrogen receptor a (Iafrati *et al.*, 1998). Additionally, we have found that absence of this estrogen receptor in both males and females does not influence blood pressures, heart rates, heart or kidney weights, or lipid profiles (J. Moyer *et al.*, unpublished data). These results indicate that the cardiovascular effects of estrogen are not mediated by estrogen receptor α and that the gene for this receptor is an example of a class III gene. A candidate receptor for mediating the cardiovascular effects of estrogen is the recently reported estrogen receptor β (Kuiper *et al.*, 1996). We have generated mice lacking estrogen receptor b and found that male mice lacking this receptor reproduce normally and female mice lacking this receptor have subtle reproductive abnormalities (Krege *et al.*, 1998). Investigations are underway to determine the role of estrogen receptor β in cardiovascular homeostasis. If the cardiovascular and reproductive effects of estrogen are mediated by different receptors, it may be possible to develop drugs that confer the beneficial cardiovascular effects of estrogen without conferring its adverse reproductive effects such as increased risk for breast and uterine cancer.

The proposed classification of candidate genes according to their effects in gene targeting experiments may be useful in thinking about the complex effects of mutations on whole animal biology and also in making predictions of how and under which circumstances human genetic variants are likely to affect blood pressures. Specifically, it is reasonable to predict that class I genes such as the angiotensinogen gene are likely to directly affect resting blood pressures in the population, class II genes such as the ACE gene are likely to affect blood pressures only in the presence of additional genetic or environmental factors, and class III genes such as the estrogen receptor α gene are unlikely to affect blood pressures under any circumstances.

5. Summary and the future

Our approach to studying blood pressure regulation is to analyze the blood pressure effects of mutating genes in the mouse. With carefully planned experimental designs and controls, it is possible to study the effects of these mutations as single variables. The first mutation usually involves inactivation of the target gene for determination of whether the gene is essential for normal homeostasis. Mice lacking the genes for angiotensinogen, angiotensin-converting enzyme, and the angiotensin type 1A receptor, for examples, all have significantly reduced blood pressures, indicating that these genes are essential for normal blood pressure regulation.

Common variants of blood pressure determining genes are likely to influence the level of gene expression; to test whether the quantitative level of gene expression directly influences blood pressure, we use the gene titration approach in which mice are generated having decreased, normal, and increased expression of the target gene for development of a blood pressure dose–response curve. Using this approach, we have observed that angiotensinogen gene copy number directly influences blood pressures, while angiotensin-converting enzyme gene copy number results in compensatory changes that likely normalize the blood pressures.

We propose a classification of candidate genes based on the experimental effects of quantitatively varying their level of expression in mice. Class I genes are those genes that directly affect blood pressures; class II genes are those genes in which compensatory adaptations normalize the blood pressures; class III genes are those genes that result in no measurable compensatory adaptations or changes in blood pressures.

Although most of the initial work in the field of gene targeting has been directed at mutating individual genes, second generation experiments in which more than one mutation are combined through breeding will likely shed light on how individual genetic variants interact with other genetic factors. For example, it will be important to test the hypothesis that the angiotensin-converting enzyme gene, a class II gene, directly affects blood pressures when combined with other mutations or environmental factors.

It is anticipated that the widespread application of these sorts of genetic approaches in combination with careful physiological analyses will help to clearly resolve the complexities of the genetics of blood pressure control.

Acknowledgments

I thank Simon John for reading the manuscript. I thank Oliver Smithies, Guilford Medical Associates, and my family for their support. I was previously a Howard Hughes Medical Institute Physician Research Fellow and was then funded by an N.I.H. Clinical Investigator Development Award (HL03470).

References

Abel, K.J. and Gross, K.W. (1990) Physical characterization of genetic rearrangements at the mouse renin loci. *Genetics* **124**: 937–947.

Bernstein, K.E., Martin, B.M., Edwards, A.S. and Bernstein, E.A. (1989) Mouse angiotensin-converting enzyme is a protein composed of two homologous domains. *J. Biol. Chem.* **264**: 11945–11951.

Bonnardeaux, A., Davies, E., Jeunemaitre, X., *et al.* (1994) Angiotensin II type 1 receptor gene polymorphisms in human essential hypertension. *Hypertension* **24**: 63–69.

Brockway, B.P., Mills, P.A. and Azar, S.H. (1991) A new method for continuous chronic measurement and recording of blood pressure, heart rate and activity in the rat via radio-telemetry. *Clin. Exper. Hyper. Theory Pract.* **A13**: 885–895.

Bronson, S.K. and Smithies, O. (1994) Altering mice by homologous recombination using embryonic stem cells. *J. Biol. Chem.* **269**: 27155–27158.

Capecchi, M.R. (1994) Targeted gene replacement. *Scient. Amer.* **270**: 52–59.

Chen, X., Li, W., Yoshida, H., *et al.* (1997) Targeting deletion of angiotensin type 1B receptor gene in the mouse. *Am. J. Physiol.* **272**: F299-F304.

Cicilia, G.T., Rapp, J.P., Wang, J.M., St.Lezin, E., Ng, S.C. and Kurtz, T.W. (1993) Linkage of 11b-hydroxylase mutations with altered steroid biosynthesis and blood pressure in the Dahl rat. *Nature Genet.* **3**: 346–353.

Clark, A.F., Sharp, M.G.F., Morley, S.D., Fleming, S., Peters, J. and Mullins, J.J. (1997) Renin-1 is essential for normal renal juxtaglomerular cell granulation and macula densa morphology. *J. Biol. Chem.* **272**: 18185–18190.

Colledge, W.H., Ratcliff, R., Foster, D., Williamson, R. and Evans, M.J. (1992) Cystic fibrosis mouse with intestinal obstruction. *Lancet* **340**: 680

Dahl, L.K., Heine, M. and Tassinari, L. (1962) Role of genetic factors in susceptibility to experimental hypertension due to chronic excess salt ingestion. *Nature* **194**: 480–482.

Dorin, J.R., Dickinson, P., Alton, E.W.F.W., *et al.* (1992) Cystic fibrosis in the mouse by targeted insertional mutagenesis. *Nature* **359**: 211–215.

Duru, K., Farrow, S., Wang, J.M., Lockette, W. and Kurtz, T. (1994) Frequency of a deletion polymorphism in the gene for angiotensin converting enzyme is increased in African-Americans with hypertension. *Am. J. Hypertension* **7**: 759–762.

Esther, C.R., Howard, T.E., Marino, E.M., Goddard, J.M., Capecchi, M.R. and Bernstein, K.E. (1996) Mice lacking angiotensin-converting enzyme have low blood pressure, renal pathology, and reduced male fertility. *Lab. Invest.* **74**: 953–965.

Fukamizu, A., Sugimura, K., Takimoto, E., *et al.* (1993) Chimeric renin-angiotensin system demonstrates sustained increase in blood pressure of transgenic mice carrying both human renin and human angiotensinogen genes. *J. Biol. Chem.* **268**: 11617–11621.

Grady, E.F., Sechi, L.A., Griffin, C.A., Schambelan, M. and Kalinyak, J.E. (1991) Expression of AT2 receptors in the developing rat fetus. *J. Clin. Invest.* **88**: 921–933.

Hansson, J.H., Nelson-Williams, C., Suzuki, H., *et al.* (1995) Hypertension caused by a truncated epithelial sodium channel gamma subunit: genetic heterogeneity of Liddle syndrome. *Nature Genet.* **11**: 76–82.

Harrap, S.B., Davidson, H.R., Connor, J.M., Soubrier, F., Corvol, P., Fraser, R., Foy, C.J.W. and Watt, G.C.M. (1993) The angiotensin I converting enzyme gene and predisposition to high blood pressure. *Hypertension* **21**: 455–460.

Hein, L., Barsh, G.S., Pratt, R.E., Dzau, V.J. and Kobilka, B.K. (1995) Behavioural and cardiovascular effects of disrupting the angiotensin II type–2 receptor gene in mice. *Nature* **377**: 744–747.

Higashimori, K., Zhao, Y., Higaki, J., Kamitani, A., Katsuya, T., Nakura, J., Miki, T., Mikami, H. and Ogihara, T. (1993) Association analysis of a polymorphism of the angiotensin converting enzyme gene with essential hypertension in the Japanese population. *Biochem. Biophys. Res. Commun.* **191**: 399–404.

Hilbert, P., Lindpaintner, K., Beckmann, J.S., *et al.* (1991) Chromosomal mapping of two genetic loci associated with blood- pressure regulation in hereditary hypertensive rats. *Nature* **353**: 521–529.

Huang, H., John, S.W.M. and Steinhelper, M.E. (1996). Organization of the mouse cardiac natriuretic peptide locus encoding BNP and ANP. *J. Molec. Cell. Cardiol.* **28**: 1823–1828.

Iafrati, M.D., Karas, R.H., Aronovitz, M., Kim, S., Sullivan, T.R.J., Lubahn, D.B., O'Donnell, T.F.J., Korach, K.S. and Mendelsohn, M.E. (1998) Estrogen inhibits the vascular injury response in estrogen receptor alpha-deficient mice. *Nature Med.* **3**: 545–548.

Ichiki, T., Labosky, P.A., Shiota, C., *et al.* (1995) Effects on blood pressure and exploratory behaviour of mice lacking angiotensin II type–2 receptor. *Nature* **377**: 748–750.

Ito, M., Oliverio, M.I., Mannon, P.J., Best, C.F., Maeda, N., Smithies, O. and Coffman, T.M. (1995) Regulation of blood pressure by the type IA angiotensin II receptor gene. *Proc. Natl Acad. Sci. USA* **92**: 3521–3525.

Jacob, H.J., Lindpaintner, K., Lincoln, S.E., Kusumi, K., Bunker, R.K., Mao, Y.P., Ganten, D., Dzau, V.J. and Lander, E.S. (1991) Genetic mapping of a gene causing hypertension in the stroke-prone spontaneously hypertensive rat. *Cell* **67**: 213–224.

Jeunemaitre, X., Lifton, R.P., Hunt, S.C., Williams, R.R. and Lalouel, J.M. (1992a) Absence of linkage between the angiotensin converting enzyme locus and human essential hypertension. *Nature Genet.* **1**: 72–75.

Jeunemaitre, X., Soubrier, F., Kotelevtsev, Y.V., *et al.* (1992b) Molecular basis of human hypertension: role of angiotensinogen. *Cell* **71**: 169–180.

John, S.W.M., Krege, J.H., Oliver, P.M., Hagaman, J.R., Hodgin, J.B., Pang, S.C., Flynn, T.G. and Smithies, O. (1995) Genetically decreases in atrial natriuretic peptide and salt-sensitive hypertension. *Science* **267**: 679–681.

Jung, S., Rajewsky, K. and Radbruch, A. (1993) Shutdown of class switch recombination by deletion of a switch region control element. *Science* **259**: 984–987.

Kim, H.-S., Krege, J.H., Kluckman, K.D., *et al.* (1995) Genetic control of blood pressure and the angiotensinogen locus. *Proc. Natl Acad. Sci. USA* **92**: 2735–2739.

Koller, B.H. and Smithies, O. (1992) Altering genes in animals by gene targeting. *Annu. Rev. Immunol.* **10**: 705–730.

Krege, J.H. (1996) Mouse systems for studying the genetics of hypertension and related disorders. *Trends Cardiovasc. Med.* **6**: 232–238.

Krege, J.H., Hodgin, J.B., Hagaman, J.R. and Smithies, O. (1995a) A noninvasive computerized tail-cuff system for measuring blood pressure in mice. *Hypertension* **25**: 1111–1115.

Krege, J.H., John, S.W.M., Kim, H.-S., Hagaman, J.R., Langenbach, L.L., Peng, L. and Smithies, O. (1995b). Renin genotypes and blood pressures in mice. *Hypertension* **26**: 576.

Krege, J.H., John, S.W.M., Langenbach, L.L., Hodgin, J.B., Hagaman, J.R., Bachman, E.S., Jennette, J.C., O'Brien, D.A. and Smithies, O. (1995c) Male–female differences in fertility and blood pressure in ACE-deficient mice. *Nature* **375**: 146–148.

Krege, J.H., Kurtz, T.W. and Smithies, O. (1995d) Using animal models to dissect the genetics of complex traits. In: *Molecular Genetics and Gene Therapy of Cardiovascular Diseases* (ed. S.C. Mockrin). Marcel Dekker, New York, pp. 271–292.

Krege, J.H., Moyer, J.S., Langenbach, L.L., Peng, L., Zhang, S.H., Maeda, N., Reddick, R.L. and Smithies, O. (1996) Angiotensin-converting enzyme gene and atherosclerosis. *Arterio. Thromb. Vasc. Biol.* **17**: 1245–1250.

Krege, J.H., Kim, H.-S., Moyer, J.S., Jennette, J.C., Peng, L., Hiller, S.K. and Smithies, O. (1997). Angiotensin converting enzyme gene mutations, blood pressures and cardiovascular homeostasis. *Hypertension* **29**: 216–221.

Krege, J.H., Hodgin, J.B., Couse, J.F., *et al.* (1998) Generation and reproductive phenotypes of mice lacking estrogen receptor β. *Proc. Natl Acad. Sci. USA* **95**: 15677–15682.

Kuiper, G.G.J.M., Enmark, E., Pelto-Huikko, M., Nilsson, S. and Gustafsson, J.-A. (1996) Cloning of a novel estrogen receptor expressed in rat prostate and ovary. *Proc. Natl Acad. Sci. USA* **93**: 5925–5930.

Lubahn, D.B., Moyer, J.S., Golding, T.S., Couse, J.F., Korach, K.S. and Smithies, O. (1993) Alteration of reproductive function but not prenatal sexual development after insertional disruption of the mouse estrogen receptor gene. *Proc. Natl Acad. Sci. USA* **90**: 11162–11166.

Morise, T., Takeuchi, Y. and Takeda, R. (1994) Angiotensin-converting enzyme polymorphism and essential hypertension. *Lancet* **343**: 125

Mullins, J.J., Peters, J. and Ganten, D. (1990) Fulminant hypertension in transgenic rats harbouring the mouse Ren–2 gene. *Nature* **344**: 541–544.

Niimura, F., Labosky, P.A., Kakuchi, J., *et al.* (1995) Gene targeting in mice reveals a requirement for angiotensin in the development and maintenance of kidney morphology and growth factor regulation. *J. Clin. Invest.* **96**: 2947–2954.

Olson, E.N., Arnold, H.-H., Rigby, P.W.J. and Wold, B.J. (1996) Know your neighbors: Three phenotypes in null mutants of the myogenic bHLH gene MRF4. *Cell* **85**: 1–4.

Rapp, J.P. (1983) A paradigm for identification of primary genetic causes of hypertension in rats. *Hypertension* **5(suppl. I)**: I–198–I–203.

Schmidt, S., van Hooft, I.M.S., Grobbee, D.E., Ganten, D. and Ritz, E. (1993) Polymorphism of the angiotensin I converting enzyme gene is apparently not related to high blood pressure: Dutch Hypertension and Offspring Study. *J. Hypertension* **11**: 345–348.

Sharp, M.G.F., Fettes, D., Brooker, G., Clark, A.F., Peters, J., Fleming, S. and Mullins, J.J. (1996) Targeted inactivation of the Ren-2 gene in mice. *Hypertension* **28**: 1126–1131.

Shesely, E.G., Maeda, N., Deshi, C., Krege, J.H., Kim, H.-S., Laubach, V.E., Sherman, P.A., Sessa, W.C. and Smithies, O. (1996). Elevated blood pressure in mice lacking endothelial nitric oxide synthase. *Proc. Natl Acad. Sci. USA* **93**: 13176–13181.

Sibony, M., Gasc, J.-M., Soubrier, F., Alhenc-Gelas, F. and Corvol, P. (1993) Gene expression and tissue localization of the two' isoforms of angiotensin I converting enzyme. *Hypertension* **21**: 827–835.

Sigmund, C.D. and Gross, K.W. (1991) Structure, expression, and regulation of the murine renin genes. *Hypertension* **18**: 446–457.

Smithies, O. (1993) Animal models of human genetic diseases. *Trends Genet.* **9**: 112–116.

Smithies, O. and Kim, H.S. (1994) Targeted gene duplication and disruption for analyzing quantitative genetic traits in mice. *Proc. Natl Acad. Sci. USA* **91**: 3612–3615.

Smithies, O. and Maeda, N. (1995) Gene targeting approaches to complex genetic diseases: atherosclerosis and essential hypertension. *Proc. Natl Acad. Sci. USA* **92**: 5266–5272.

Snouwaert, J.N., Brigman, K.K., Latour, A.M., Malouf, N.N., Boucher, R.C., Smithies, O. and Koller, B.H. (1992) An animal model for cystic fibrosis made by gene targeting. *Science* **257**: 1083–1088.

Soubrier, F., Alhenc-Gelas, F., Hubert, C., Allegrini, J., John, M., Tregear, G. and Corvol, P. (1988) Two putative active centers in human angiotensin I-converting enzyme revealed by molecular cloning. *Proc. Natl Acad. Sci. USA* **85**: 9386–9390.

St. Lezin, E.M. and Kurtz, T.W. (1993) The renin gene and hypertension. *Seminars Nephrol.* **13**: 581–585.

Sugaya, T., Nishimatsu, S.-I., Tanimoto, K., *et al.* (1995) Angiotensin II type 1a receptor-deficient mice with hypotension and hyperreninemia. *J. Biol. Chem.* **270**: 18719–18722.

Tanimoto, K., Sugiyama, F., Goto, Y., Ishida, J., Takimoto, E., Yagami, K.-I., Fukamizu, A. and Murakami, K. (1994) Angiotensinogen-deficient mice with hypotension. *J. Biol. Chem.* **269**: 31334–31337.

Tiret, L., Rigat, B., Viskis, S., Breda, C., Corvol, P., Cambien, F. and Soubrier, F. (1992) Evidence, from combined segregation and linkage analysis, that a variant of the angiotensin I-converting enzyme (ACE) gene controls plasma ACE levels. *Am. J. Hum. Genet.* **51**: 197–205.

Tsuchida, S., Matsusaka, T., Chen, X., *et al.* (1998) Murine double nullizygotes of the angiotensin type 1A and 1B receptor genes duplicate severe abnormal phenotypes of angiotensinogen nullizygotes. *J. Clin. Invest.* **101**: 755–760.

Tsutsumi, K. and Saavedra, J.M. (1991) Characterization and development of angiotensin II receptor subtypes (AT1 and AT2) in rat brain. *Am. J. Physiol.* **261**: R209-R216.

Wagner, D., Metzger, R., Paul, M., Ludwig, G., Suzuki, F., Takahashi, S., Murakami, K. and Ganten, D. (1990) Androgen dependence and tissue specificity of renin messenger RNA expression in mice. *J. Hypertension* **8**: 45–52.

Wilson, C.M., Cherry, M., Taylor, B.A. and Wilson, J.D. (1981) Genetic and endocrine control of renin activity in the submaxillary gland of the mouse. *Biochem. Genet.* **19**: 509–523.

Zee, R.Y.L., Lou, Y.K., Griffiths, L.R. and Morris, B.J. (1992) Association of a polymorphism of the angiotensin I-converting enzyme gene with essential hypertension. *Biochem. Biophys. Res. Commun.* **184**: 9–15.

Zhang, S.H., Reddick, R., Burkey, B. and Maeda, N. (1994) Diet-induced atherosclerosis in mice heterozygous and homozygous for apolipoprotein E gene disruption. *J. Clin. Invest.* **94**: 937–945.

7

Monogenic forms of mineralocorticoid hypertension

Perrin C. White

1. Introduction

A useful way to analyse the contribution of genetic factors to the development of human hypertension is the study of hypertensive syndromes inherited as monogenic traits. The syndromes that have been identified and studied thus far are all caused by direct or indirect dysregulation of the kidneys' system for controlling intravascular volume through sodium resorption from the urine.

The regulated step in renal sodium resorption is passive diffusion through a sodium-permeable channel in the apical membranes of epithelial cells lining the distal convoluted tubule and collecting duct. Activating mutations in regulatory subunits of this channel lead to Liddle's syndrome which is characterized by excessive resorption of sodium, increased intravascular volume and hypertension. The normal regulator of this channel is a steroid, aldosterone, secreted by the adrenal cortex. Because aldosterone regulates sodium resorption and potassium excretion, it is referred to as a 'mineralocorticoid' hormone.

Aldosterone increases resorption of sodium from the urine through at least two mechanisms (*Figure 1*, reviewed in White, 1994). It increases the apparent number of epithelial sodium channels. This may reflect an increase in the percentage of time that each channel stays open, possibly mediated by methylation of the channel (Duchatelle *et al.*, 1992), and/or an increase in the actual number of channels (Palmer and Frindt, 1992). Aldosterone also increases synthesis of a sodium/potassium ATPase located in the basolateral cell membrane which generates the electrochemical gradient that drives diffusion through the sodium channels (Horisberger and Rossier, 1992). These actions are the result of transcriptional effects mediated by a specific nuclear receptor referred to as the mineralocorticoid or 'type 1 steroid' receptor. These receptors are expressed in renal distal tubules and cortical collecting ducts and also in other mineralocorticoid target tissues, including salivary glands and the colon. Mineralocorticoid

Molecular Genetics of Hypertension, edited by A.F. Dominiczak, J.M.C. Connell and F. Soubrier.

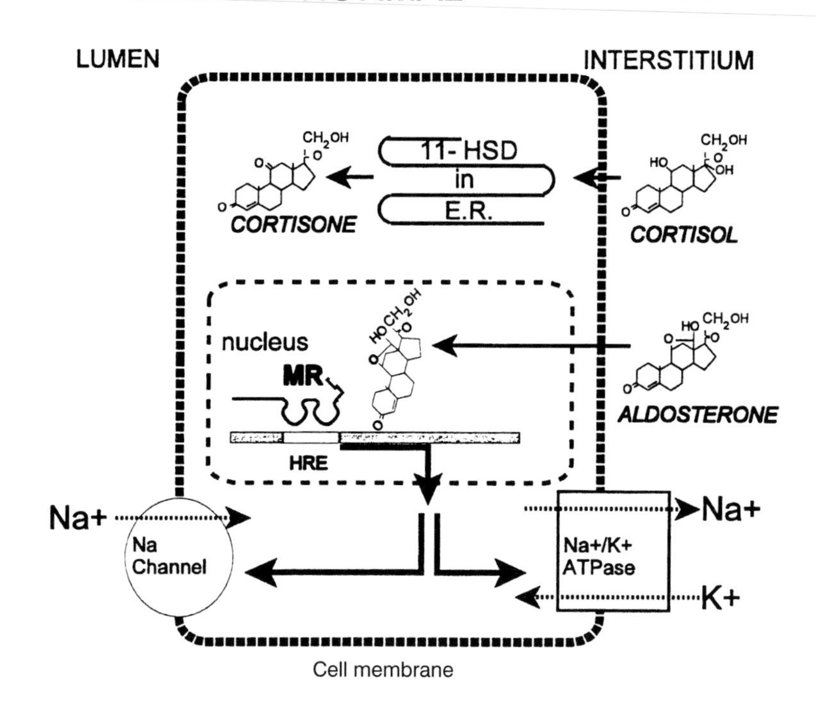
NORMAL
LUMEN
INTERSTITIUM
11-HSD
in
E.R.
CORTISONE
CORTISOL
nucleus
MR
HRE
ALDOSTERONE
Na+
Na
Channel
Na+
Na+/K+
ATPase
K+
Cell membrane

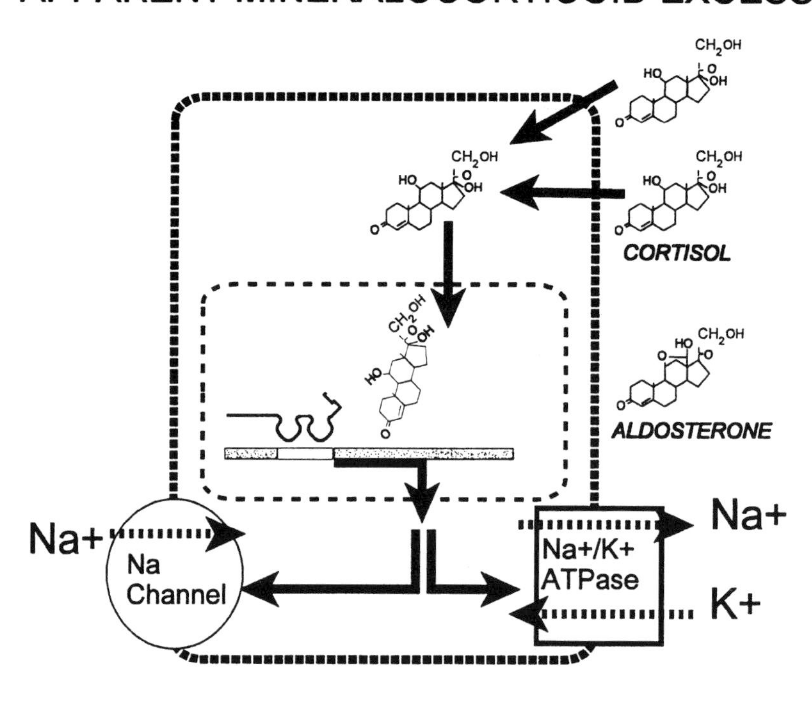
APPARENT MINERALOCORTICOID EXCESS
CORTISOL
ALDOSTERONE
Na+
Na
Channel
Na+/K+
ATPase
Na+
K+

receptors are also expressed in the myocardium (Lombes *et al.*, 1995) and aldosterone has direct actions on the heart in animal studies including induction of myocardial hypertrophy and fibrosis (Brilla *et al.*, 1993; Young *et al.*, 1994).

Excess secretion of aldosterone or other mineralocorticoids, or abnormal sensitivity to mineralocorticoids, may result in hypokalemia, suppressed plasma renin activity and hypertension. This chapter reviews studies of several such conditions. In order to understand the pathophysiology of these disorders, it is first necessary to review several aspects of aldosterone biosynthesis and action.

2. Normal synthesis and actions of aldosterone

2.1 Enzymes required for synthesis of aldosterone and cortisol

Cortisol is synthesized from cholesterol in the zona fasciculata of the adrenal cortex. This process requires five enzymatic conversions (*Figure 2*): cleavage of the cholesterol side-chain to yield pregnenolone, 17α-hydroxylation and 3β-dehydrogenation to 17-hydroxyprogesterone, and successive hydroxylations at the 21 and 11β positions. A '17-deoxy' pathway is also active in the zona fasciculata, in which 17α-hydroxylation does not occur and the final product is normally corticosterone.

Aldosterone is synthesized via a similar 17-deoxy pathway in the adrenal zona glomerulosa. However, corticosterone is not the final product in the zona glomerulosa; instead, corticosterone is successively hydroxylated and oxidized at the 18 position to yield aldosterone.

The same enzymes catalyze cholesterol side-chain cleavage, 3β-dehydrogenation and 21-hydroxylation, respectively, in the zona fasciculata. Although cortisol and aldosterone syntheses both require 11β-hydroxylation of steroid intermediates, these steps are catalyzed by different isozymes, respectively termed steroid 11β-hydroxylase and aldosterone synthase (reviewed in White *et al.*, 1994). The latter isozyme catalyzes the subsequent 18-hydroxylation and 18-oxidation steps required for aldosterone synthesis as well as 11β-hydroxylation, and thus itself converts deoxycorticosterone to aldosterone (Curnow *et al.*, 1991; Kawamoto *et al.*, 1990a; Kawamoto *et al.*, 1992; Ogishima *et al.*, 1991).

Figure 1. Schematic of mineralocorticoid action. *Top*: a normal mineralocorticoid target cell in a renal cortical collecting duct. Aldosterone occupies nuclear receptors (MR) that bind to hormone response elements, increasing transcription of genes and directly or indirectly increasing activities of apical sodium (Na) channels and the basolateral sodium–potassium (Na/K) ATPase. This increases resorption of sodium from and excretion of potassium into the tubular lumen. Cortisol, which circulates at higher levels than aldosterone, cannot occupy the receptor because it is oxidized to cortisone by 11β-hydroxysteroid dehydrogenase (11-HSD). *Bottom*: a cell from a patient with the syndrome of apparent mineralocorticoid excess. Because 11-HSD is absent, cortisol inappropriately occupies mineralocorticoid receptors, leading to increased gene transcription, increased activity of sodium channels and the Na/K ATPase, increased resorption of sodium and excretion of potassium, and hypertension.

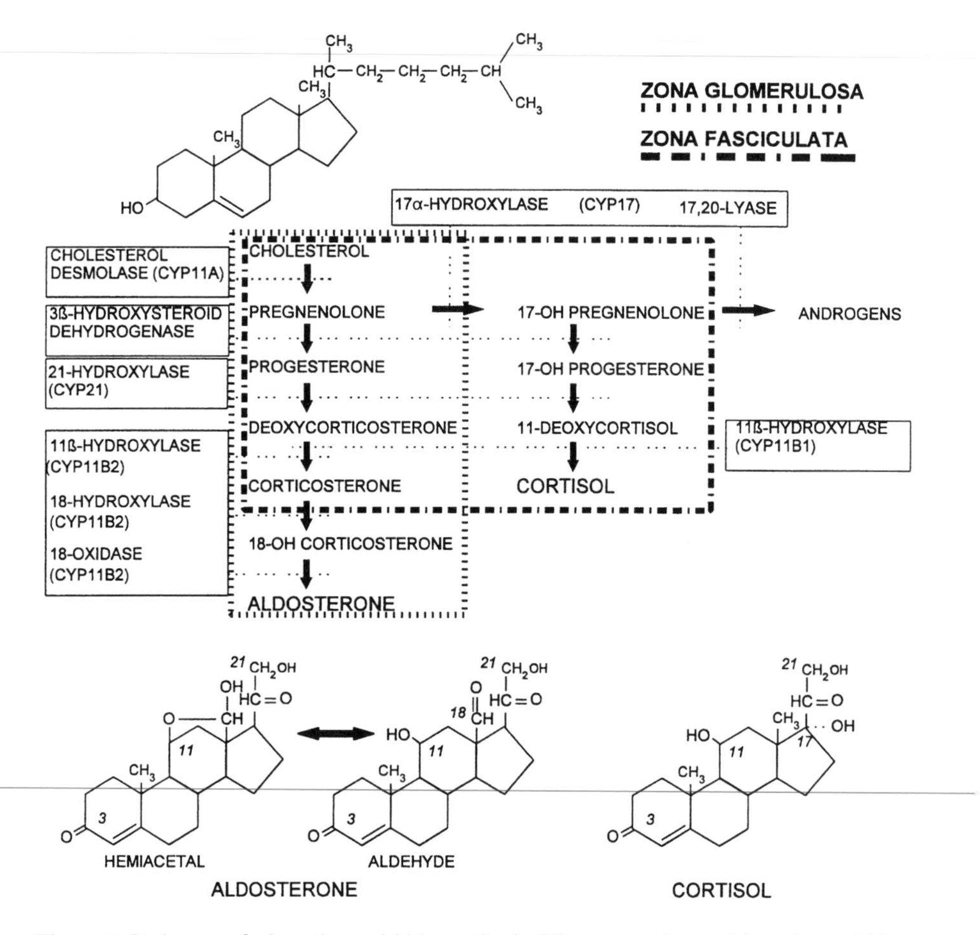

Figure 2. Pathways of adrenal steroid biosynthesis. The conversions taking place within the zonae glomerulosa and fasciculata are marked by broken rectangles; note that several conversions take place in both zones. The pathways of biosynthesis of aldosterone and cortisol from cholesterol are shown with the planar structures of these substances, respectively, at the bottom and top of the figure. Aldosterone exists in two conformations (18-aldehyde and hemiacetal) that are freely interconvertible; the hemiacetal predominates under physiological conditions. The enzymes responsible for each biosynthetic step are listed in boxes on the left; the last three steps of aldosterone biosynthesis are mediated by a single enzyme, aldosterone synthase (CYP11B2). Deficiencies of the enzymes listed in bold boxes cause congenital adrenal hyperplasia.

With the exception of 3β-hydroxysteroid dehydrogenase, which is a 'short chain dehydrogenase' (see below) all of the enzymes required for cortisol and aldosterone biosynthesis are cytochromes P450, heme-containing enzymes with molecular weights of about 50 000. P450s comprise a gene superfamily and are as little as 15–20% identical in their amino acid sequences. They are systematically designated by 'CYP' followed by a number and, if there are closely related isozymes, by additional letters and numbers. Schenkman and Greim (1993) is an extensive review of cytochrome P450 enzymes. Steroid 17α-hydroxylase (CYP17) and 21-hydroxylase (CYP21) are located in the endoplasmic reticulum ('microsomal' enzymes) whereas cholesterol desmolase (CYP11A), 11β-hydroxylase (CYP11B1) and aldosterone synthase (CYP11B2) are found in the inner membrane of mitochondria.

P450s are 'mixed function oxidases.' They carry out oxidative conversions utilizing molecular oxygen and reducing equivalents (i.e. electrons) provided by NADPH via accessory electron transport proteins. Microsomal P450s utilize a single accessory protein, NADPH-dependent cytochrome P450 reductase. This is a flavoprotein containing one molecule each of FAD and FMN (Porter and Kasper, 1985). Mitochondrial P450s require two proteins; NADPH-dependent adrenodoxin reductase donates electrons to adrenodoxin which in turn transfers them to the P450. Adrenodoxin (or ferredoxin) reductase is also a flavoprotein but contains only a molecule of FMN (Solish *et al.*, 1988). Adrenodoxin (or ferredoxin) contains nonheme iron complexed with sulfur (Picado-Leonard *et al.*, 1988). Only one isoform of each accessory protein has been documented in mammals.

2.2 Normal regulation of cortisol and aldosterone biosynthesis

Cortisol synthesis is primarily controlled by ACTH (corticotropin) (Waterman and Simpson, 1989). ACTH secretion is stimulated mainly by corticotropin releasing hormone (CRH) from the paraventricular nucleus of the hypothalamus and is subject to feedback inhibition by cortisol and other glucocorticoids. ACTH acts through a specific G protein-coupled receptor on the surface of cells of the adrenal cortex (Mountjoy *et al.*, 1992) to increase levels of cAMP (cyclic adenosine 3′,5′ monophosphate). cAMP has short term (minutes to hours) effects on transport of cholesterol into mitochondria through increasing the synthesis of a short lived protein, steroidogenic acute regulatory (STAR) protein (Clark *et al.*, 1995; Clark *et al.*, 1994). It has longer term (hours to days) effects on transcription of genes encoding the enzymes required to synthesize cortisol (John *et al.*, 1986). The transcriptional effects occur at least in part through increased activity of protein kinase A (Wong *et al.*, 1989) which phosphorylates transcriptional regulatory factors, not all of which have been identified.

The rate of aldosterone synthesis, which is normally 100–1000 fold less than that of cortisol synthesis, is regulated mainly by the renin–angiotensin system and by potassium levels with ACTH having only a short term effect (Quinn and Williams, 1988). Renin is a proteolytic enzyme secreted by the juxtaglomerular apparatus of the nephron in response to decreased volume, as sensed by stretch receptors in the afferent arteriole. Renin digests angiotensinogen to angiotensin I, a decapeptide which is converted by angiotensin converting enzyme to an octapeptide, angiotensin II. Angiotensin II occupies a G protein-coupled receptor (Curnow *et al.*, 1992; Murphy *et al.*, 1991; Sasaki *et al.*, 1991), activating phospholipase C. The latter protein hydrolyses phosphatidylinositol bisphosphate to produce inositol triphosphate and diacylglycerol, which raise intracellular calcium levels and activate protein kinase C. Phosphorylation of an unidentified factor(s) is presumed to increase transcription of enzymes required for aldosterone synthesis.

Because the necessary precursors for aldosterone biosynthesis (in particular, deoxycorticosterone) are also synthesized in the much larger zona fasciculata, it is apparent that there must be unique regulated steps in aldosterone biosynthesis in the zona glomerulosa or aldosterone biosynthesis would simply be regulated by ACTH. These regulated steps seem to be those mediated by aldosterone synthase (CYP11B2) (Adler *et al.*, 1993).

The *CYP11B1* and *CYP11B2* genes encoding the two 11β-hydroxylase isozymes are regulated in a manner consistent with their respective roles in cortisol and aldosterone biosynthesis. *CYP11B1* is expressed at high levels in normal adrenal glands (Mornet *et al.*, 1989), and transcription of this gene is appropriately regulated by cAMP (Kawamoto *et al.*, 1990b). *CYP11B2* transcripts cannot be detected by hybridization to Northern blots of normal adrenal RNA (Mornet *et al.*, 1989), but such transcripts have been detected in normal adrenal RNA using a more sensitive assay wherein RNA is reverse-transcribed (RT) and then amplified using the polymerase chain reaction (PCR; RT–PCR) (Curnow *et al.*, 1991). *CYP11B2* transcripts are present at increased levels in aldosterone secreting tumors (Curnow *et al.*, 1991; Kawamoto *et al.*, 1990a). In primary cultures of human zona glomerulosa cells, angiotensin II markedly increases levels of both *CYP11B1* and *CYP11B2* transcripts. ACTH increases *CYP11B1* mRNA levels more effectively than angiotensin II does, but it has no effect on *CYP11B2* transcription in these cells (Curnow *et al.*, 1991). In NCI-H295 human adrenocortical carcinoma cells, which synthesize both cortisol and aldosterone, angiotensin II increases CYP11B2 mRNA levels and aldosterone synthesis (Bird *et al.*, 1993). These cells are apparently unresponsive to ACTH, but both CYP11B1 and CYP11B2 mRNA levels are increased by administration of a cAMP analog (Clyne *et al.*, 1997; Staels *et al.*, 1993).

Although studies with primary cultures of zona glomerulosa cells suggests that *CYP11B1* is expressed in both the zonae fasciculata and glomerulosa, these two genes seem to be expressed in a mutually exclusive manner when examined by *in situ* hybridization of normal human adrenal cortex (Pascoe *et al.*, 1995). The difference in regulation of *CYP11B1* and *CYP11B2* is presumably due to the extensive divergence between the 5′ regions flanking these genes (Honda *et al.*, 1993; Lala *et al.*, 1992).

Both human genes include (Mornet *et al.*, 1989) a TATA box variant, a palindromic cAMP response element, and several recognition sites for steroidogenic factor-1 (SF-1). SF-1 sites appear in the regulatory regions of all steroid hydroxylase genes expressed in the adrenal cortex and the gonads (Morohashi *et al.*, 1992; Rice *et al.*, 1991). SF-1 (also called Ad4BP) is an orphan nuclear receptor, that is, it is a member of the steroid and thyroid hormone receptor superfamily but its ligand is not known. Intriguingly, it is closely related in its predicted amino acid sequence to FTZ-F1, a *Drosophila* protein that regulates transcription of a homeobox gene, *fushi tarazu* (Honda *et al.*, 1993; Lala *et al.*, 1992).

All of these sequences are required for normal transcription of murine and bovine *CYP11B* genes (Hashimoto *et al.*, 1992; Honda *et al.*, 1990; Mouw *et al.*, 1989; Rice *et al.*, 1989). Whereas *CYP11B1* expression has not yet been studied in detail (Kawamoto *et al.*, 1992), 5′ flanking elements have been identified that are required for basal and hormone induced expression of *CYP11B2* reporter constructs in H295R human adrenocortical carcinoma cells (Clyne *et al.*, 1997).

These elements include the putative cAMP response element and an element located between 129 and 114 upstream from the start of translation that binds both SF-1 and COUP-TF, another orphan nuclear receptor (Wang *et al.*, 1989).

As yet, the genetic elements responsible for the differential regulation of human *CYP11B1* and *CYP11B2* in the zonae fasciculata and glomerulosa have not been identified.

2.3 Mechanisms conferring specificity on the mineralocorticoid receptor

Although aldosterone has actions that are distinct from those of cortisol, the mineralocorticoid ('type 1') receptor has a high degree of sequence identity with the glucocorticoid or 'type 2' receptor (Arriza *et al.*, 1987). The central portion of the polypeptide contains a DNA binding domain consisting of two 'zinc fingers'; this region is also involved in dimerization of liganded receptors. The amino acid sequences of the mineralocorticoid and glucocorticoid receptors are 94% identical in this region, and the two receptors interact with similar hormone response elements. The carboxyl terminus is a ligand binding domain that is 57–60% identical in amino acid sequence in the two receptors. In fact, the mineralocorticoid receptor has very similar *in vitro* binding affinities for aldosterone and for glucocorticoids such as corticosterone and cortisol (Arriza *et al.*, 1987; Krozowski and Funder, 1983). If the mineralocorticoid receptor can bind glucocorticoids and can interact with glucocorticoid response elements in transcriptional regulatory regions of genes, it is not obvious how mineralocorticoids can have distinct physiological effects.

The mechanism by which the glucocorticoid and mineralocorticoid receptors influence transcription of distinct sets of genes probably involves the amino terminal domains of these receptors, which are less than 15% identical in amino acid sequence. In the glucocorticoid receptor, this region is known to interact with other nuclear transcription factors (Pearce and Yamamoto, 1993). Presumably the mineralocorticoid receptor interacts with a different set of accessory transcription factors, permitting it to have distinct transcriptional effects.

Whereas the mineralocorticoid receptor has similar *in vitro* affinities for aldosterone, cortisol and corticosterone, the latter two steroids are weak mineralocorticoids *in vivo*. It has been proposed (Edwards *et al.*, 1988; Funder *et al.*, 1988; Stewart *et al.*, 1987) that the physiological mechanism conferring specificity for aldosterone upon the mineralocorticoid receptor is oxidation of cortisol and corticosterone to cortisone or 11-dehydrocorticosterone, respectively, by 11β-hydroxysteroid dehydrogenase (11-HSD) (*Figure 1*). Whereas cortisone and 11-dehydrocorticosterone are poor agonists for the mineralocorticoid receptor, aldosterone is a poor substrate for 11-HSD because, in solution, its 11β-hydroxyl group is normally in a hemiacetal conformation with the 18-aldehyde group and is not accessible to the enzyme. Evidence supporting this hypothesis comes from studies of a form of hypertension, the syndrome of apparent mineralocorticoid excess (AME, see below).

3. Hypertensive forms of congenital adrenal hyperplasia

Congenital adrenal hyperplasia (CAH), the inherited inability to synthesize cortisol, can be caused by mutations in any of the latter four enzymes required for cortisol biosynthesis (*Figure 2*) or by mutations in the STAR protein required for cholesterol transport into mitochondria (Lin *et al.*, 1995). More than 90% of cases are caused by 21-hydroxylase deficiency. This usually affects both aldosterone and cortisol biosynthesis, leading to signs of aldosterone deficiency including hyponatremia, hyperkalemia and hypovolemia that may, if untreated, progress to shock and death within weeks after birth. In contrast, most remaining cases of CAH are associated with hypertension. Most of these are due to 11β-hydroxylase deficiency, and a lesser number to 17α-hydroxylase deficiency. These autosomal recessive disorders represent the first Mendelian forms of hypertension in which the affected genes were cloned and causative mutations identified (White *et al.*, 1994 and Yanase *et al.*, 1991 are detailed reviews).

3.1 11β-Hydroxylase deficiency

Clinical presentation. In most populations, 11β-hydroxylase deficiency comprises approximately 5–8% of cases of CAH (Zachmann *et al.*, 1983) and thus it occurs in approximately 1 in 200 000 births. A large number of cases of 11β-hydroxylase deficiency has been reported in Israel among Jewish immigrants from Morocco; the incidence in this group is currently estimated to be 1/5000–1/7000 births (Rosler *et al.*, 1992).

In 11β-hydroxylase deficiency, 11-deoxycortisol and deoxycorticosterone are not efficiently converted to cortisol and corticosterone, respectively. Decreased production of glucocorticoids reduces their feedback inhibition on the hypothalamus and anterior pituitary, increasing secretion of ACTH. This stimulates the zona fasciculata of the adrenal cortex to overproduce steroid precursors proximal to the blocked 11β-hydroxylase step. Thus, 11β-hydroxylase deficiency can be diagnosed by detecting high basal or ACTH-stimulated levels of deoxycorticosterone and/or 11-deoxycortisol in the serum, or increased excretion of the tetrahydro metabolites of these compounds in a 24-hour urine collection. Obligate heterozygous carriers of 11β-hydroxylase deficiency alleles (e.g. parents) have no consistent biochemical abnormalities detectable even after stimulation of the adrenal cortex with intravenous ACTH (Pang *et al.*, 1980), consistent with an autosomal recessive mode of inheritance.

Approximately two-thirds of patients with the severe, 'classic' form of 11β-hydroxylase deficiency have high blood pressure (Rosler *et al.*, 1992; Rosler *et al.*, 1982), often beginning in the first few years of life (Mimouni *et al.*, 1985). Although the hypertension is usually of mild to moderate severity, left ventricular hypertrophy and/or retinopathy have been observed in up to one-third of patients, and deaths from cerebrovascular accidents have been reported (Hague and Honour, 1983; Rosler *et al.*, 1992). Other signs of mineralocorticoid excess such as hypokalemia and muscle weakness or cramping occur in a minority of patients and are not well correlated with blood pressure. Plasma renin activity is usually suppressed in older children and levels of aldosterone are consequently low even though the ability to synthesize aldosterone is actually unimpaired.

The cause of hypertension in 11β-hydroxylase deficiency is not well understood. It might be assumed that it is caused by elevated serum levels of deoxycorticosterone but blood pressure and deoxycorticosterone levels are poorly correlated in patients (Rosler *et al.*, 1982; Zachmann *et al.*, 1983). In addition, this steroid has only weak mineralocorticoid activity when administered to humans or other animals and it is possible that other metabolites of deoxycorticosterone are responsible for the development of hypertension. The 18-hydroxy and 19-nor metabolites of deoxycorticosterone are thought to be more potent mineralocorticoids (Griffing *et al.*, 1983), but consistent elevation of these steroids in 11β-hydroxylase deficiency has not been documented. Moreover, synthesis of these steroids requires hydroxylations within the adrenal (19-nor-deoxycorticosterone is synthesized via 19-hydroxy and 19-oic intermediates) that are probably mediated primarily by CYP11B1 (Ohta *et al.*, 1988). This is unlikely to take place efficiently in 11β-hydroxylase deficiency.

In addition to hypertension, patients with 11β-hydroxylase deficiency often exhibit signs of androgen excess. This occurs because accumulated cortisol precursors in the adrenal cortex are shunted (through the activity of 17α-hydroxylase/17,20-lyase) into the pathway of androgen biosynthesis, which is active in the human adrenal in both sexes. Affected females are born with some degree of masculinization of their external genitalia. This includes clitoromegaly and partial or complete fusion of the labioscrotal folds. Such ambiguous genitalia can be difficult to distinguish from those of a normal cryptorchid male (Bistritzer *et al.*, 1984; Harinarayan *et al.*, 1992; Rosler *et al.*, 1992). In contrast to the external genitalia, the gonads and the internal genital structures (Fallopian tubes, uterus and cervix) arising from the Muellerian ducts are normal and affected females have intact reproductive potential if their external genital abnormalities are corrected surgically.

Other signs of androgen excess that occur postnatally in both sexes include rapid somatic growth in childhood and accelerated skeletal maturation leading to premature closure of the epiphyses and short adult stature. Additionally, patients may have premature development of sexual and body hair (premature adrenarche) and acne. Androgens may affect the hypothalamic–pituitary–gonadal axis leading to amenorrhea or oligomenorrhea in females and true precocious puberty or, conversely, poor spermatogenesis in males (Hochberg *et al.*, 1985).

Glucocorticoid administration (usually with hydrocortisone) replaces deficient cortisol and thus reduces ACTH secretion, suppressing excessive adrenal androgen production and preventing further virilization. Such therapy should also suppress ACTH-dependent production of mineralocorticoid agonists and ameliorate hypertension. If hypertension has been of long standing prior to treatment, additional anti-hypertensive drugs may be required to lower blood pressure into the normal range. These may include potassium sparing diuretics such as spironolactone or amiloride and/or a calcium channel blocker such as nifedipine (Nadler *et al.*, 1985). Because the renin–angiotensin system is suppressed in these patients, angiotensin converting enzyme inhibitors are unlikely to be effective. Thiazide diuretics should not be used except in combination with a potassium sparing diuretic because they will otherwise cause hypokalemia in patients with mineralocorticoid excess.

A mild, 'nonclassic' form of 11β-hydroxylase has been described (Zachmann *et al.*, 1983) in which patients have relatively mild signs of androgen excess; they are not characteristically hypertensive. This disorder appears to be quite rare as compared with nonclassic 21-hydroxylase deficiency (Speiser *et al.*, 1985).

Genetic analysis. In humans, CYP11B1 and CYP11B2 are encoded by two genes (Mornet *et al.*, 1989) on chromosome 8q21–q22 (Chua *et al.*, 1987; Wagner *et al.*, 1991; *Figure 3*). Each contains nine exons spread over approximately 7000 base pairs (7 kb) of DNA. The encoded proteins are 93% identical in predicted amino acid sequence. They are each synthesized with 503 amino acid residues, but a signal peptide is cleaved in mitochondria to yield the mature protein of 479 residues. The nucleotide sequences of these genes are 95% identical in coding regions and about 90% identical in introns. The genes are approximately 40 kb apart (Lifton *et al.*, 1992b; Pascoe *et al.*, 1992a), and *CYP11B2* is on the left if the genes are pictured as being transcribed left to right (Lifton *et al.*, 1992a; Pascoe *et al.*, 1992b).

Deficiency of 11β-hydroxylase results from mutations in *CYP11B1* (*Figure 3*). At this time, 20 mutations have been identified in patients with classic 11β-hydroxylase deficiency (Curnow *et al.*, 1993; Geley *et al.*, 1996; Helmberg *et al.*, 1992; Naiki *et al.*, 1993). In Moroccan Jews, a group that has a high prevalence of 11β-hydroxylase deficiency, almost all affected alleles carry the same mutation, Arg-448 to His (R448H) (White *et al.*, 1991). This probably represents a founder effect, but this mutation has also occurred independently in other ethnic groups, and another mutation of the same residue (R448C) has also been reported (Geley *et al.*, 1996). This apparent mutational 'hotspot' contains a CpG dinucleotide. Such dinucleotides are prone to methylation of the cytosine followed by deamidation to TpG; several other mutations in CYP11B1 (T318M, R374Q, R384Q) are of this type.

These and almost all other missense mutations identified thus far are in regions of known functional importance (Nelson and Strobel, 1988; Poulos, 1991; Ravichandran *et al.*, 1993) and abolish enzymatic activity (Curnow *et al.*, 1993).

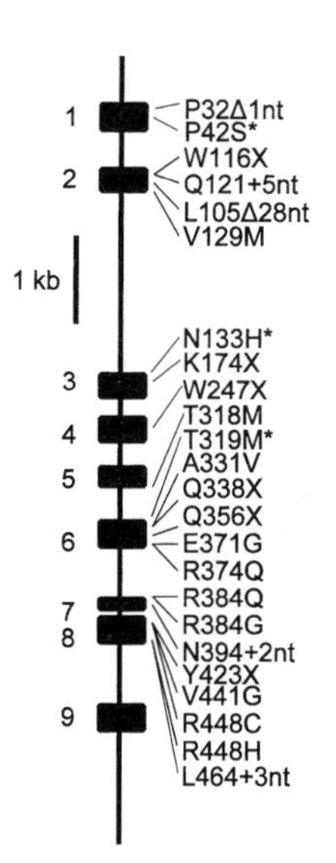

Figure 3. Diagram of the *CYP11B1* gene, showing locations of mutations causing congenital adrenal hyperplasia due to 11β-hydroxylase deficiency. The gene is drawn to scale as marked. Exons are represented by numbered boxes. Mutations are in single letter code: A, alanine; C, cysteine; D, aspartic acid; E, glutamic acid; F, phenylalanine; G, glycine; H, histidine; I, isoleucine; K, lysine; L, leucine; M, methionine; N, asparagine; P, proline; Q, glutamine; R, arginine; S, serine; T, threonine; V, valine; W, tryptophan; X, nonsense (stop) mutation; Y, tyrosine; +, insertion of nucleotides (nt); Δ, deletion of nucleotides. For example, W116X represents a nonsense mutation of Trp–116. Mutations yielding partially active enzymes are marked with asterisks.

For example, Arg-448 is adjacent to Cys-450 which is a ligand of the heme iron atom of this cytochrome P450 enzyme. T318M modifies an absolutely conserved residue that is thought to be critical for proton transfer to the bound oxygen molecule (Ravichandran *et al.*, 1993). E371G and R374Q also mutate highly conserved residues and may affect binding of adrenodoxin. R384Q is in a region that may form part of the substrate binding pocket (Ravichandran *et al.*, 1993). Almost all P450s have a basic residue (H or R) at this or the immediately adjacent position (Nelson and Strobel, 1988). Finally, V441G is adjacent to the highly conserved heme binding region, and this mutation may change the secondary structure of the protein.

Other mutations found in patients with the classic form of the disease are nonsense or frameshift mutations that also abolish enzymatic activity. One, a nonsense mutation of Trp-247 (W247X) has been identified in several unrelated kindreds in Austria and also probably represents a founder effect (Geley *et al.*, 1996).

Each patient with mild, nonclassic disease carries at least one mutation that reduces but does not destroy activity; the other mutation may be either mild or severe (Johrer *et al.*, 1997).

Although classic patients apparently completely lack 11β-hydroxylase activity, they differ significantly in the severity of the various signs and symptoms of their disease. There is not a strong correlation between severity of hypertension and biochemical parameters such as plasma levels of the 11β-hydroxylase substrates, deoxycortisol and deoxycorticosterone, and urinary excretion of tetrahydrodeoxycortisol (Rosler *et al.*, 1992; White *et al.*, 1991). Moreover, there is no consistent correlation between the severity of hypertension and degree of virilization. These phenotypic variations must be governed by factors outside the *CYP11B1* locus.

3.2 17α-Hydroxylase deficiency

Clinical presentation. Because cortisol cannot be synthesized, 17-deoxy steroids are synthesized in excessive quantities. Corticosterone, a glucocorticoid agonist, can be synthesized in the affected gland so that affected individuals do not suffer from adrenal insufficiency. However, adequate levels of corticosterone are synthesized only at the expense of excessive secretion of deoxycorticosterone, so that patients with 17α-hydroxylase deficiency tend to develop hypertension similar to that seen in 11β-hydroxylase deficiency (reviewed in Kater and Biglieri, 1994 and Yanase *et al.*, 1991). However, in contrast to individuals with 11β-hydroxylase deficiency who often have prominent signs of androgen excess, patients with severe 17α-hydroxylase deficiency are unable to synthesize sex steroids. Males are born with female-appearing external genitalia. Females appear normal at birth but remain sexually infantile. The ovaries have poor follicular development and in rare cases appear as streak gonads. This condition is often not diagnosed in either sex until the age at which puberty is expected.

Some patients with mild 17α-hydroxylase deficiency are able to synthesize at least some sex steroids so that males have partially virilized (ambiguous) genitalia and females develop at least some signs of puberty. Some (mostly female) patients have been reported to have isolated 17α-hydroxylase deficiency with normal 17,20-lyase activity. Conversely, a few patients have been thought to have isolated 17,20-lyase deficiency (Yanase *et al.*, 1992).

Genetic analysis. The affected enzyme, a microsomal cytochrome P450 with 508 amino acids, is encoded by the *CYP17* gene on chromosome 10q24–25 (Kagimoto *et al.*, 1988; Picado-Leonard and Miller, 1987; Sparkes *et al.*, 1991). It consists of eight exons spread over 6.7 kb. At this time, at least 18 different mutations have been identified in 27 individuals (*Figure 4*, reviewed in Yanase, 1995, and Yanase *et al.*, 1991). All known mutations are in coding regions. A 4 bp duplication in the last exon causing a shift in the reading frame has been found in 10 patients in the Netherlands or of Dutch Mennonite ancestry; this presumably represents a founder effect (Imai *et al.*, 1992). Complete deficiency is associated with frameshifts or nonsense mutations. In one kindred, two exons were replaced by a segment of the *E. coli* lac operon; the mechanism by which this rearrangement occurred is not known but it may have involved viral integration into the genome (Biason *et al.*, 1991).

Partial deficiency is associated with missense mutations in *CYP17* or, in one case, deletion of a single codon (ΔF53 or 54) maintaining the reading frame of translation (Ahlgren *et al.*, 1992; Yanase *et al.*, 1989). These mutations have been introduced into cDNA and expressed in cultured cells. These studies suggest that greater 20% of normal activity is required to synthesize sufficient androgens to permit normal male sexual development; 50% of normal activity must be sufficient for normal development because obligate heterozygous males are asymptomatic. Patients with isolated 17,20 lyase deficiency have mutations in CYP17 that interfere with electron transfer from cytochrome P450 reductase (Geller *et al.*, 1997).

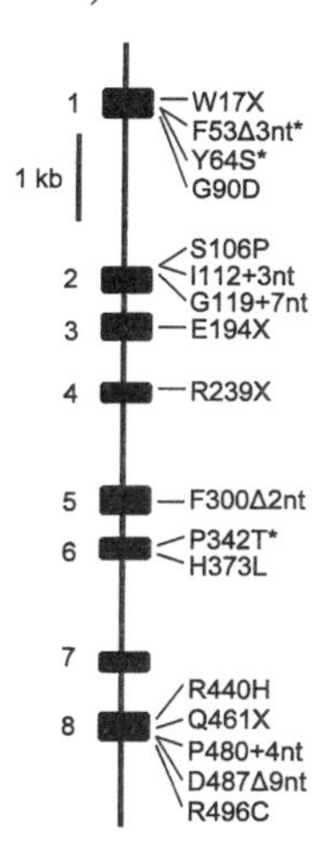

Figure 4. Diagram of the *CYP17* gene, showing locations of mutations causing congenital adrenal hyperplasia due to 17α-hydroxylase deficiency.

4. Glucocorticoid-suppressible hyperaldosteronism

4.1 Clinical presentation

Glucocorticoid-suppressible hyperaldosteronism (also called dexamethasone-suppressible hyperaldosteronism or glucocorticoid-remediable aldosteronism) is a form of hypertension inherited in an autosomal dominant manner with high penetrance (New and Peterson, 1967; Sutherland *et al.*, 1966). It is characterized by moderate hypersecretion of aldosterone, suppressed plasma renin activity and rapid reversal of these abnormalities after administration of glucocorticoids. It is clearly a rare disorder but until several years ago the absence of reliable biochemical or genetic markers made it difficult to ascertain.

Hypokalemia is usually mild and may be absent. Absolute levels of aldosterone secretion are usually moderately elevated in the untreated state but may be within normal limits. Plasma renin activity is strongly suppressed, so that the ratio of aldosterone secretion to renin activity is always abnormally high. Levels of 18-hydroxycortisol and 18-oxocortisol are elevated to 20–30 times normal (Connell *et al.*, 1986; Gomez-Sanchez *et al.*, 1988; Stockigt and Scoggins, 1987). The ratio of urinary excretion of tetrahydro metabolites of 18-oxocortisol to those of aldosterone exceeds 2.0 whereas this ratio averages 0.2 in normal individuals. Elevation of 18-oxocortisol is the most consistent and reliable biochemical marker of the disease, although it may also be elevated in cases of primary aldosteronism (Hamlet *et al.*, 1988). This steroid may be of pathophysiological significance; it is an agonist for the mineralocorticoid receptor and has been shown to raise blood pressure in animal studies (Hall and Gomez-Sanchez, 1986).

Once an affected individual has been identified in a kindred, additional cases may be ascertained within that kindred using biochemical (18-oxocortisol levels) or genetic (see below) markers (Rich *et al.*, 1992). It is apparent from these studies that affected individuals have blood pressures that are markedly elevated as compared to unaffected individuals in the same kindred, although some patients may in fact have normal blood pressures. Even young children typically have blood pressures greater than the 95th percentile for age, and most are frankly hypertensive before the age of 20. The hypertension is often of only moderate severity and blood pressures exceeding 180/120 are unusual. Associated signs of hypertension are frequent including left ventricular hypertrophy on the electrocardiogram and retinopathy. Some affected kindreds have remarkable histories of early (before age 45 years) death from strokes in many family members (O'Mahony *et al.*, 1989; Rich *et al.*, 1992).

Steroid biosynthesis is otherwise normal so that affected individuals have normal growth and sexual development.

Most laboratory and clinical abnormalities are suppressed by treatment with glucocorticoids, whereas infusion of ACTH exacerbates these problems (Ganguly *et al.*, 1984; Oberfield *et al.*, 1981). This suggests that aldosterone is being inappropriately synthesized in the zona fasciculata and is being regulated by ACTH. Moreover, 18-hydroxycortisol and 18-oxocortisol, the steroids that are characteristically elevated in this disorder, are 17α-hydroxylated analogues of 18-hydroxycorticosterone and aldosterone, respectively. Because 17α-hydroxylase is not expressed in the zona glomerulosa, the presence of large amounts of a 17α-hydroxy, 18-oxo-steroid suggests that an enzyme with 18-oxidase activity (i.e. aldosterone synthase, CYP11B2) is abnormally expressed in the zona fasciculata (White, 1991).

The initial treatment of choice in adults is dexamethasone (1–2 mg/day). Within 2–4 days of initiating therapy, oversecretion of aldosterone should be completely suppressed and plasma renin activity and potassium (if low) should increase into the normal range. Blood pressure usually also decreases into the normal range. Children with this condition should be treated cautiously because of potential adverse effects of glucocorticoid therapy on growth. If therapy is indicated, children should be treated with the lowest effective dose of hydrocortisone. If hypertension is of long standing, it may not completely respond to glucocorticoids. This problem is similar to that observed in patients with 11β-hydroxylase deficiency and the choice of adjunctive therapy is governed by the same considerations. Patients with this disorder usually respond poorly to conventional antihypertensive medications unless they are also treated with glucocorticoids.

It is important to distinguish glucocorticoid-suppressible hyperaldosteronism from aldosterone-producing adenomas, considering that the latter condition is best treated by surgical removal of the affected adrenal gland (Melby, 1991). Secretion of 18-hydroxy- and 18-oxocortisol may be increased in patients with adenomas, but the ratio of urinary excretion of tetrahydro metabolites of 18-oxocortisol and aldosterone is rarely greater than 1.0 (Hamlet *et al.*, 1988; Ulick *et al.*, 1990a). Suppression of aldosterone secretion with glucocorticoids (Hamlet *et al.*, 1988) and familial aggregation (Gordon *et al.*, 1992) are both unusual findings in adenomas but have been reported. However, presentation of an adenoma during childhood is exceedingly rare. Conversely, rare patients with glucocorticoid suppressible hyperaldosteronism eventually become resistant to glucocorticoids and are then indistinguishable from patients with primary aldosteronism (Stockigt and Scoggins, 1987). The mechanism by which this occurs is not known.

4.2 Genetic analysis

Chimeric CYP11B1/B2 *genes cause glucocorticoid-suppressible hyperaldosteronism.* All patients with glucocorticoid-suppressible hyperaldosteronism have the same type of mutation, a chromosome that carries three *CYP11B* genes instead of the normal two (*Figure 5*; Lifton *et al.*, 1992a, 1992b; Pascoe *et al.*, 1992a). The middle gene on this chromosome is a chimera with 5′ and 3′ ends corresponding to *CYP11B1* and *CYP11B2* respectively. The chimeric gene is flanked by presumably normal *CYP11B2* and *CYP11B1* genes. In all kindreds analysed thus far, the breakpoints (the points of transition between *CYP11B1* and *CYP11B2* sequences) are located between intron 2 and exon 4. As the breakpoints are not identical in different kindreds, these must represent independent mutations.

The chromosomes carrying chimeric genes are presumably generated by unequal crossing over. The high homology and proximity of the *CYP11B1* and *CYP11B2* genes makes it possible for them to become misaligned during meiosis. If this occurs, crossing over between the misaligned genes creates two chromosomes, one of which carries one *CYP11B* gene (i.e. a deletion) whereas the other carries three *CYP11B* genes.

The invariable presence of a chimeric gene in patients with this disorder suggests that this gene is regulated like *CYP11B1* (expressed at high levels in the zona fasciculata and regulated primarily by ACTH) because it has transcriptional regulatory sequences identical to those of *CYP11B1*. If the chimeric gene has

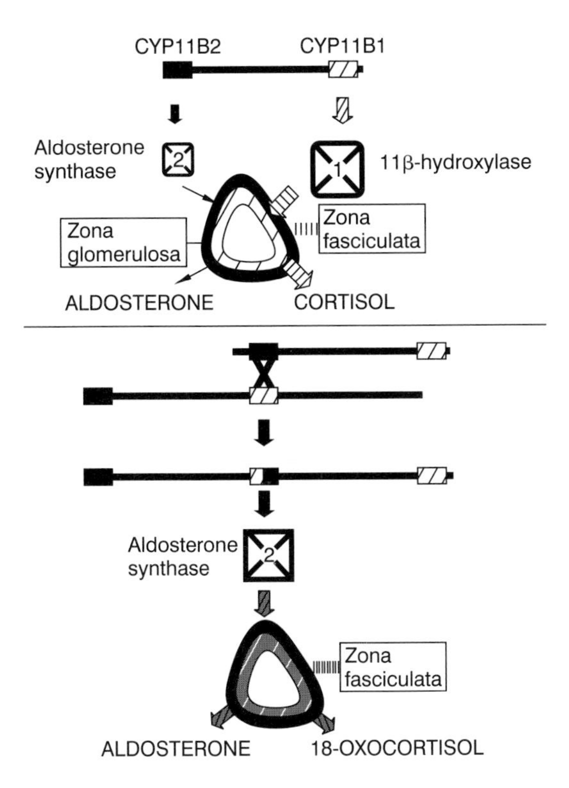

Figure 5. *Top:* schematic of *CYP11B1* and *CYP11B2* genes, showing patterns of expression of each. *Bottom:* unequal crossing over generating a chimeric *CYP11B1/2* gene that has aldosterone synthase activity but is expressed in the zona fasciculata. This causes glucocorticoid-suppressible hyperaldosteronism.

enzymatic activity similar to that of *CYP11B2,* a single copy of such an abnormally regulated gene should be sufficient to cause the disorder, consistent with the known autosomal dominant mode of inheritance of this syndrome. Recently, abnormal expression of the chimeric gene in the zona fasciculata was directly demonstrated by *in situ* hybridization studies of an adrenal gland from a patient with this disorder (Pascoe *et al.,* 1995).

The chimeric genes causing glucocorticoid-suppressible hyperaldosteronism may be readily detected by hybridization to Southern blots of genomic DNA, or they may be specifically amplified using the PCR (Jonsson *et al.,* 1995). As these techniques are widely used in molecular genetics laboratories, direct molecular genetic diagnosis may be more practical in many cases than assays of 18-oxocortisol levels, which are not routinely available (Dluhy and Lifton, 1995).

The limited region in which crossover breakpoints have been observed in glucocorticoid-suppressible hyperaldosteronism alleles suggests that there are functional constraints on the structures of chimeric genes able to cause this disorder.

One obvious constraint is that sufficient *CYP11B2* coding sequences must be present in the chimeric gene so that the encoded enzyme actually has aldosterone synthase (i.e. 18-hydroxylase and 18-oxidase) activity. As determined by expressing chimeric cDNAs in cultured cells, chimeric enzymes with amino termini from CYP11B1 and carboxyl termini from CYP11B2 have 18-oxidase activity only if at least the region encoded by exons 5–9 corresponds to CYP11B2. If the sequence of exon 5 instead corresponds to CYP11B1, the enzyme has 11β-hydroxylase but no 18-oxidase activity (Pascoe *et al.*, 1992a). This is entirely consistent with the observation that no breakpoints in glucocorticoid suppressible hyperaldosteronism alleles occur after exon 4. The chimeric enzymes either have strong 18-oxidase activity or none detectable and there does not appear to be any location of crossover that yields an enzyme with an intermediate level of 18-oxidase activity. Thus, there is no evidence for allelic variation in this disorder (i.e. variations in clinical severity are unlikely to be the result of different crossover locations).

Presumably the transcriptional regulatory region of the chimeric gene must correspond completely to that of *CYP11B1* or the chimeric gene will not be expressed at sufficiently high levels in the zona fasciculata to cause the disorder. Although transcriptional regulatory elements in the *CYP11B* genes have not been completely defined, the fact that no breakpoints have been detected before intron 2 in glucocorticoid-suppressible hyperaldosteronism alleles suggests that there is a transcriptional enhancer in exon 1-intron 2 of *CYP11B1* or, conversely, a silencer in this region of *CYP11B2*.

Other factors such as kallikrein levels may affect the development of hypertension in this disorder (Dluhy and Lifton, 1995). One study found that blood pressure in persons with glucocorticoid-suppressible hyperaldosteronism is higher when the disease is inherited from the mother than when it is paternally inherited (Jamieson *et al.*, 1995). It is theoretically possible that the gene is imprinted (i.e. the maternal and paternal copies are expressed differently), but it seems more likely that exposure of the fetus to elevated levels of maternal aldosterone subsequently exacerbates the hypertension.

Allelic variation in CYP11B2. Whereas it originally seemed possible that a 'mild' form of glucocorticoid-suppressible hyperaldosteronism might be a common etiology of essential hypertension, the lack of allelic variation in this disorder makes this unlikely. However, other polymorphisms in the 5′ flanking region of *CYP11B2* have been documented (Lifton *et al.*, 1992a; White and Slutsker, 1995), although none has been shown to affect expression of the gene. If any does influence regulation of *CYP11B2*, it might be a risk factor for the development of hypertension. It has also been suggested that polymorphisms in the coding sequence of *CYP11B2* might increase the aldosterone synthase activity of the enzyme and thus might be a risk factor for hypertension (Fardella *et al.*, 1995). One coding sequence polymorphism, K173R, has been characterized in humans. It is not associated with any significant differences in enzymatic activity, but is associated with low renin hypertension in a small sample of patients from Chile (36).

Thus far, the most extensively studied polymorphism is located 344 nucleotides 5′ of the start of translation; this position may be either a C or a T (-344C/T)(White and Slutsker, 1995). These alleles are present at approximately equal frequencies in Caucasian populations (White and Slutsker, 1995; Kupari *et al.*, 1998). This position comprises part of a binding site for the SF-1 transcription factor (see above), and the C allele binds SF-1 approximately four times as strongly as the T allele (unpublished observations). The functional significance of this is obscure, because this site may be deleted from reporter constructs without affecting expression (Clyne *et al.*, 1997). Inconsistent associations have been observed between this polymorphism and serum aldosterone levels or aldosterone excretion, with the C allele associated with higher aldosterone levels in some studies (Pojoga *et al.*, 1998), but lower levels in others (Brand *et al.*, 1998; Hautanen *et al.*, 1998a). Associations with blood pressure have also been inconsistent (Brand *et al.*, 1998; Hautanen *et al.*, 1998a; Kupari *et al.*, 1998; Pojoga *et al.*, 1998). However, associations have also been noted between -344C and other cardiovascular parameters including increased left ventricular diameter and mass, increased response of left ventricular mass to increases in dietary salt (Kupari *et al.*, 1998) and decreased baroreflex sensitivity (Ylitalo *et al.*, 1997). These associations all appear to be strongest in young adults. Both left ventricular hypertrophy (Levy *et al.*, 1990) and decreased baroreflex sensitivity (La Rovere *et al.*, 1998) are well established predictors of morbidity and mortality from myocardial infarction, raising the possibility that the -344C/T polymorphism may represent an independent cardiovascular risk factor. Preliminary studies suggest that this may in fact be true, at least in high risk individuals with other risk factors such as dysplipidemias and smoking (Hautanen *et al.*, 1998b). At this time, it remains possible that the -344C/T polymorphism is merely a marker for an associated polymorphism that is affecting gene expression directly. To answer this question, additional populations should be studied and the vicinity of *CYP11B2* sequenced more completely in different individuals.

It is notable that many kindreds with glucocorticoid-suppressible hyperaldosteronism are of Anglo-Irish extraction (Lifton *et al.*, 1992b; Pascoe *et al.*, 1992a). Moreover, the chromosomes carrying chimeric genes tend to occur in association with specific polymorphisms in the *CYP11B* genes (Lifton *et al.*, 1992b), even though the duplications generating the chimeric genes are apparently independent events. This suggests that one of these polymorphisms is, or is in linkage disequilibrium with, a structural polymorphism that predisposes to unequal crossing over during meiosis. Such features might include sequences similar to *chi* sites in bacteriophage lambda; this type of sequence has been postulated to increase the frequency of recombination in the *CYP21* genes (Amor *et al.*, 1988). Additionally, in approximately 40% of alleles in Caucasians, the second intron of *CYP11B2* has a sequence almost identical to that of *CYP11B1* (White and Slutsker, 1995). This region could promote misalignment of chromosomal segments during meiosis and thus increase the risk of unequal crossing over.

5. Loss of specificity of the mineralocorticoid receptor: the syndrome of apparent mineralocorticoid excess

5.1 Clinical features

AME is an inherited syndrome in which children present with hypertension, hypokalemia and low plasma renin activity. Other clinical features include moderate intrauterine growth retardation and postnatal failure to thrive. Consequences of the often severe hypokalemia include nephrocalcinosis, nephrogenic diabetes insipidus and rhabdomyolysis. Complications of hypertension have included cerebrovascular accidents, and several patients have died during infancy or adolescence. Several affected sibling pairs have been reported but parents have usually been asymptomatic, suggesting that AME is a genetic disorder with an autosomal recessive mode of inheritance.

A low salt diet or blockade of mineralocorticoid receptors with spironolactone ameliorate the hypertension whereas ACTH and hydrocortisone exacerbate it. Levels of all known mineralocorticoids are low (Oberfield *et al.*, 1983; Ulick *et al.*, 1979). These findings suggest that cortisol (i.e. hydrocortisone) acts as a stronger mineralocorticoid than is normally the case. Indeed, patients with AME have abnormal cortisol metabolism. Cortisol half-life in plasma is prolonged from approximately 80 to 120–190 min (Ulick *et al.*, 1979). Very low levels of cortisone metabolites are excreted in the urine as compared with cortisol metabolites, indicating a marked deficiency in 11-HSD, the enzyme catalysing the conversion of cortisol to cortisone. This has been assayed directly by administering 11α[^{3}H]cortisol to subjects and measuring the appearance of tritiated water. Most often it is measured as an increase in the sum of the urinary concentrations of tetrahydrocortisol and allo-tetrahydrocortisol, divided by the concentration of tetrahydrocortisone, abbreviated (THF+aTHF)/THE. However, 11-reduction is unimpaired; labeled cortisone administered to patients is excreted entirely as cortisol and other 11β-reduced metabolites (Shackleton *et al.*, 1985).

Similar but milder abnormalities occur with licorice intoxication (Stewart *et al.*, 1987). The active component of licorice, glycyrrhetinic acid, inhibits 11-HSD in isolated rat kidney microsomes (Monder *et al.*, 1989). Thus, it appears that licorice intoxication is a reversible pharmacological counterpart to the inherited syndrome of apparent mineralocorticoid excess.

Juvenile hypertension, marked hypokalemia and suppressed plasma renin activity are also found in Liddle's syndrome caused by activating mutations in the regulatory subunits of the sodium channel. However, Liddle's syndrome has an autosomal dominant mode of inheritance, and whereas it can be treated by blockade of the cortical collecting duct's sodium channel with amiloride or triamterine, blockade of the mineralocorticoid receptor with spironolactone is not effective.

5.2 Isozymes of 11β-hydroxysteroid dehydrogenase (11-HSD)

There are two distinct isozymes of 11-HSD. Both are members of the 'short chain dehydrogenase' family. These enzymes all have a highly conserved nucleotide cofactor binding domain near the amino terminus; the cofactor functions as an electron acceptor for dehydrogenation (NAD^+ or $NADP^+$) and as an electron donor for reduction (NADH or NADPH). Completely conserved tyrosine and lysine residues toward the carboxyl terminus function in catalysis. Conservative substitutions of either of these residues destroy enzymatic activity in a number of related enzymes (Chen *et al.*, 1993; Ensor and Tai, 1993; Ghosh *et al.*, 1991; Obeid and White, 1992).

X-ray crystallographic studies of a related enzyme, 3α,20β-hydroxysteroid dehydrogenase from *S. hydrogenans*, demonstrated that the conserved tyrosine and lysine residues are located near the pyridine ring of the cofactor in a cleft presumed to be the substrate binding site (Ghosh *et al.*, 1991). In 11-HSD, these two residues may facilitate a hydride ion (a proton plus two electrons) transfer from the 11α position to $NADP^+$ or NAD^+. It is hypothesized that the ε-amino group of the lysine deprotonates the phenolic group of the tyrosine. Deprotonation of a phenolic group in aqueous solution normally has a pK_a of about 10, but the local alkaline milieu provided by lysine lowers the apparent pK_a of the phenolic group of tyrosine into the physiological range. The deprotonated phenolic group then removes a proton from the 11β-hydroxyl group of the steroid, leaving a negative charge on the 11 position of the steroid nucleus. This allows transfer of the 11α hydrogen (as a hydride) to the pyridine group of the cofactor.

The first isozyme of 11-HSD that was characterized, termed the liver or 11-HSD1 isozyme, was originally isolated from rat liver microsomes (Lakshmi and Monder, 1988) and the corresponding cDNA was cloned (Agarwal *et al.*, 1989). It requires $NADP^+$ as a cofactor and has an affinity for steroids in the micromolar range. Although the enzyme purified from rat liver functions only as a dehydrogenase, the recombinant enzyme expressed from cloned cDNA exhibits both 11β-dehydrogenase and the reverse oxoreductase activity (conversion of 11-dehydrocorticosterone to corticosterone) when expressed in mammalian cells (Agarwal *et al.*, 1989) suggesting that the reductase activity is destroyed during purification from the liver.

Several lines of evidence suggest that this isozyme does not play a significant role in conferring ligand specificity on the mineralocorticoid receptor. It is expressed at highest levels in the liver, which does not respond to mineralocorticoids, and although it is expressed at high levels in the rat kidney (Agarwal *et al.*, 1989), it is expressed at much lower levels in human (Tannin *et al.*, 1991) kidneys. Even in rat kidney, immunoreactivity to the protein is observed primarily in proximal tubules and not in distal tubules and collecting ducts, the sites of mineralocorticoid action (Rundle *et al.*, 1989), Finally, when the *HSD11L* (*HSD11B1*) gene encoding this isozyme was cloned (Tannin *et al.*, 1991) and examined for mutations in patients with AME, none were found (Nikkila *et al.*, 1993).

Accordingly, a second isozyme was sought in mineralocorticoid target tissues. Evidence for an NAD^+ dependent isozyme was obtained from histochemical studies of rat kidney (Mercer and Krozowski, 1992). In isolated rabbit kidney cortical collecting duct cells, 11-HSD was detected in the microsomal fraction (Rusvai and Naray-Fejes-Toth, 1993) This activity was almost exclusively NAD^+ dependent and had a very high affinity for steroids (K_m for corticosterone of 26 nM). There was almost no reduction of 11-dehydrocorticosterone to corticosterone, suggesting that, unlike the liver isozyme, the kidney or 11-HSD2 isozyme only catalyzed dehydrogenation. The enzyme in the human placenta had similar characteristics (Brown *et al.*, 1993); it was NAD^+ dependent and had K_m values for steroids in the nanomolar range. Similar activities were noted in sheep kidney (Yang and Yu, 1994) and in many human fetal tissues (Stewart *et al.*, 1994).

Thus far, 11-HSD2 has not been purified to homogeneity in active form from any source. This rendered the cloning of the corresponding cDNA more difficult. It was eventually accomplished by expression screening strategies in which pools of clones were assayed for their ability to confer NAD^+ dependent 11-HSD activity on *Xenopus* oocytes or cultured mammalian cells. Positive pools were divided into smaller pools and rescreened until a single positive clone was identified. Both sheep (Agarwal *et al.*, 1994) and human (Albiston *et al.*, 1994) cDNA encoding this isoform were isolated in this manner. Recombinant 11-HSD2 has properties that are virtually identical to the activity found in mineralocorticoid target tissues. The recombinant enzyme functions exclusively as a dehydrogenase; no reductase activity is detectable with either NADH or NADPH as a cofactor (Agarwal *et al.*, 1994; Albiston *et al.*, 1994). It has an almost exclusive preference for NAD^+ as a cofactor and a very high affinity for glucocorticoids. This isozyme is expressed in mineralocorticoid target tissues, particularly the kidney, and in human placenta, whereas it is not detected in the liver.

The predicted amino acid sequence of 11-HSD2 is only 21% identical to that of 11-HSD1. It consists of 404 amino acid residues. The corresponding gene, termed *HSD11K* or *HSD11B2*, is located on chromosome 16q22 (Agarwal *et al.*, 1995). It consists of five exons spaced over approximately 6 kb (*Figure* 6). This organization differs from *HSD11B1 (HSD11L)*, suggesting that the two isozymes are only distantly related.

5.3 Detection of mutations in HSD11β2 in patients with AME

Thus far, 18 different mutations in the *HSD11B2* gene have been published involving 21 kindreds with AME (*Figure* 6). These mutations all affect enzymatic activity or pre-mRNA splicing, thus confirming in its entirety the hypothesis that 11-HSD protects the mineralocorticoid receptor from high concentrations of cortisol (Dave-Sharma *et al.*, 1998; Kitanaka *et al.*, 1997; Li *et al.*, 1997; Li *et al.*, 1998; Mune *et al.*, 1995; Mune and White, 1996; Stewart *et al.*, 1996; Wilson *et al.*, 1998; Wilson *et al.*, 1995a; Wilson *et al.*, 1995b). Most patients are homozygous for single mutations, with only three published patients being compound heterozygotes for two different mutations. This suggests that the prevalence of AME mutations in the general population is low, so that the disease is found mostly in limited populations in which inbreeding is relatively high. Six kindreds are of Native American origin. Three from Minnesota or Canada carry the same mutation (L250S, L251P), consistent with a founder effect, but the others are each homozygous for a different mutation. The reason for the relatively high prevalence of this very rare disease among Native Americans is not immediately apparent.

Of the mutations identified thus far, three shift the reading frame of translation, a third deletes three amino acids including the catalytic tyrosine residue (Y232),

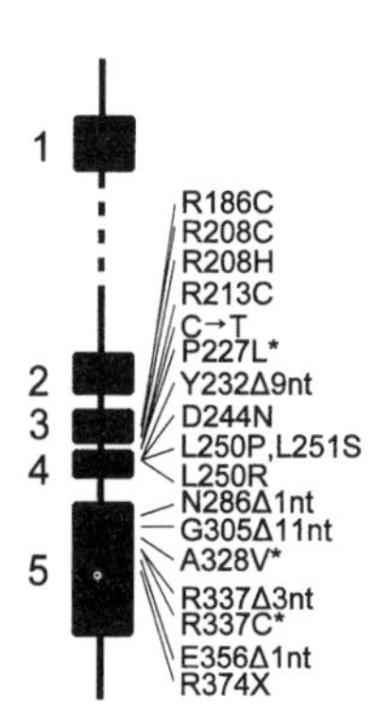

Figure 6. Diagram of the *HSD11K* (*HSD11B2*) gene, showing locations of mutations causing apparent mineralocorticoid excess.

and one is a nonsense mutation. These mutations are all presumed to completely destroy enzymatic activity. One mutation in the third intron leads to skipping of the fourth exon during processing of pre-mRNA (Mune *et al.*, 1995). As the fourth exon encodes the catalytic site, the resulting enzyme is again presumably inactive. The remaining mutations each affect no more than two amino acids. Most of these have been introduced into cDNA and expressed in cultured cells to determine their effects. Several completely inactivate the enzyme, whereas others have almost unimpaired activity in whole cells (Li *et al.*, 1998; Mune and White, 1996; Wilson *et al.*, 1998). However, all mutations are associated with decreased or absent activity in lysed cells, suggesting that they adversely affect protein stability once cells are lysed; this has been confirmed by Western blots for several mutations (Mune and White, 1996).

Both the wild type enzyme and most mutants are concentrated in the nucleus as determined by Western blots of cell fractions. This may reflect the enzyme's function in protecting the nuclear mineralocorticoid receptor from excessive concentrations of cortisol.

Genotype–phenotype correlation. Although the number of patients with AME is small, sufficient data now exist to demonstrate a statistically significant correlation (R^2=0.65, P<0.0001) between degree of enzymatic impairment and biochemical severity as measured by the product: precursor ratio, THE/(THF+aTHF) (Mune and White, 1996, and unpublished observations) This correlation is most obvious for the partially active mutants. It is remarkable that trivial impairment of enzymatic activity (as measured in transfected intact cells) is apparently sufficient to compromise metabolism of cortisol in the kidney, suggesting that there is very little excess capacity to metabolize cortisol in this organ. This seems to raise a paradox, because AME is a recessive disorder and heterozygous carriers, who would be expected to have 50% of normal activity, are asymptomatic. Altered stability or kinetic properties of the partially active mutants may be important, including alterations in enzyme inhibition by end product (i.e. cortisone or corticosterone) or by other circulating steroids.

Because of the small numbers of patients, and the possible confounding effects of prior antihypertensive therapy, it is difficult to correlate biochemical severity with measures of clinical severity, except that serum potassium levels do tend to be lower in individuals carrying more severe mutations. Although correlations between genotype and blood pressure levels are not statistically significant, anecdotal reports suggest that mutations that do not destroy activity are indeed associated with milder disease (Li *et al.*, 1998; Mune *et al.*, 1995; Wilson *et al.*, 1998; Wilson *et al.*, 1995b). In particular, the so-called 'type II' variant of AME, which is associated with only slightly abnormal precursor: product ratios (Mantero *et al.*, 1994; Ulick *et al.*, 1990b), is caused by mutations in HSD11B2 that affect enzymatic activity very mildly (Li *et al.*, 1998).

HSD11B2 *as a candidate locus for essential hypertension.* Whereas apparent 11-HSD deficiency causes severe hypertension, it is reasonable to hypothesize that milder decreases in enzymatic activity might be associated with common 'essential' hypertension. Patients with AME are often born with a mild to moderate degree of intrauterine growth retardation. Although the reason for this is not known, it seems likely that deficiency of 11-HSD in the placenta permits excessive quantities of maternal glucocorticoids to cross the placenta and thus inhibit fetal growth (Reinisch *et al.*, 1978). Thus, a hypothetical mild form of 11-HSD deficiency might also present with low birth weight and subsequent hypertension (Edwards *et al.*, 1993). In rats, placental 11-HSD activity is inversely correlated with placental weight and directly correlated with term fetal weight (Benediktsson *et al.*, 1993). In human population studies, most of which are retrospective, low birth weight and increased placental weight are indeed risk factors for subsequent development of adult hypertension (Barker *et al.*, 1990). Although variations in 11-HSD might in principle be responsible for this correlation, studies in humans (Rogerson *et al.*, 1997; Stewart *et al.*, 1995) have not found such a correlation between placental 11-HSD activity and placental weight. A weak but significant positive correlation was observed between 11-HSD activity and fetal birth weight in the first study (Stewart *et al.*, 1995), but the subsequent larger study of the identical population (Rogerson *et al.*, 1997) was unable to confirm this. Thus, the currently available data do not support the idea that low 11-HSD activity is a risk factor for low birth weight in humans who do not suffer from AME. Of course, this does not rule out a possible effect of genetically determined mild variations in 11-HSD activity upon blood pressure or more specifically on salt sensitivity. Molecular studies of *HSD11B2* should unambiguously determine if this gene is frequently involved in the development of hypertension. These might include linkage studies (see Chapter 1) and a search for frequent polymorphisms in *HSD11B2* that might be associated with the development of hypertension. Additional insights into the physiology of this enzyme might be obtained by 'knocking out' the corresponding gene in mice (see Chapter 6).

6. Summary

Aldosterone, the most important mineralocorticoid, regulates electrolyte excretion and intravascular volume mainly through its effects on renal distal convoluted tubules and cortical collecting ducts. Excess secretion of aldosterone or other mineralocorticoids, or abnormal sensitivity to mineralocorticoids, may result in hypertension, suppressed plasma renin activity and hypokalemia. Such conditions often have a genetic basis, and studies of these conditions have provided valuable insights into normal and abnormal physiology of mineralocorticoid action. Deficiencies of steroid 11β-hydroxylase or 17α-hydroxylase are types of congenital adrenal hyperplasia, the autosomal recessive inability to synthesize cortisol. These two defects often cause hypertension due to overproduction of cortisol precursors that are, or are metabolized to, mineralocorticoid agonists. These disorders result from mutations in the *CYP11B1* and *CYP17* genes encoding the corresponding enzymes. Glucocorticoid-suppressible hyperaldosteronism is an autosomal dominant form of hypertension in which aldosterone secretion is abnormally regulated by ACTH. It is caused by recombinations between linked genes encoding closely related isozymes, 11β-hydroxylase (*CYP11B1*) and aldosterone synthase (*CYP11B2*), generating a dysregulated chimeric gene with aldosterone synthase activity. Apparent mineralocorticoid excess is a loss of functional ligand specificity of the mineralocorticoid receptor caused by deficiency of the kidney isozyme of 11β-hydroxysteroid dehydrogenase, an enzyme that normally metabolizes cortisol to cortisone to prevent it from occupying the receptor. This autosomal recessive form of severe hypertension results from mutations in the *HSD11K* (*HSD11B2*) gene.

Acknowledgments

I thank my present and former colleagues in my laboratory whose work I have cited in this chapter, including Phyllis Speiser, Leigh Pascoe, Kathleen Curnow, Anil Agarwal, Grace Tannin, Jihad Obeid, Heli Nikkila, Tomoatsu Mune and Fraser Rogerson. Work from my laboratory discussed in this chapter was supported by grants DK37867 and DK42169 from the National Institutes of Health.

References

Adler, G.K., Chen, R., Menachery, A.I., Braley, L.M. and Williams, G.H. (1993) Sodium restriction increases aldosterone biosynthesis by increasing late pathway, but not early pathway, messenger ribonucleic acid levels and enzyme activity in normotensive rats. *Endocrinology* **133**: 2235–2240.

Agarwal, A.K., Monder, C., Eckstein, B. and White, P.C. (1989) Cloning and expression of rat cDNA encoding corticosteroid 11 beta-dehydrogenase. *J. Biol. Chem.* **264**: 18939–18943.

Agarwal, A.K., Mune, T., Monder, C. and White, P.C. (1994) NAD+-dependent isoform of 11 beta hydroxysteroid dehydrogenase: cloning and characterization of cDNA from sheep kidney. *J. Biol. Chem.* **269**: 25959–25962.

Agarwal, A.K., Rogerson, F.M., Mune, T. and White, P.C. (1995) Gene structure and chromosomal localization of the human HSD11K gene encoding the kidney (type 2) isozyme of 11b-hydroxysteroid dehydrogenase. *Genomics* **29**: 195–199.

Ahlgren, R., Yanase, T., Simpson, E.R., Winter, J.S. and Waterman, M.R. (1992) Compound heterozygous mutations (Arg 239—stop, Pro 342—Thr) in the CYP17 (P45017 alpha) gene lead to ambiguous external genitalia in a male patient with partial combined 17 alpha- hydroxylase/17,20-lyase deficiency. *J. Clin. Endocrinol. Metab.* **74**: 667–672.

Albiston, A.L., Obeyesekere, V.R., Smith, R.E. and Krozowski, Z.S. (1994) Cloning and tissue distribution of the human 11-HSD type 2 enzyme. *Mol. Cell Endocrinol.* **105**: R11–R17.

Amor, M., Parker, K.L., Globerman, H., New, M.I. and White, P.C. (1988) Mutation in the CYP21B gene (Ile-172—Asn) causes steroid 21-hydroxylase deficiency. *Proc. Natl Acad. Sci. U.S.A.* **85**: 1600–1604.

Arriza, J.L., Weinberger, C., Cerelli, G., Glaser, T.M., Handelin, B.L., Housman, D.E. and Evans, R.M. (1987) Cloning of human mineralocorticoid receptor complementary DNA: structural and functional kinship with the glucocorticoid receptor. *Science* **237**: 268–275.

Barker, D.J., Bull, A.R., Osmond, C. and Simmonds, S.J. (1990) Fetal and placental size and risk of hypertension in adult life. *Br. Med. J.* **301**: 259–262.

Benediktsson, R., Lindsay, R.S., Noble, J., Seckl, J.R. and Edwards, C.R. (1993) Glucocorticoid exposure in utero: new model for adult hypertension. *Lancet* **341**: 339–341.

Biason, A., Mantero, F., Scaroni, C., Simpson, E.R. and Waterman, M.R. (1991) Deletion within the CYP17 gene together with insertion of foreign DNA is the cause of combined complete 17 alpha- hydroxylase/17,20-lyase deficiency in an Italian patient. *Mol. Endocrinol.* 5: 2037–2045.

Bird, I.M., Hanley, N.A., Word, R.A., Mathis, J.M., McCarthy, J.L., Mason, J.I. and Rainey, W.E. (1993) Human NCI-H295 adrenocortical carcinoma cells: a model for angiotensin-II-responsive aldosterone secretion. *Endocrinology* **133**: 1555–1561.

Bistritzer, T., Sack, J., Eshkol, A., Zur, H. and Katznelson, D. (1984) Sex reassignment in a girl with 11 beta-hydroxylase deficiency. *Isr. J. Med. Sci.* **20**: 55–58.

Brand, E., Chatelain, N., Mulatero, P., Fery, I., Curnow, K.M., Jeunemaitre, X., Corvol, P., Pascoe, L. and Soubrier, F. (1998) Structural analysis and evaluation of the aldosterone synthase gene in hypertension. *Hypertension* **32**: 198–204.

Brilla, C.G., Matsubara, L.S. and Weber, K.T. (1993) Anti-aldosterone treatment and the prevention of myocardial fibrosis in primary and secondary hyperaldosteronism. *J. Mol. Cell. Cardiol.* **25**: 563–575.

Brown, R.W., Chapman, K.E., Edwards, C.R. and Seckl, J.R. (1993) Human placental 11 beta-hydroxysteroid dehydrogenase: evidence for and partial purification of a distinct NAD-dependent isoform. *Endocrinology* **132**: 2614–2621.

Chen, Z., Jiang, J.C., Lin, Z.G., Lee, W.R., Baker, M.E. and Chang, S.H. (1993) Site-specific mutagenesis of *Drosophila* alcohol dehydrogenase: evidence for involvement of tyrosine-152 and lysine–156 in catalysis. *Biochemistry* **32**: 3342–3346.

Chua, S.C., Szabo, P., Vitek, A., Grzeschik, K.H., John, M. and White, P.C. (1987) Cloning of cDNA encoding steroid 11 beta-hydroxylase (P450c11). *Proc. Natl Acad. Sci. U.S.A.* **84**: 7193–7197.

Clark, B.J., Wells, J., King, S.R. and Stocco, D.M. (1994) The purification, cloning, and expression of a novel luteinizing hormone-induced mitochondrial protein in MA-10 mouse Leydig tumor cells. Characterization of the steroidogenic acute regulatory protein (StAR). *J. Biol. Chem.* **269**: 28314–28322.

Clark, B.J., Soo, S.C., Caron, K.M., Ikeda, Y., Parker, K.L. and Stocco, D.M. (1995) Hormonal and developmental regulation of the steroidogenic acute regulatory protein. *Mol. Endocrinol.* **9**: 1346–1355.

Clyne, C.D., Zhang, Y., Slutsker, L., Mathis, J.M., White, P.C. and Rainey, W.E. (1997) Angiotensin II and potassium regulate human CYP11B2 transcription through common cis elements. *Mol. Endocrinol.* **11**: 638–649.

Connell, J.M., Kenyon, C.J., Corrie, J.E., Fraser, R., Watt, R. and Lever, A.F. (1986) Dexamethasone-suppressible hyperaldosteronism. Adrenal transition cell hyperplasia? *Hypertension* **8**: 669–676.

Curnow, K.M., Tusie-Luna, M.T., Pascoe, L., Natarajan, R., Gu, J.L., Nadler, J.L. and White, P.C. (1991) The product of the CYP11B2 gene is required for aldosterone biosynthesis in the human adrenal cortex. *Mol. Endocrinol.* **5**: 1513–1522.

Curnow, K.M., Pascoe, L. and White, P.C. (1992) Genetic analysis of the human type-1 angiotensin II receptor. *Mol. Endocrinol.* **6**: 1113–1118.

Curnow, K.M., Slutsker, L., Vitek, J., Cole, T., Speiser, P.W., New, M.I., White, P.C. and Pascoe, L. (1993) Mutations in the CYP11B1 gene causing congenital adrenal hyperplasia and hypertension cluster in exons 6, 7, and 8. *Proc. Natl Acad. Sci. U.S.A.* **90**: 4552–4556.

Dave-Sharma, S., Wilson, R.C., Harbison, M.D., *et al.* (1998) Examination of genotype and phenotype relationships in 14 patients with apparent mineralocorticoid excess. *J. Clin. Endocrinol. Metab.* **83**: 2244–2254.

Dluhy, R.G. and Lifton, R.P. (1995) Glucocorticoid-remediable aldosteronism (GRA): diagnosis, variability of phenotype and regulation of potassium homeostasis. *Steroids* **60**: 48–51.

Duchatelle, P., Ohara, A., Ling, B.N., Kemendy, A.E., Kokko, K.E., Matsumoto, P.S. and Eaton, D.C. (1992) Regulation of renal epithelial sodium channels. *Mol. Cell Biochem.* **114**: 27–34.

Edwards, C.R., Stewart, P.M., Burt, D., Brett, L., McIntyre, M.A., Sutanto, W.S., de Kloet, E.R. and Monder, C. (1988) Localisation of 11 beta-hydroxysteroid dehydrogenase—tissue specific protector of the mineralocorticoid receptor. *Lancet* **2**: 986–989.

Edwards, C.R., Benediktsson, R., Lindsay, R.S. and Seckl, J.R. (1993) Dysfunction of placental glucocorticoid barrier: link between fetal environment and adult hypertension? *Lancet* **341**: 355–357.

Ensor, C.M. and Tai, H.H. (1993) Site-directed mutagenesis of the conserved tyrosine-151 of human placental NAD^+-dependent 15-hydroxyprostaglandin dehydrogenase yields a catalytically inactive enzyme. *Biochem. Biophys. Res. Commun.* **176**: 840–845.

Fardella, C.E., Rodriguez, H., Hum, D.W., Mellon, S.H. and Miller, W.L. (1995) Artificial mutations in P450c11AS (aldosterone synthase) can increase enzymatic activity: a model for low-renin hypertension? *J. Clin. Endocrinol. Metab.* **80**: 1040–1043.

Funder, J.W., Pearce, P.T., Smith, R. and Smith, A.I. (1988) Mineralocorticoid action: target tissue specificity is enzyme, not receptor, mediated. *Science* **242**: 583–585.

Ganguly, A., Weinberger, M.H., Guthrie, G.P. and Fineberg, N.S. (1984) Adrenal steroid responses to ACTH in glucocorticoid-suppressible aldosteronism. *Hypertension* **6**: 563–567.

Geley, S., Kapelari, K., Johrer, K., Peter, M., Glatzl, J., Vierhapper, H., Sippell, W.G., White, P.C. and Kofler, R. (1996) CYP11B1 mutations causing congenital adrenal hyperplasia due to 11b-hydroxylase deficiency. *J. Clin. Endocrinol. Metab.* **81**: 2896–2901.

Geller, D.H., Auchus, R.J., Mendonca, B.B. and Miller, W.L. (1997) The genetic and functional basis of isolated 17,20-lyase deficiency. *Nature Genet.* **17**: 201–205.

Ghosh, D., Weeks, C.M., Grochulski, P., Duax, W.L., Erman, M., Rimsay, R.L. and Orr, J.C. (1991) Three-dimensional structure of holo 3 alpha,20 beta- hydroxysteroid dehydrogenase: a member of a short-chain dehydrogenase family. *Proc. Natl Acad. Sci. USA* **88**: 10064–10068.

Gomez-Sanchez, C.E., Gill, J.R., Jr., Ganguly, A. and Gordon, R.D. (1988) Glucocorticoid-suppressible aldosteronism: a disorder of the adrenal transitional zone. *J. Clin. Endocrinol. Metab.* **67**: 444–448.

Gordon, R.D., Klemm, S.A., Tunny, T.J. and Stowasser, M. (1992) Primary aldosteronism: hypertension with a genetic basis. *Lancet* **340**: 159–161.

Griffing, G.T., Dale, S.L., Holbrook, M.M. and Melby, J.C. (1983) 19-nor-deoxycorticosterone excretion in primary aldosteronism and low renin hypertension. *J. Clin. Endocrinol. Metab.* **56**: 218–221.

Hague, W.M. and Honour, J.W. (1983) Malignant hypertension in congenital adrenal hyperplasia due to 11 beta-hydroxylase deficiency. *Clin. Endocrinol.(Oxf).* **18**: 505–510.

Hall, C.E. and Gomez-Sanchez, C.E. (1986) Hypertensive potency of 18-oxocortisol in the rat. *Hypertension* **8**: 317–322.

Hamlet, S.M., Gordon, R.D., Gomez-Sanchez, C.E., Tunny, T.J. and Klemm, S.A. (1988) Adrenal transitional zone steroids, 18-oxo and 18- hydroxycortisol, useful in the diagnosis of primary aldosteronism, are ACTH-dependent. *Clin. Exp. Pharmacol. Physiol.* **15**: 317–322.

Harinarayan, C.V., Ammini, A.C., Karmarkar, M.G., Prakash, V., Gupta, R., Taneja, N., Mohapatra, I., Kucheria, K. and Ahuja, M.M. (1992) Congenital adrenal hyperplasia and complete masculinization masquarading as sexual precocity and cryptorchidism. *Indian Pediatr.* **29**: 103–106.

Hashimoto, T., Morohashi, K., Takayama, K., Honda, S., Wada, T., Handa, H. and Omura, T. (1992) Cooperative transcription activation between Ad1, a CRE-like element, and other elements in the CYP11B gene promoter. *J. Biochem. (Tokyo)* **112**: 573–575.

Hautanen, A., Lankinen, L., Kupari, M., Janne, O.A., Adlercreutz, H., Nikkila, H. and White, P.C. (1998a) Associations between aldosterone synthase gene polymorphism and the adrenocortical function in males. *J. Intern. Med.* **244**: 11–18.

Hautanen, A., Toivanen, P., Manttari, M., Tenkanen, L., Manninen, V., Kayes, K.M., Rosenfeld, S. and White, P.C. (1998b) Variants of the aldosterone synthase gene and the risk of coronary heart disease in dyslipidemic middle-aged men. *Circulation* **98**: I-531. 1998b. Abstract.

Helmberg, A., Ausserer, B. and Kofler, R. (1992) Frame shift by insertion of 2 basepairs in codon 394 of CYP11B1 causes congenital adrenal hyperplasia due to steroid 11 beta-hydroxylase deficiency. *J. Clin. Endocrinol. Metab.* **75**: 1278–1281.

Hochberg, Z., Schechter, J., Benderly, A., Leiberman, E. and Rosler, A. (1985) Growth and pubertal development in patients with congenital adrenal hyperplasia due to 11-beta-hydroxylase deficiency. *Am. J. Dis. Child* **139**: 771–776.

Honda, S., Morohashi, K. and Omura, T. (1990) Novel cAMP regulatory elements in the promoter region of bovine P-450(11 beta) gene. *J. Biochem. (Tokyo)* **108**: 1042–1049.

Honda, S., Morohashi, K., Nomura, M., Takeya, H., Kitajima, M. and Omura, T. (1993) Ad4BP regulating steroidogenic P-450 gene is a member of steroid hormone receptor superfamily. *J. Biol. Chem.* **268**: 7494–7502.

Horisberger, J.D. and Rossier, B.C. (1992) Aldosterone regulation of gene transcription leading to control of ion transport. *Hypertension* **19**: 221–227.

Imai, T., Yanase, T., Waterman, M.R., Simpson, E.R. and Pratt, J.J. (1992) Canadian Mennonites and individuals residing in the Friesland region of The Netherlands share the same molecular basis of 17 alpha-hydroxylase deficiency. *Hum. Genet.* **89**: 95–96.

Jamieson, A., Slutsker, L., Inglis, G.C., Fraser, R., White, P.C. and Connell, J.M. (1995) Glucocorticoid-suppressible hyperaldosteronism: effects of crossover site and parental origin of chimaeric gene on phenotypic expression. *Clin. Sci.* **88**: 563–570.

John, M.E., John, M.C., Boggaram, V., Simpson, E.R. and Waterman, M.R. (1986) Transcriptional regulation of steroid hydroxylase genes by corticotropin. *Proc. Natl Acad. Sci. USA* **83**: 4715–4719.

Johrer, K., Geley, S., Strasser-Wozak, E.M., Azziz, R., Wollmann, H.A., Schmitt, K., Kofler, R. and White, P.C. (1997) CYP11B1 mutations causing nonclassic adrenal hyperplasia due to 11b-hydroxylase deficiency. *Hum. Mol. Genet.* **6**: 1829–1834.

Jonsson, J.R., Klemm, S.A., Tunny, T.J., Stowasser, M. and Gordon, R.D. (1995) A new genetic test for familial hyperaldosteronism type I aids in the detection of curable hypertension. *Biochem. Biophys. Res. Commun.* **207**: 565–571.

Kagimoto, M., Winter, J.S., Kagimoto, K., Simpson, E.R. and Waterman, M.R. (1988) Structural characterization of normal and mutant human steroid 17 alpha-hydroxylase genes: molecular basis of one example of combined 17 alpha-hydroxylase/17,20 lyase deficiency. *Mol. Endocrinol.* **2**: 564–570.

Kater, C.E. and Biglieri, E.G. (1994) Disorders of steroid 17 alpha-hydroxylase deficiency. *Endocrinol. Metab. Clin. North Am.* **23**: 341–357.

Kawamoto, T., Mitsuuchi, Y., Ohnishi, T., *et al.* (1990a) Cloning and expression of a cDNA for human cytochrome P–450aldo as related to primary aldosteronism. *Biochem. Biophys. Res. Commun.* **173**: 309–316.

Kawamoto, T., Mitsuuchi, Y., Toda, K., *et al.* (1990b) Cloning of cDNA and genomic DNA for human cytochrome P–45011 beta. *FEBS Lett.* **269**: 345–349.

Kawamoto, T., Mitsuuchi, Y., Toda, K., *et al.* (1992) Role of steroid 11 beta-hydroxylase and steroid 18-hydroxylase in the biosynthesis of glucocorticoids and mineralocorticoids in humans. *Proc. Natl Acad. Sci. USA* **89**: 1458–1462.

Kitanaka, S., Katsumata, N., Tanae, A., Hibi, I., Takeyama, K., Fuse, H., Kato, S. and Tanaka, T. (1997) A new compound heterozygous mutation in the 11 beta-hydroxysteroid dehydrogenase type 2 gene in a case of apparent mineralocorticoid excess. *J. Clin. Endocrinol. Metab.* **82**: 4054–4058.

Krozowski, Z.S. and Funder, J.W. (1983) Renal mineralocorticoid receptors and hippocampal corticosterone binding species have identical intrinsic steroid specificity. *Proc. Natl Acad. Sci. U.S.A.* **80**: 6056–6060.

Kupari, M., Hautanen, A., Lankinen, L., Koskinen, P., Virolainen, J., Nikkila, H. and White, P.C. (1998) Associations between human aldosterone synthase (CYP11B2) gene polymorphisms and left ventricular size, mass and function. *Circulation* **97**: 569–575.

La Rovere, M.T., Bigger, J.T., Marcus, F.I., Mortara, A. and Schwartz, P.J. (1998) Baroreflex sensitivity and heart-rate variability in prediction of total cardiac mortality after myocardial infarction. *Lancet* **351**: 478–484.

Lakshmi, V. and Monder, C. (1988) Purification and characterization of the corticosteroid 11 beta-dehydrogenase component of the rat liver 11 beta-hydroxysteroid dehydrogenase complex. *Endocrinology* **123**: 2390–2398.

Lala, D.S., Rice, D.A. and Parker, K.L. (1992) Steroidogenic factor I, a key regulator of steroidogenic enzyme expression, is the mouse homolog of fushi tarazu-factor I. *Mol. Endocrinol.* **6**: 1249–1258.

Levy, D., Garrison, R.J., Savage, D.D., Kannel, W.B. and Castelli, W.P. (1990) Prognostic implications of echocardiographically determined left ventricular mass in the Framingham Heart Study. *N. Engl. J. Med.* **322**: 1561–1566.

Li, A., Li, K.X., Marui, S., *et al.* (1997) Apparent mineralocorticoid excess in a Brazilian kindred: hypertension in the heterozygote state. *J. Hypertension* **15**: 1397–1402.

Li, A., Tedde, R., Krozowski, Z.S., Pala, A., Li, K.X., Shackleton, C.H., Mantero, F., Palermo, M. and Stewart, P.M. (1998) Molecular basis for hypertension in the 'type II variant' of apparent mineralocorticoid excess. *Am. J. Hum. Genet.* **63**: 370–379.

Lifton, R.P., Dluhy, R.G., Powers, M., Rich, G.M., Cook, S., Ulick, S. and Lalouel, J.M. (1992a) A chimaeric 11 beta-hydroxylase/aldosterone synthase gene causes glucocorticoid-remediable aldosteronism and human hypertension. *Nature* **355**: 262–265.

Lifton, R.P., Dluhy, R.G., Powers, M., *et al.* (1992b) Hereditary hypertension caused by chimaeric gene duplications and ectopic expression of aldosterone synthase. *Nature Genet.* **2**: 66–74.

Lin, D., Sugawara, T., Strauss, J.F., Clark, B.J., Stocco, D.M., Saenger, P., Rogol, A. and Miller, W.L. (1995) Role of steroidogenic acute regulatory protein in adrenal and gonadal steroidogenesis. *Science* **267**: 1828–1831.

Lombes, M., Alfaidy, N., Eugene, E., Lessana, A., Farman, N. and Bonvalet, J.P. (1995) Prerequisite for cardiac aldosterone action. Mineralocorticoid receptor and 11 beta-hydroxysteroid dehydrogenase in the human heart. *Circulation* **92**: 175–182.

Mantero, F., Tedde, R., Opocher, G., Dessi Fulgheri, P., Arnaldi, G. and Ulick, S. (1994) Apparent mineralocorticoid excess type II. *Steroids* **59**: 80–83.

Melby, J.C. (1991) Diagnosis of hyperaldosteronism. *Endocrinol. Metab. Clin. North Am.* **20**: 247–255.

Mercer, W.R. and Krozowski, Z.S. (1992) Localization of an 11 beta hydroxysteroid dehydrogenase activity to the distal nephron. Evidence for the existence of two species of dehydrogenase in the rat kidney. *Endocrinology* **130**: 540–543.

Mimouni, M., Kaufman, H., Roitman, A., Morag, C. and Sadan, N. (1985) Hypertension in a neonate with 11 beta-hydroxylase deficiency. *Eur. J. Pediatr.* **143**: 231–233.

Monder, C., Stewart, P.M., Lakshmi, V., Valentino, R., Burt, D. and Edwards, C.R. (1989) Licorice inhibits corticosteroid 11 beta-dehydrogenase of rat kidney and liver: *in vivo* and *in vitro* studies. *Endocrinology* **125**: 1046–1053.

Mornet, E., Dupont, J., Vitek, A. and White, P.C. (1989) Characterization of two genes encoding human steroid 11 beta-hydroxylase (P-450(11) beta). *J. Biol. Chem.* **264**: 20961–20967.

Morohashi, K., Honda, S., Inomata, Y., Handa, H. and Omura, T. (1992) A common transacting factor, Ad4-binding protein, to the promoters of steroidogenic P-450s. *J. Biol. Chem.* **267**: 17913–17919.

Mountjoy, K.G., Robbins, L.S., Mortrud, M.T. and Cone, R.D. (1992) The cloning of a family of genes that encode the melanocortin receptors. *Science* **257**: 1248–1251.

Mouw, A.R., Rice, D.A., Meade, J.C., Chua, S.C., White, P.C., Schimmer, B.P. and Parker, K.L. (1989) Structural and functional analysis of the promoter region of the gene encoding mouse steroid 11 beta-hydroxylase. *J. Biol. Chem.* **264**: 1305–1309.

Mune, T. and White, P.C. (1996) Apparent mineralocorticoid excess: genotype is correlated with biochemical phenotype. *Hypertension* **27**: 1193–1199.

Mune, T., Rogerson, F.M., Nikkila, H., Agarwal, A.K. and White, P.C. (1995) Human hypertension caused by mutations in the kidney isozyme of 11 beta-hydroxysteroid dehydrogenase. *Nature Genet.* **10**: 394–399.

Murphy, T.J., Alexander, R.W., Griendling, K.K., Runge, M.S. and Bernstein, K.E. (1991) Isolation of a cDNA encoding the vascular type-1 angiotensin II receptor. *Nature* **351**: 233–236.

Nadler, J.L., Hsueh, W. and Horton, R. (1985) Therapeutic effect of calcium channel blockade in primary aldosteronism. *J. Clin. Endocrinol. Metab.* **60**: 896–899.

Naiki, Y., Kawamoto, T., Mitsuuchi, Y., Miyahara, K., Toda, K., Orii, T., Imura, H. and Shizuta, Y. (1993) A nonsense mutation (TGG [Trp116]-TAG [Stop]) in CYP11B1 causes steroid 11beta-hydroxylase deficiency. *J. Clin. Endocrinol. Metab.* **77**: 1677–1682.

Nelson, D.R. and Strobel, H.W. (1988) On the membrane topology of vertebrate cytochrome P-450 proteins. *J. Biol. Chem.* **263**: 6038–6050.

New, M.I. and Peterson, R.E. (1967) A new form of congenital adrenal hyperplasia. *J. Clin. Endocrinol. Metab.* **27**: 300–305.

Nikkila, H., Tannin, G.M., New, M.I., Taylor, N.F., Kalaitzoglou, G., Monder, C. and White, P.C. (1993) Defects in the HSD11 gene encoding 11b-hydroxysteroid dehydrogenase are not found in patients with apparent mineralocorticoid excess or 11-oxoreductase deficiency. *J. Clin. Endocrinol. Metab.* **77**: 687–691.

O'Mahony, S., Burns, A. and Murnaghan, D.J. (1989) Dexamethasone-suppressible hyperaldosteronism: a large new kindred. *J. Hum. Hypertension* **3**: 255–258.

Obeid, J. and White, P.C. (1992) Tyr-179 and Lys-183 are essential for enzymatic activity of 11 beta-hydroxysteroid dehydrogenase. *Biochem. Biophys. Res. Commun.* **188**: 222–227.

Oberfield, S.E., Levine, L.S., Stoner, E., *et al.* (1981) Adrenal glomerulosa function in patients with dexamethasone- suppressible hyperaldosteronism. *J. Clin. Endocrinol. Metab.* **53**: 158–164.

Oberfield, S.E., Levine, L.S., Carey, R.M., Greig, F., Ulick, S. and New, M.I. (1983) Metabolic and blood pressure responses to hydrocortisone in the syndrome of apparent mineralocorticoid excess. *J. Clin. Endocrinol. Metab.* **56**: 332–339.

Ogishima, T., Shibata, H., Shimada, H., Mitani, F., Suzuki, H., Saruta, T. and Ishimura, Y. (1991) Aldosterone synthase cytochrome P-450 expressed in the adrenals of patients with primary aldosteronism. *J. Biol. Chem.* **266**: 10731–10734.

Ohta, M., Fujii, S., Ohnishi, T. and Okamoto, M. (1988) Production of 19-oic-11-deoxycorticosterone from 19-oxo-11- deoxycorticosterone by cytochrome P-450(11)beta and nonenzymatic production of 19-nor-11-deoxycorticosterone from 19-oic-11-deoxycorticosterone. *J. Steroid Biochem.* **29**: 699–707.

Palmer, L.G. and Frindt, G. (1992) Regulation of apical membrane Na and K channels in rat renal collecting tubules by aldosterone. *Semin. Nephrol.* **12**: 37–43.

Pang, S., Levine, L.S., Lorenzen, F., Chow, D., Pollack, M., Dupont, B., Genel, M. and New, M.I. (1980) Hormonal studies in obligate heterozygotes and siblings of patients with 11 beta-hydroxylase deficiency congenital adrenal hyperplasia. *J. Clin. Endocrinol. Metab.* **50**: 586–589.

Pascoe, L., Curnow, K.M., Slutsker, L., Connell, J.M., Speiser, P.W., New, M.I. and White, P.C. (1992a) Glucocorticoid-suppressible hyperaldosteronism results from hybrid genes created by unequal crossovers between CYP11B1 and CYP11B2. *Proc. Natl Acad. Sci. U.S.A.* **89**: 8327–8331.

Pascoe, L., Curnow, K.M., Slutsker, L., Rosler, A. and White, P.C. (1992b) Mutations in the human CYP11B2 (aldosterone synthase) gene causing corticosterone methyloxidase II deficiency. *Proc. Natl Acad. Sci. U.S.A.* **89**: 4996–5000.

Pascoe, L., Jeunemaitre, X., Lebrethon, M.C., Curnow, K.M., Gomez-Sanchez, C.E., Gasc, J.M., Saez, J.M. and Corvol, P. (1995) Glucocorticoid-suppressible hyperaldosteronism and adrenal tumors occurring in a single French pedigree. *J. Clin. Invest.* **96**: 2236–2246.

Pearce, D. and Yamamoto, K.R. (1993) Mineralocorticoid and glucocorticoid receptor activities distinguished by nonreceptor factors at a composite response element. *Science* **259**: 1161–1165.

Picado-Leonard, J. and Miller, W.L. (1987) Cloning and sequence of the human gene for P450c17 (steroid 17 alpha-hydroxylase/17,20 lyase): similarity with the gene for P450c21. *DNA* **6**: 439–448.

Picado-Leonard, J., Voutilainen, R., Kao, L.C., Chung, B.C., Strauss, J.F. and Miller, W.L. (1988) Human adrenodoxin: cloning of three cDNAs and cycloheximide enhancement in JEG-3 cells *J. Biol. Chem.* **263**: 3240–3244.

Pojoga, L., Gautier, S., Blanc, H., Guyene, T.T., Poirier, O., Cambien, F. and Benetos, A. (1998) Genetic determination of plasma aldosterone levels in essential hypertension. *Am. J. Hypertension* **11**: 856–860.

Porter, T.D. and Kasper, C.B. (1985) Coding nucleotide sequence of rat NADPH-cytochrome P-450 oxidoreductase cDNA and identification of flavin-binding domains. *Proc. Natl Acad. Sci. U.S.A.* **82**: 973–977.

Poulos, T.L. (1991) Modeling of mammalian P450s on basis of P450cam X-ray structure. *Methods Enzymol.* **206**: 11–30.

Quinn, S.J. and Williams, G.H. (1988) Regulation of aldosterone secretion. *Annu. Rev. Physiol.* **50**: 409–426.

Ravichandran, K.G., Boddupalli, S.S., Hasemann, C.A., Peterson, J.A. and Deisenhofer, J. (1993) Crystal structure of hemoprotein domain of P450BM-3, a prototype for microsomal P450's. *Science* **261**: 731–736.

Reinisch, J., Simon, N.G. and Karwo, W.G. (1978) Prenatal exposure to prednisone in humans and animals retards intrauterine growth. *Science* **202**: 436–438.

Rice, D.A., Aitken, L.D., Vandenbark, G.R., Mouw, A.R., Franklin, A., Schimmer, B.P. and Parker, K.L. (1989) A cAMP-responsive element regulates expression of the mouse steroid 11 beta-hydroxylase gene. *J. Biol. Chem.* **264**: 14011–14015.

Rice, D.A., Mouw, A.R., Bogerd, A.M. and Parker, K.L. (1991) A shared promoter element regulates the expression of three steroidogenic enzymes. *Mol. Endocrinol.* **5**: 1552–1561.

Rich, G.M., Ulick, S., Cook, S., Wang, J.Z., Lifton, R.P. and Dluhy, R.G. (1992) Glucocorticoid-remediable aldosteronism in a large kindred: clinical spectrum and diagnosis using a characteristic biochemical phenotype. *Ann. Intern. Med.* **116**: 813–820.

Rogerson, F.M., Kayes, K.M. and White, P.C. (1997) Variation in placental type 2 11β-hydroxysteroid dehydrogenase activity is not related to birth weight or placental weight. *Mol. Cell Endocrinol.* **128**: 103–109.

Rosler, A., Leiberman, E., Sack, J., Landau, H., Benderly, A., Moses, S.W. and Cohen, T. (1982) Clinical variability of congenital adrenal hyperplasia due to 11 beta-hydroxylase deficiency. *Hormone Res.* **16**: 133–141.

Rosler, A., Leiberman, E. and Cohen, T. (1992) High frequency of congenital adrenal hyperplasia (classic 11 beta-hydroxylase deficiency) among Jews from Morocco. *Am. J. Med. Genet.* **42**: 827–834.

Rundle, S.E., Funder, J.W., Lakshmi, V. and Monder, C. (1989) The intrarenal localization of mineralocorticoid receptors and 11 beta-dehydrogenase: immunocytochemical studies. *Endocrinology* **125**: 1700–1704.

Rusvai, E. and Naray-Fejes-Toth, A. (1993) A new isoform of 11 beta-hydroxysteroid dehydrogenase in aldosterone target cells. *J. Biol. Chem.* **268**: 10717–10720.

Sasaki, K., Yamamo, Y., Bardhan, S., Iwai, N., Murray, J.J., Hasegawa, M., Matsuda, Y. and Inagami, T. (1991) Cloning and expression of a complementary cDNA encoding a bovine adrenal angiotensin II type-1 receptor. *Nature* **351**: 230–233.

Schenkman, J.B. and Greim, H. (1993) *Cytochrome P450* (eds J.B. Schenkman and H. Greim, H). Berlin, Springer-Verlag.

Shackleton, C.H., Rodriguez, J., Arteaga, E., Lopez, J.M. and Winter, J.S. (1985) Congenital 11 beta-hydroxysteroid dehydrogenase deficiency associated with juvenile hypertension: corticosteroid metabolite profiles of four patients and their families. *Clin. Endocrinol.(Oxf).* **22**: 701–712.

Solish, S.B., Picado-Leonard, J., Morel, Y., Kuhn, R.W., Mohandas, T.K., Hanukoglu, I. and Miller, W.L. (1988) Human adrenodoxin reductase: two mRNAs encoded by a single gene on chromosome 17cen–q25 are expressed in steroidogenic tissues. *Proc. Natl Acad. Sci. U.S.A.* **85**: 7104–7108.

Sparkes, R.S., Klisak, I. and Miller, W.L. (1991) Regional mapping of genes encoding human steroidogenic enzymes: P450scc to 15q23–q24, adrenodoxin to 11q22; adrenodoxin reductase to 17q24–q25; and P450c17 to 10q24–q25. *DNA Cell Biol.* **10**: 359–365.

Speiser, P.W., Dupont, B., Rubinstein, P., Piazza, A., Kastelan, A. and New, M.I. (1985) High frequency of nonclassical steroid 21-hydroxylase deficiency. *Am. J. Hum. Genet.* **37**: 650–667.

Staels, B., Hum, D.W. and Miller, W.L. (1993) Regulation of steroidogenesis in NCI-H295 cells: a cellular model of the human fetal adrenal. *Mol. Endocrinol.* **7**: 423–433.

Stewart, P.M., Wallace, A.M., Valentino, R., Burt, D., Shackleton, C.H. and Edwards, C.R. (1987) Mineralocorticoid activity of liquorice: 11-beta-hydroxysteroid dehydrogenase deficiency comes of age. *Lancet* **2**: 821–824.

Stewart, P.M., Murry, B.A. and Mason, J.I. (1994) Type 2 11beta-hydroxysteroid dehydrogenase in human fetal tissues. *J. Clin. Endocrinol. Metab.* **78**: 1529–1532.

Stewart, P.M., Rogerson, F.M. and Mason, J.I. (1995) Type 2 11 beta-hydroxysteroid dehydrogenase messenger ribonucleic acid and activity in human placenta and fetal membranes: its relationship to birth weight and putative role in fetal adrenal steroidogenesis. *J. Clin. Endocrinol. Metab.* **80**: 885–890.

Stewart, P.M., Krozowski, Z.S., Gupta, A., Milford, D.V., Howie, A.J., Sheppard, M.C. and Whorwood, C.B. (1996) Hypertension in the syndrome of apparent mineralocorticoid excess due to mutation of the 11b-hydroxysteroid dehydrogenase type 2 gene. *Lancet* **347**: 88–91.

Stockigt, J.R. and Scoggins, B.A. (1987) Long term evolution of glucocorticoid-suppressible hyperaldosteronism. *J. Clin. Endocrinol. Metab.* **64**: 22–26.

Sutherland, D.J., Ruse, J.L. and Laidlaw, J.C. (1966) Hypertension, increased aldosterone secretion and low plasma renin activity relieved by dexamethasone. *Can. Med. Assoc. J.* **95**: 1109–1119.

Tannin, G.M., Agarwal, A.K., Monder, C., New, M.I. and White, P.C. (1991) The human gene for 11 beta-hydroxysteroid dehydrogenase. Structure, tissue distribution, and chromosomal localization. *J. Biol. Chem.* **266**: 16653–16658.

Ulick, S., Levine, L.S., Gunczler, P., Zanconato, G., Ramirez, L.C., Rauh, W., Rosler, A., Bradlow, H.L. and New, M.I. (1979) A syndrome of apparent mineralocorticoid excess associated with defects in the peripheral metabolism of cortisol. *J. Clin. Endocrinol. Metab.* **49**: 757–764.

Ulick, S., Chan, C.K., Gill, J.R., Jr., Gutkin, M., Letcher, L., Mantero, F. and New, M.I. (1990a) Defective fasciculata zone function as the mechanism of glucocorticoid-remediable aldosteronism. *J. Clin. Endocrinol. Metab.* **71**: 1151–1157.

Ulick, S., Tedde, R. and Mantero, F. (1990b) Pathogenesis of the type 2 variant of the syndrome of apparent mineralocorticoid excess. *J. Clin. Endocrinol. Metab.* **70**: 200–206.

Wagner, M.J., Ge, Y., Siciliano, M. and Wells, D.E. (1991) A hybrid cell mapping panel for regional localization of probes to human chromosome 8. *Genomics* **10**: 114–125.

Wang, L.H., Tsai, S.Y., Cook, R.G., Beattie, W.G., Tsai, M.J. and O'Malley, B.W. (1989) COUP transcription factor is a member of the steroid receptor superfamily. *Nature* **340**: 163–166.

Waterman, M.R. and Simpson, E.R. (1989) Regulation of steroid hydroxylase gene expression is multifactorial in nature. *Recent Prog. Horm. Res.* **45**: 533–563.

White, P.C. (1991) Defects in cortisol metabolism causing low-renin hypertension. *Endocrinol. Res.* **17**: 85–107.

White, P.C. (1994) Disorders of aldosterone biosynthesis and action. *N. Engl. J. Med.* **331**: 250–258.

White, P.C. and Slutsker, L. (1995) Haplotype analysis of CYP11B2. *Endocrinol. Res.* **21**: 437–442.

White, P.C., Dupont, J., New, M.I., Leiberman, E., Hochberg, Z. and Rosler, A. (1991) A mutation in CYP11B1 (Arg–448-His) associated with steroid 11 beta-hydroxylase deficiency in Jews of Moroccan origin. *J. Clin. Invest.* **87**: 1664–1667.

White, P.C., Curnow, K.M. and Pascoe, L. (1994) Disorders of steroid 11 beta hydroxylase isozymes. *Endocrinol. Rev.* **15**: 421–438.

Wilson, R.C., Dave-Sharma, S., Wei, J.Q., *et al.* (1998) A genetic defect resulting in mild low-renin hypertension. *Proc. Natl Acad. Sci. U.S.A.* **95**: 10200–10205.

Wilson, R.C., Harbison, M.D., Krozowski, Z.S., *et al.* (1995a) Several homozygous mutations in the gene for 11b-hydroxysteroid dehydrogenase type 2 in patients with apparent mineralocorticoid excess. *J. Clin. Endocrinol. Metab.* **80**: 3145–3150.

Wilson, R.C., Krozowski, Z.S., Li, K., *et al.* (1995b) A mutation in the HSD11B2 gene in a family with apparent mineralocorticoid excess. *J. Clin. Endocrinol. Metab.* **80**: 2263–2266.

Wong, M., Rice, D.A., Parker, K.L. and Schimmer, B.P. (1989) The roles of cAMP and cAMP-dependent protein kinase in the expression of cholesterol side chain cleavage and steroid 11 beta-hydroxylase genes in mouse adrenocortical tumor cells. *J. Biol. Chem.* **264**: 12867–12871.

Yanase, T. (1995) 17 alpha-Hydroxylase/17,20-lyase defects. *J. Steroid Biochem. Mol. Biol.* **53**: 153–157.

Yanase, T., Kagimoto, M., Suzuki, S., Hashiba, K., Simpson, E.R. and Waterman, M.R. (1989) Deletion of a phenylalanine in the N-terminal region of human cytochrome P-450(17 alpha) results in partial combined 17 alpha-hydroxylase/17,20-lyase deficiency. *J. Biol. Chem.* **264**: 18076–18082.

Yanase, T., Simpson, E.R. and Waterman, M.R. (1991) 17 alpha-hydroxylase/17,20-lyase deficiency: from clinical investigation to molecular definition. *Endocrinol. Rev.* **12**: 91–108.

Yanase, T., Waterman, M.R., Zachmann, M., Winter, J.S., Simpson, E.R. and Kagimoto, M. (1992) Molecular basis of apparent isolated 17,20-lyase deficiency: compound heterozygous mutations in the C-terminal region (Arg(496)–Cys, Gln(461)—Stop) actually cause combined 17 alpha-hydroxylase/17,20-lyase deficiency. *Biochim. Biophys. Acta* **1139**: 275–279.

Yang, K. and Yu, M. (1994) Evidence for distinct isoforms of 11-beta-hydroxysteroid dehydrogenase in the ovine liver and kidney. *J. Steroid Biochem. Mol. Biol.* **49**: 245–250.

Ylitalo, A., Hautanen, A., Airaksinen, K.E.J., Savolainen, M.J., Kauma, H., Kupari, M., Kesaniemi, A. and Huikari, H. (1997) Baroreflex sensitivity and aldosterone synthase gene polymorphism in middle aged subjects. *Circulation* **96(suppl. 1)**: 227. Abstract.

Young, M., Fullerton, M.J., Dilley, R. and Funder, J.W. (1994) Mineralocorticoids, hypertension, and cardiac fibrosis. *J. Clin. Invest.* **93**: 2578–2583.

Zachmann, M., Tassinari, D. and Prader, A. (1983) Clinical and biochemical variability of congenital adrenal hyperplasia due to 11 beta-hydroxylase deficiency. A study of 25 patients. *J. Clin. Endocrinol. Metab.* **56**: 222–229.

8

The amiloride-sensitive epithelial Na^+ channel: structure and function of a key molecule for salt homeostasis

Pascal Barbry

1. Introduction

High resistance epithelia can reabsorb external sodium against large negative concentration gradients. Ussing and co-workers have explained the phenomenon by the existence of transcellular sodium reabsorption through high-resistance epithelia: coupling between passive electrodiffusion of sodium through the apical membrane, and active extrusion of intracellular sodium by basolateral Na^+/K^+/ATPase generates a vectorial transcellular sodium transport (Koefoed-Johnsen and Ussing, 1958; *Figure 1*). In distal colon and distal segments of the nephron, the same mechanism is involved in the control of sodium excretion. An ion channel mediates the electrodiffusion of Na^+ across the apical membrane. This passive diffusion corresponds to the limiting step of the transcellular transport. The channel associated with that apical diffusion is characterized by a high selectivity for sodium and lithium over potassium. It can be blocked by low concentration of the diuretics amiloride and triamterene. Hormones, such as aldosterone and vasopressin, are involved in the fine control of the transcellular Na^+ reabsorption. Molecular identification of the amiloride-sensitive sodium channel proteins has been achieved (Canessa *et al.*, 1993; Lingueglia *et al.*, 1993); three homologous subunits, entitled αENaC, βENaC, and γENaC (for epithelial Na^+ channel), correspond to the pore-forming subunits (Canessa *et al.*, 1994b). They are distinct from voltage-dependent Na^+ channels (Noda *et al.*, 1986). Instead, they constitute with more than 30 homologous proteins, a new gene super-family of ionic channels. This super-family can be divided into four subfamilies (Barbry and Hofman, 1997). The first subfamily, found in mammals,

Molecular Genetics of Hypertension, edited by A.F. Dominiczak, J.M.C. Connell and F. Soubrier.

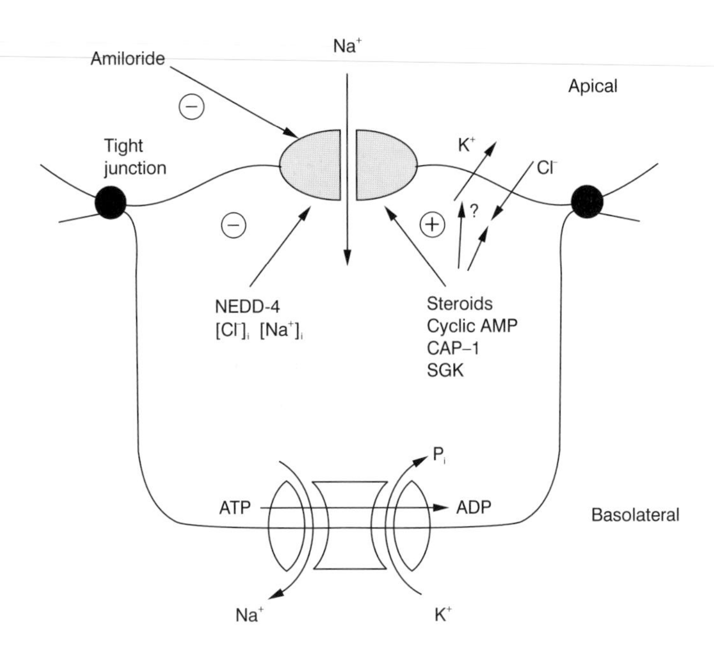

Figure 1. Mechanism of active sodium reabsorption across tight epithelia. Passive electrodiffusion of sodium through the apically expressed ENaC complex is coupled to an active extrusion of sodium through the basolateral sodium/potassium ATPase, resulting in a vectorial transport of sodium. Positive (Berthiaume *et al.*, 1987; Chen *et al.*, 1999; Chraïbi *et al.*, 1998; Renard *et al.*, 1995) and negative (Komwatana *et al.*, 1996; Staub *et al.*, 1996) effectors of sodium absorption are indicated. Sodium transport is coupled to chloride and/or potassium transport, which can also be affected by some of these effectors.

contains the constitutively activated channels involved in vectorial transport of electrolytes. It includes αENaC, βENaC, γENaC, the epithelial Na^+ channel subunits, and δENaC, a human α-like subunit (Waldmann *et al.*, 1995a).

The other three subfamilies are associated with diverse physiological functions in a range of tissues and species. This paper reviews some recent data about functional and molecular properties of the epithelial Na^+ channel. It presents the properties of some related ionic channels, when they have some relevant implications for understanding the function of ENaC. It then analyses the cell-specific expression of ENaC, and the mechanisms of its regulation by steroids.

2. Pharmacological and biophysical properties of the epithelial Na^+ channel

2.1 Pharmacology

The diuretic properties of amiloride and triamterene are explained by the blockade of a Na^+ permeability at the apical membrane of epithelial tissues (Bentley, 1968; Eigler and Crabbe, 1969; Gross and Kokko, 1977). Amiloride was discovered in 1964 after screening for nonsteroidal saliuretic agents with antikaliuretic properties in rat (Cragoe, 1979). Amiloride analogs can inhibit distinct Na^+ transport systems: (i) the epithelial Na^+ channel, (ii) the Na^+/H^+ exchange system, (iii) the Na^+/Ca^{2+} exchange system, which have usually distinct sensitivities for amiloride derivatives. Phenamil, or benzamil, substituted on the guanidino moiety of the amiloride molecule, are potent inhibitors of the epithelial Na^+ channel, but poor inhibitors of the Na^+/H^+ antiporter (Barbry and Lazdunski, 1996). Conversely, 5-*N*-disubstituted derivatives of amiloride, such as ethylisopropylamiloride (EIPA), are the most potent inhibitors of the ubiquitous isoform of the Na^+/H^+ exchanger, while the Na^+ channel is not blocked by these derivatives (Wakabayashi *et al.*, 1997). The Na^+/Ca^{2+} exchange system is poorly inhibited by amiloride, but some amiloride derivatives that are substituted on the guanidino moiety, such as dichlorobenzamil, inhibit it, although with a low affinity (Kaczorowski *et al.*, 1985). Importantly, the pharmacological characterization of a Na^+ transport system is not always sufficient to identify the correct Na^+ pathway, since some isoforms of Na^+/H^+ antiporters are not sensitive to amiloride and to EIPA (Wakabayashi *et al.*, 1997). EIPA-sensitive electrogenic Na^+ transport, presumably through Na^+ permeant channels, has also been described in cultured alveolar type II cells (Matalon *et al.*, 1996).

2.2 Electrophysiology

Noise analysis induced by submaximal concentrations of amiloride allowed Lindemann and Van Driessche to identify the unitary properties of the epithelial Na^+ channel (Lindemann and Van Driessche, 1977). Later, the same channel was characterized by the patch clamp technique in A6 cells [derived from amphibian kidney (Hamilton and Eaton, 1985)], in apical membrane of cortical collecting tubule (Letz *et al.*, 1995; Palmer and Frindt, 1986), in intact epithelium of the toad urinary bladder (Frings *et al.*, 1988) and in primary culture of fetal rat lung epithelial cells (Voilley *et al.*, 1994). The channel is characterized by a low unitary conductance (~4 pS in 140 m*M* NaCl), and a high selectivity for sodium and lithium over potassium. Whole-cell patch clamp or noise analysis show that the same channel is present in distal segments of colon (Clauss *et al.*, 1987; Zeiske *et al.*, 1982), in airway epithelium (Chinet *et al.*, 1993), in granular duct cells of mouse mandibular gland (Dinudom *et al.*, 1995), in stria vascularis marginal cells from the cochlea (Iwasa *et al.*, 1994), and in many other high resistance epithelia (Barbry and Lazdunski, 1996; Palmer, 1992).

3. Primary structure

The epithelial Na^+ channel can be expressed in *Xenopus laevis* oocytes by injection of mRNAs derived from Na^+-reabsorbing high resistance epithelia (Asher *et al.*, 1992b; George *et al.*, 1989; Hinton and Eaton, 1989; Kroll *et al.*, 1989). The highest expression was observed after injection of mRNAs prepared from rat distal colon (i.e. 5–10 times higher than with any other tissue) recovered from animals treated for at least 10 days with a low-sodium diet, or injected with high doses of dexamethasone. Dexamethasone-treated (Lingueglia *et al.*, 1993) or aldosterone-treated (Canessa *et al.*, 1993) rat distal colon mRNAs were used to build a *Xenopus* oocyte cDNA expression library. A clone was characterized after five steps of purification (Lingueglia *et al.*, 1993). It is 3081 nucleotides long and encodes a 699 amino acid protein. This protein was called RCNaCh, for rat colon Na^+ channel (Lingueglia *et al.*, 1993), or αENaC, for epithelial Na^+ channel (Canessa *et al.*, 1993).

The level of amiloride-sensitive current generated in *Xenopus* oocyte after expression of the cloned protein was much lower than after injection of total mRNA (Canessa *et al.*, 1993; Lingueglia *et al.*, 1993; Voilley *et al.*, 1994). This suggested that a factor was lacking to permit full expression of the channel activity. Canessa *et al.* successfully complemented the Na^+ channel activity induced by expression of αENaC by coexpressing different pools of a rat colon cDNA library (Canessa *et al.*, 1994b). This functional complementation strategy led to identification of two homologous cDNAs, called βENaC and γENaC, sharing ~35% identity with αENaC. Coexpression of the three subunits increases the amplitude of the current by two orders of magnitude. The ionic selectivity, gating properties, and pharmacological profile of the channel formed after coexpression of the three subunits in oocyte are similar to those of the native channel. A unitary conductance of 4.5 pS was measured in 140 mM NaCl, of 6.5 pS in 140 mM LiCl. The channel has no conductance for K^+ (Canessa *et al.*, 1994b). Independently, Lingueglia *et al.* designed degenerate oligonucleotides deduced from an alignment between the cloned subunit of the Na^+ channel and three homologous proteins, called degenerins, involved in mechanosensitivity in the nematode *Caenorhabditis elegans* (Lingueglia *et al.*, 1994). Amplification by the polymerase chain reaction (PCR) of reverse transcribed (RT) RNA from distal colon allowed the characterization of a second subunit, corresponding to the γENaC cloned by Canessa *et al.* (Canessa *et al.*, 1994b). Coexpression with the first subunit increased channel activity 18±5-fold (Lingueglia *et al.*, 1994).

The human αENaC, βENaC, γENaC have been subsequently characterized from lung (Voilley *et al.*, 1994, 1995) and kidney (McDonald *et al.*, 1994, 1995) cDNA libraries. Human αENaC gene maps to chromosome 12p13 (Voilley *et al.*, 1994), while βENaC and γENaC genes are co-localized within a common 400 kb fragment on chromosome 16p12–13 (Voilley *et al.*, 1995). A βENaC subunit variant has been described in human, where the last 42 amino acid residues are modified, in a region implicated in the regulation of channel activity (Voilley *et al.*, 1995). This modification increases the Na^+ channel activity in *Xenopus* oocytes (Jeunemaitre *et al.*, 1997). When human ENaCs are expressed in the oocyte, the Na^+ channel properties are identical to those of the rat: a voltage-independent channel selective for Na^+ was characterized by a unitary conductance of 5.6 pS in 140 mM NaCl and by a unitary conductance of 7.8 pS in 140 mM LiCl (Waldmann *et al.*, 1995a). The rat and human subunits can be readily exchanged without affecting the functional properties of the channels expressed into oocyte (McDonald *et al.*, 1995). Nearly identical results have been obtained after expression of *Xenopus* EnaCs (Puoti *et al.*, 1995).

The bovine and chicken αENaC sequences have also been established (Fuller *et al.*, 1995; Goldstein *et al.*, 1997; Killick and Richardson, 1997). The bovine sequence shows marked divergence in the last 113 COOH terminal residues, which alters the channel kinetics (Fuller *et al.*, 1995). In chicken cochlea, three αENaC splice variants have been characterized (Killick and Richardson, 1997). They result in a truncated nonfunctional protein.

4. The epithelial Na$^+$ channel superfamily

Epithelial amiloride-sensitive Na$^+$ channels are expressed in many multi-cellular organisms: in *Hirudo medicinalis* integument, their activity controls leech volume body (Weber *et al.*, 1995); in *Lumbricus terrestris* intestine, amiloride-sensitive sodium transport displays seasonal changes, via an unknown hormonal regulatory mechanism (Cornell, 1984). Identification of the ENaC subunits has also revealed the existence of similitude with a family of proteins previously identified in the nematode *Caenorhabditis elegans*. While αENaC, βENaC, and γENaC share no significant identity with previously cloned ionic channels, a ~12% amino acid identity was observed between them and the proteins MEC-4, MEC-10 and DEG-1, also called degenerins (Canessa *et al.*, 1993; Chalfie *et al.*, 1993; Lingueglia *et al.*, 1993).

In *Helix aspersa* neurons, as well as in *Aplysia californica* bursting and motor neurons, Phe–Met–Arg–Phe–NH_2 (FMRFamide) and structurally related peptides induce a fast excitatory depolarizing response. This response is due to direct activation of an amiloride-sensitive Na$^+$ channel (Belkin and Abrams, 1993; Green *et al.*, 1994). This current is carried through amiloride-sensitive, but tetrodotoxin- or lidocainein-sensitive, Na$^+$-selective channels, with a 6.6 pS unitary conductance in 100 mM Na$^+$. A cDNA has been isolated from *Helix aspersa* nervous tissue (Lingueglia *et al.*, 1995). It encodes a FMRFamide-activated Na$^+$ channel (FaNaCh) that can be blocked by amiloride. FaNaCh displays the structural organization of epithelial Na$^+$ channel subunits and of *C. elegans* degenerins. It corresponds to the first example of an ionotropic receptor for a peptide.

Comparison of αENaC, βENaC and γENaC, FaNaCh or degenerin sequences with databases of expressed sequence tags (ESTs) has revealed the existence of homologs in mammals. The first mammalian protein identified by this approach was derived from a human testis EST. The messenger RNA, detected in brain, pancreas, testis, and ovary, encodes a protein of 638 amino acids (Waldmann *et al.*, 1995a). The highest sequence identity (37%) was observed with αENaC. When expressed alone in *Xenopus* oocytes, this protein generates a small amiloride-sensitive Na$^+$ current, as does αENaC alone. When expressed with βENaC and γENaC, it generates a large amiloride-sensitive Na$^+$ current, with biophysical and pharmacological properties slightly distinct from those observed after expression of αENaC, βENaC, and γENaC (higher unitary conductance, lower sensitivity to amiloride derivatives, different selectivity sequence and gating). Taken together, these results suggests that this protein, called δENaC, but which is not a fourth subunit of the epithelial Na$^+$ channel, rather corresponds to an 'αENaC-like' subunit. Human δENaC gene maps to chromosome 1p36.3–p36.2 (Waldmann *et al.*, 1996a).

Acid Sensing Ion Channels (ASIC, also called BNC, Brain Na$^+$ Channel) were subsequently characterized (Chen *et al.*, 1998; García-Añoveros *et al.*, 1997; Ishibashi and Marumo, 1998; Lingueglia *et al.*, 1997; Price *et al.*, 1996; Waldmann *et al.*, 1996b,1997a, 1997b). The ASICs participate in the formation of cationic channels in the central nervous system and in sensory neurons (Waldmann *et al.*, 1996b). These channels are gated by acidification of the external solution. Each ASIC is characterized by specific ionic selectivity, kinetics of opening and closure, and sensitivity to the external pH or to amiloride.

Identification of degenerins, ENaC, FaNaCh, and ASICs clearly defines a large gene super-family. This family contains not only constitutively activated ionic channels, involved in vectorial transport of electrolytes, as the amiloride-sensitive Na^+ channel, but also proteins involved in sensory perception, especially mechanosensation, as *C. elegans* MEC-4 and MEC-10, and ligand-gated ionotropic receptors, such as FaNaCh and ASIC. Some channels can also participate in vectorial transport of electrolytes and in sensory perception, as evidenced by the role of ENaC in mediating salt perception in rat taste bud cells (Kretz *et al.*, 1999). Many other mammalian members of this super-family remain to be identified. For instance, the molecular relationship between ENaC and moderately or nonselective Na^+ channels, characterized by p_{Na^+}/p_{K^+} ratios below 6 has to be clarified, although they are usually classified within the same family of ionic channels (see for instance Palmer, 1992). Although most of the members of the ENaC-degenerin super-family which have been expressed so far are characterized by a p_{Na^+}/p_{K^+} ratio greater than 10, mutant forms of ASIC2 and of UNC105 and some heteromultimeric complex of ASICs exhibit a lower Na^+ selectivity. It has been proposed that degradation of the highly selective Na^+ channel by extracellular proteases might alter the p_{Na^+}/p_{K^+} ratio, and change channel biophysical properties (Lewis and Alles, 1986), (Chraïbi *et al.*, 1998). It has also been proposed that these low selective channels would correspond to different mixtures of αENaC, βENaC and γ (Fyfe and Canessa, 1998). Finally, this type of nonselective channels might also very well correspond to new isoforms of nucleotide-gated cationic channel, which are also sensitive to amiloride (Schwiebert *et al.*, 1997). Additional work is clearly needed to clarify this issue.

5. Quaternary structure

Functional expression into *Xenopus* oocytes has clearly demonstrated that the functional Na^+ channel contains at least one copy of each ENaC subunit, implying a minimal stoichiometry of three. In order to identify the exact number of subunits into a complex, biochemical and functional experiments have been performed with FaNaCh, the FMRFamide receptor (Coscoy *et al.*, 1998), as well as with ENaC (Berdiev *et al.*, 1998; Firsov *et al.*, 1998; Kosari *et al.*, 1998; Snyder *et al.*, 1998).

FaNaCh biochemical properties were analyzed after stable transfection of the human embryonic kidney cells HEK-293 (Coscoy *et al.*, 1998). FaNaCh was used in these biochemical studies because of two useful properties. First, and unlike ENaC, active FMRFamide-activated Na^+ channels are formed by multimerization of only one protein. Second, and unlike ENaC, the FaNaCh channel is totally silent in the absence of FMRFamide, and is therefore not toxic for cells expressing it. In cells expressing functional channels, a protein with an apparent molecular mass of 82 kDa was detected at the cell surface. The 82-kDa form was derived from an incompletely glycosylated form of 74 kDa found in the endoplasmic reticulum. Covalent bonding by bifunctional cross-linkers resulted in the formation of covalent multimers that contained up to four subunits. Hydrodynamic properties of the solubilized FaNaCh complex, corrected for the contribution of the detergent to the size of the solubilized complex, also indicated a stoichiometry of four subunits per complex. Similar results were obtained after solubilization with Triton-X100 and with CHAPS. This stoichiometry clearly differs from that proposed by Cheng *et al.*, who used sucrose gradient sedimentation analysis to determine the sedimentation properties of ENaC subunits complexes (Cheng *et al.*, 1998). After solubilization of transfected cells with digitonin, these authors identified a complex of αENaC, βENaC and γENaC at a sedimentation coefficient of 25 S. This high sedimentation coefficient suggested to these authors that the channel might contain up to nine subunits. This stoichiometry of nine was also derived from functional measurements by the same group (Snyder *et al.*, 1998). However, the specific volume of digitonin (v = 0.73 cm^3g^{-1}) is very near the specific volume of proteins. The contribution of digitonin to the mass of the solubilized complex was therefore impossible to determine, leading to a large imprecision in the molecular weight of the glycoprotein complex (Lichtenberg *et al.*, 1992). Using an independent biochemical method, Firsov quantified the number of subunits present into a functional complex expressed at the cell surface (Firsov *et al.*, 1998). It was shown that the three ENaC subunits assemble according to a fixed stoichiometry, αENaC being more abundant than βENaC and γENaC (Firsov *et al.*, 1998).

Functional measurements have also been used in order to define the stoichiometry of these channels (Firsov *et al.*, 1998; Kosari *et al.*, 1998; Snyder *et al.*, 1998). They are based on a methodology first described by McKinnon for Shaker K^+ channel (McKinnon, 1991). In these experiments, wild type and mutant ENaC were mixed at different ratios. Mutations affecting αENaC, βENaC or γENaC introduced differential sensitivities to amiloride (Firsov *et al.*, 1998; Kosari *et al.*, 1998), to zinc (Firsov *et al.*, 1998), or to methanethiosulfonates (Kosari *et al.*, 1998; Snyder *et al.*, 1998). Analysis of the specific currents associated with the different complexes led Firsov *et al.* and Kosari *et al.* to suggest a stoichiometry of four subunits per complex, with two α, one β and one γ (Firsov *et al.*, 1998; Kosari *et al.*, 1998). However, using mutant forms of the three human ENaC subunits with altered channel inhibition by methanethiosulfonates, Snyder *et al.* suggested a stoichiometry of three α, three β, and three γENaC per complex (Snyder *et al.*, 1998). This discrepancy can probably be explained if the main hypothesis made by MacKinnon, i.e. the equivalence between mutant and wild type proteins, is violated. In order to check whether the mutations introduced into a channel subunit were changing the association of the complex, Firsov *et al.* quantified the expression of the cell surface of the wild type and of their mutants. The mutant forms used in their study did not modify the cell surface expression, suggesting equivalence between mutant and wild type proteins (Firsov *et al.*, 1998).

A different functional methodology was developed by Berdiev *et al.* (Berdiev *et al.*, 1998). It took advantage of the different unitary properties of a wild type

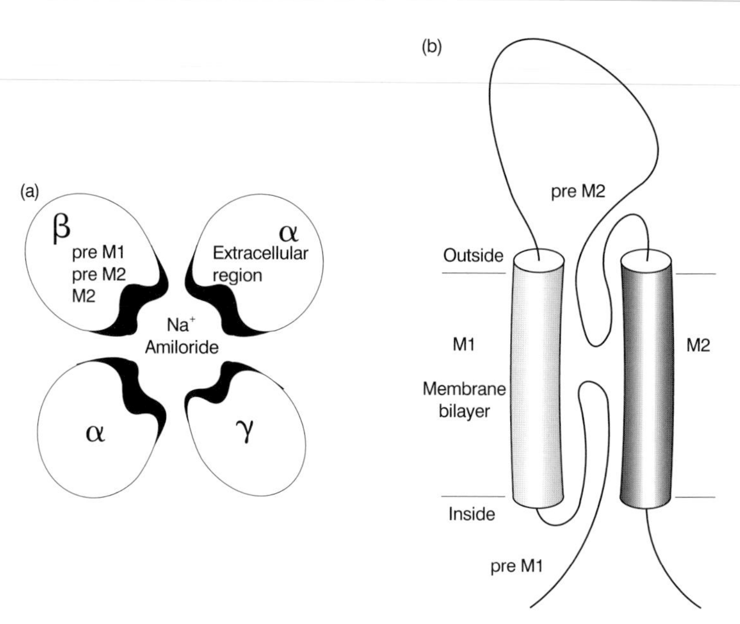

Figure 2. (a) View from above of the ENaC complex, derived from experiments performed on ENaC by Berdiev *et al.* (1998), Firsov *et al.* (1998) and Kosari *et al.* (1998) and on FaNaCh, a related protein, by Coscoy *et al.* (1998). An alternative model containing nine subunits has been proposed by Snyder *et al.* (1998). See text for discussion. Regions involved in the formation of the ionic pore and of the amiloride binding site are indicated. (b) Membrane topology of ENaC, deduced from experiments performed on ENaC (Renard *et al.*, 1994; Schild *et al.*, 1997; Waldmann *et al.*, 1995b), ASIC (Coscoy *et al.*, 1999) and degenerins (García-Añoveros *et al.*, 1995; Lai *et al.*, 1996).

αENaC and of a mutant channel, $\alpha_{\Delta 278-283}$-rENaC, characterized by distinct amiloride binding properties, after expression in planar lipid (Berdiev *et al.*, 1998). Five channel subtypes with distinct sensitivities to amiloride were found in a 1 αENaC: 1 $\alpha_{\Delta 278-283}$ENaC protein mixture. Their relative abundance is consistent with a tetrameric organization.

Despite some conflicting data (Snyder *et al.*, 1998), the tetrameric architecture therefore emerges as a classical motif for cationic channels (*Figure 2a*). Inward-rectifier K^+ channel (Yang *et al.*, 1995) and P_{2x} receptor, an ionotropic receptor for ATP (Kim *et al.*, 1997; but see also Nicke *et al.*, 1998), characterized by the presence of only two transmembrane α-helices, also form tetramers. This organization is similar to that reported for other cation selective channels, such as voltage-dependent K^+ channels (McKinnon, 1991), voltage-dependent Na^+ and Ca^{2+} channels (Tanabe *et al.*, 1987), and cyclic-nucleotide gated channels (Liu *et al.*, 1996a), that all have four-fold internal symmetry.

6. Secondary structure

Analysis of ENaCs primary structure clearly indicates the presence of two large hydrophobic domains. These two hydrophobic domains divide the proteins into five distinct domains:

(i) a 50–100 residues long NH_2-terminal segment (presumably cytoplasmic, due to the absence of a signal peptide), and a 20–100 residues long COOH-terminal segment,
(ii) two hydrophobic domains: the second one being more conserved over the family than the first one (~30% identity against ~20%),
(iii) a large domain located between the two hydrophobic domains, that represents more than 50% of the total mass of the protein. It contains several short motifs that are highly conserved among the family, such as one (two in degenerins) cysteine-rich region(s).

The same transmembrane organization is found for all members of the family (*Figure 2b*). It has been demonstrated for αENaC by a biochemical approach (Renard *et al.*, 1994) then confirmed with βENaC, γENaC, and MEC-4 (Renard, unpublished data; Lai *et al.*, 1996; Snyder *et al.*, 1994).

The large domain between the two transmembrane segments is extracellular. The NH_2 and COOH terminal domains of the proteins are cytoplasmic, and contain putative regulatory sequences, such as consensus sites for phosphorylation by protein kinases, and for interaction with cytoplasmic proteins. Slight differences between theoretical and experimental molecular weights of the extracellular domain were noticed by Renard *et al.* (Renard *et al.*, 1994). They proposed that the structure of the two transmembrane domains might be in fact more complex than classical hydrophobic α-helices (Renard *et al.*, 1994), and probably involves structures similar to the pore loop found in several voltage-dependent ionic channels (Guy and Durell, 1996).

Several mutations affecting the function of the proteins of the family have been characterized. Gain-of-function mutations are associated with neurodegenerescence (Chalfie and Wolinski, 1990) or hypercontraction (Liu *et al.*, 1996b) in nematodes, and with hereditary hypertension (Liddle's syndrome) in humans (Shimkets *et al.*, 1994); loss-of-function mutations are associated with mechano-insensitivity in nematodes, and with pseudohypoaldosteronism (PHA1) in humans (Chang *et al.*, 1996). Mutations in humans and nematodes often affect homologous regions, and sometimes equivalent residues. Such mutations clearly identify conserved domains that are important for all the proteins of the family.

6.1 Second transmembrane domain

Alignment of the different proteins of the ENaC-degenerin super-family reveals the highest level of similitude in the region around the second transmembrane region. Many experimental observations confer a high functional importance to this region (Driscoll and Chalfie, 1991; García-Añoveros *et al.*, 1995; Huang and Chalfie, 1994; Renard *et al.*, 1994; Schild *et al.*, 1997; Waldmann *et al.*, 1995b). They are consistent with a structural model where the second transmembrane domain is divided into two distinct parts. The COOH-terminal segment would correspond to a classical transmembrane α-helix. One side of this helix would interact with ions and with the amiloride molecule. The NH_2-terminal segment, where two putative β-strand structures would be linked by a coil-region that contains one (or two) conserved glycine(s), would participate in the formation of the ionic pore (Canessa *et al.*, 1994a; García-Añoveros *et al.*, 1995; Renard *et al.*, 1994).

Other residues, located in the second half of the hydrophobic domain (transmembrane α-helix), can also affect pharmacological and biophysical properties of the channel. Structure–function analysis of rat αENaC has shown that $Ser^{589}{}_{\alpha ENaC}$ is important in conferring to ENaC correct pharmacology, selectivity, conductance and gating, while $Ser^{593}{}_{\alpha ENaC}$ is important for correct selectivity, conductance and gating, but does not interact with amiloride. S589F shifts the inhibition constant for amiloride from 0.19 to 8.9 μM. When $Ile^{544}{}_{FaNaCh}$ from FaNaCh is mutated in a serine (I544S, i.e. the residue equivalent to $Ser^{589}{}_{\alpha ENaC}$), the mutant becomes totally insensitive to amiloride, and exhibits altered biophysical properties (Lingueglia *et al.*, 1995). Ser^{726} Thr^{729} and $Glu^{732}{}_{MEC\text{-}4}$ from MEC-4 are affected by loss-of-function mutations (Waldmann *et al.*, 1995b). All these residues are located on the same side of the predicted transmembrane α-helix, and are certainly oriented towards the ionic pore (Waldmann *et al.*, 1995b). Accordingly, modifications of residues located on the opposite side of the α-helix have no functional effect.

6.2 First transmembrane domain

The first hydrophobic domain is less conserved than the second one. However, when the first hydrophobic α-helix from αENaC is replaced by corresponding sequences of MEC-4, the affinity for amiloride decreases by a factor of 3, single channel conductance slightly increases, and mean open time decreases by three orders of magnitude (Waldmann *et al.*, 1995b).

Several point mutations have been identified in a highly conserved region located just before the first hydrophobic domain. $S105F_{MEC\text{-}10}$ eliminates MEC-10 function (as does an equivalent mutation in MEC-4 for MEC-4 function), but enhances $A673V_{MEC\text{-}10}$-induced degeneration (Huang and Chalfie, 1994). A direct interaction between these two residues is suggested by a recent study (Coscoy *et al.*, 1999) that shows the pre M1 region of ASIC2 participates in the formation of the ionic pore. A second mutation in the same region causes pseudohypoaldosteronism type-1 (PHA-1), an inherited disease characterized by severe neonatal salt-wasting, hyperkalemia, metabolic acidosis and unresponsiveness to mineralocorticoid hormones (Chang *et al.*, 1996; Strautnieks *et al.*, 1996). When expressed together with wild type αENaC and γENaC, $G37S_{\beta ENaC}$ is properly expressed at the cell surface, but the activity of the mutant complex is dramatically reduced (Gründer *et al.*, 1997). When ^{95}Gly of rat αENaC is mutated into a Ser, or when ^{40}Gly of rat γENaC is mutated into a Ser, the amiloride-sensitive inward current is reduced, without modification of the expression of the ENaC complex at the cell surface. Reduction of the current is of similar amplitude after expression of a mutant βENaC or after expression of a mutant γENaC. Reduction of the current is more robust after expression of a mutant αENaC with wild type βENaC and γENaC than after expression of a mutant βENaC (respectively γENaC) with wild type αENaC and γENaC (respectively βENaC). Three other PHA-1 mutations have been reported (Chang *et al.*, 1996; Strautnieks *et al.*, 1996): they correspond to a two-base pair deletion at codon 68 of human αENaC, that introduces a frameshift, and to a change in codon 508, that introduces a stop codon (R508X) in αENaC, and to a splice site mutation in γENaC.

6.3 Extracellular domain

This domain is particularly important for ligand-gated Na^+ channels (FaNaCh, ASIC, MDEG), and it will be important to determine the exact binding site of FMRFamide in FaNaCh, or of H^+ in ASIC and MDEG, and the conformational change induced by binding of the ligands. In degenerins, a missense recessive mutation within an extracellular 22 amino-acid region found in degenerins, but not in the other subfamilies, causes cell death similar to that caused by the dominant mutations affecting the predicted pore-lining (García-Añoveros *et al.*, 1995).

A point mutation affects Pro^{134} (P134T or P134S) of *C. elegans* UNC-105, located in the extracellular domain, between the first transmembrane α-helix and the highly conserved motif FPAVT. Animals harboring these mutations are hyper-contracted and paralyzed (Liu *et al.*, 1996b). The underlying defect can be corrected by a point mutation in a basement membrane collagen. A distinct point mutation affects UNC105 at the level of Glu^{677} ($E677K_{UNC-105}$). It is located close, but not in, the first half of the second hydrophobic region.

6.4 NH_2- and COOH-terminal cytoplasmic segment

The orientation of the two short NH_2- and COOH-terminal segments toward the cytoplasm confers on them an important role, since they are in contact with cytosolic protein kinases, and with all the intracellular regulatory machinery. Existence of specific regulation mechanisms for each protein is consistent with the poor conservation of the cytosolic domains among the family. Analysis of ENaC cytoplasmic segments is therefore treated in the next section, with the regulation of ENaC activity by cytosolic effectors and by protein kinases, and its implication for human pathologies.

7. Localization of ENaC gene products in epithelial tissues

Rat αENaC, βENaC and γENaC mRNAs have been detected by Northern blot analysis as unique bands of 3.6, 2.6, and 3.2 kb, respectively (Voilley *et al.*, 1997). They are expressed at a high level in epithelial tissues, such as renal cortex and medulla, distal colon, urinary bladder, lung, placenta, and salivary glands (Canessa *et al.*, 1994a; Voilley *et al.*, 1994, 1997). Low levels of transcripts were also identified in proximal colon, in uterus, in thyroid, and in intestine. No signal was detected by Northern blot analysis of total RNA from liver, stomach, duodenum, muscle (smooth and striated), heart, brain, or blood–brain barrier microvessels (Voilley *et al.*, 1997). Using *in situ* hybridization, Duc *et al.* have shown that the three subunit mRNAs are specifically coexpressed in the rat renal distal convoluted tubules, connecting tubules, cortical collecting ducts, and outer medullary collecting ducts, but not in the inner medullary collecting ducts (Duc *et al.*, 1994). In rat lung, αENaC mRNA was detected in trachea, bronchi, bronchioles, and alveoli (Farman *et al.*, 1997; Matsushita *et al.*, 1996). βENaC and γENaC mRNAs were more abundant in the bronchiolar and bronchial epithelium (Farman *et al.*, 1997; Matsushita *et al.*, 1996). In isolated taste buds from the tongue, differential

ENaC subunit mRNAs expression was noticed by an RT-PCR technique (Kretz *et al.*, 1999): in the anterior tongue, where salt taste is transduced by apical amiloride-sensitive sodium channels, mRNA of all three subunits was found. In isolated taste buds of the vallate papilla, where salt taste is not amiloride-blockable, only αENaC mRNA was detected.

α, β and γENaC expression was also analyzed at a protein level using immunohistochemistry with specific anti-peptide antibodies raised against each of the three subunits (Duc *et al.*, 1994; Renard *et al.*, 1995; Voilley *et al.*, 1997). In kidney, a majority of cells of the renal cortical collecting duct and outer medullary collecting duct express the three subunits (Duc *et al.*, 1994; Kretz *et al.*, 1999). Interestingly, Kretz *et al.* also detected some α-subunit immunoreactivity in the brush border at the level of the straight segments of the proximal nephron, consistent with the existence of functional Na^+ channels in the straight proximal segment (Willmann *et al.*, 1997). In distal colon, the three subunits were only detected in the surface epithelial cells, but not in the crypt cells (Duc *et al.*, 1994; Renard *et al.*, 1995). In rat submandibular gland, a strong apical labelling was noticed with antibodies raised against the three subunits (Duc *et al.*, 1994; Kretz *et al.*, 1999). In lung, Renard *et al.* reported a high level of expression of all three subunits in distal airways, at the level of Clara cells and of ciliated cells (Renard *et al.*, 1995). In the anterior tongue, reactivity for αENaC, βENaC, and γENaC was present in taste buds and lingual epithelium. In the posterior tongue vallate papilla, reactivity for αENaC was easily demonstrable, whereas that for βENaC and for γENaC was weaker than in the anterior tongue (Kretz *et al.*, 1999).

In human airways, Gaillard *et al.* used two different polyclonal antisera raised against β and γ ENaC to localize the channel in fetal (10–35 weeks) and adult human airways (D. Gaillard *et al.*, unpublished data). They noticed an early expression (as soon as 17 weeks of gestation) of the two subunits at the apical domain of bronchial ciliated cells, in glandular ducts and in bronchiolar ciliated and Clara cells. After 30 weeks, the distribution of the two subunits was similar in fetal and adult airways. In large airways, the two subunits were detected in ciliated cells, in cells lining glandular ducts and in the serous gland cells. In the distal bronchioles, β and γ were identified in ciliated and Clara cells. Ultrastructural immunogold labeling confirmed the identification of the two subunits in submucosal serous cells and bronchiolar Clara cells.

ENaC subunits have also been detected at a RNA level in other nonepithelial cells, such as in osteoblasts (Kizer *et al.*, 1997), or in baroreceptor nerve terminals innervating the aortic arch and carotid sinus (Drummond *et al.*, 1998). In both cases, ENaC subunits may be components of a mechanotransducer.

8. Regulation of ENaC expression

8.1 Steroids

Many factors acting in a tissue-specific manner, participate in the adaptation of ENaC activity. This explains how the same structural entity (i.e. formed by the three ENaC subunits) is able to fulfil different physiological roles, such as control of Na^+ homeostasis in kidney and in colon, correct hydration of mucus in the airways, or salt taste perception. In kidney and colon, aldosterone, which acts as the major sodium-retaining hormone, affects many cellular functions (Rossier and Palmer, 1992). It can activate transcription of specific genes after a free diffusion of the hormone across the cell membrane, binding to a cytosolic receptor, nuclear translocation of the hormone–receptor complex, and binding of the complex to steroid responsive elements located in the regulatory region of target gene promoters (Rossier and Palmer, 1992). However, an acute increase in Na^+ permeability can also occur before any transcriptional event. Both mechanisms exist in toad urinary bladder, where aldosterone increases the number of active channels (Palmer *et al.*, 1982). Asher and Garty have shown that a relatively fast increase in the apical Na^+ permeability (= 3 h) is mediated by mineralocorticoid receptors, and does not involve the transcription of new genes; however, during a later response (> 3 h) mediated by the activation of glucocorticoid receptors, an increased transcription of specific genes is observed (Asher and Garty, 1988). ENaC regulation by steroids has been more particularly studied in rat distal colon, rat kidney, and rat lung, where distinct regulatory mechanisms have been identified.

8.2 Distal colon

In rat distal colon, aldosterone induces amiloride-sensitive electrogenic Na^+ transport and inhibits electroneutral Na^+ absorption (Sandle and Binder, 1987). The Na^+ transport observed in non-stimulated control rats involves coupled transport of NaCl via a Na^+/H^+ antiporter and a Cl^-/HCO_{3^-} antiporter, and there is no, or very low, ENaC activity (Bastl and Hayslett, 1992; Charney and Feldman, 1984). The same mechanism is responsible for the Na^+ reabsorption observed in distal and proximal segments of the colon from control animals. In chicken colon, besides stimulation of electrogenic Na^+ transport, a low Na^+ diet also inhibits Na^+-driven cotransporters of amino-acids and sugars (Årnason and Skadhauge, 1991). Steroid stimulation of electrogenic amiloride-sensitive Na^+ transport in distal colon is mediated by type I mineralocorticoid receptors, as shown by the inhibitory effect of spironolactone, a specific type I mineralocorticoid receptor antagonist (Bastl and Hayslett, 1992), and by the control of electrogenic Na^+ absorption by nanomolar concentrations of aldosterone (Fromm *et al.*, 1993). Synthetic steroids with a high specificity for glucocorticoid receptors stimulate the electroneutral transport of NaCl but are inactive on the electrogenic Na^+ transport (Bastl and Hayslett, 1992; Turnamian and Binder, 1989).

The observation that colonic Na^+ channels are strongly regulated at the RNA level by dietary salt intake in rat and chicken has been one of the keys to successful expression cloning of the Na^+ channel (Canessa *et al.*, 1993, 1994b; Lingueglia *et al.*, 1993, 1994). A low Na^+ diet stimulates the renin–angiotensin system and leads to large increases in aldosterone secretion and Na^+ channel activity (Pácha *et al.*, 1993). Since messenger RNA derived from steroid-treated tissues lead to a more robust ENaC expression in *Xenopus* oocyte, transcription of specific RNAs has to be increased by the hormone (Asher *et al.*, 1992a).

Lingueglia *et al.* (1994) and Renard *et al.* (1995) have shown that aldosterone and high doses of dexamethasone stimulate the transcription of βENaC and γENaC, but not of αENaC in rat distal colon. Aldosterone-treated animals then express similar levels of αENaC, βENaC and γENaC transcripts. The expression of βENaC and γENaC RNAs was increased 5–10 fold after a chronic low-Na^+ diet (P. Barbry, unpublished data; Asher *et al.*, 1996). In normally fed animals βENaC and γENaC RNAs are still significantly expressed. Therefore post-translational mechanisms have to control the functional expression of neosynthesized channels in order to explain the low functional ENaC activity in these animals. Expression of βENaC and γENaC proteins at the apical membrane of colonocytes is largely increased by a low-Na^+ diet or by a dexamethasone-treatment, as detected by immunohistochemical stainings with specific anti-peptide antibodies (Lingueglia *et al.*, 1994; Renard *et al.*, 1995). Aldosterone treatment also increases the expression of the αENaC protein. Renard *et al.* proposed that this regulation occurs at a post-translational level, as the result of multimerization of the three subunits after steroid treatment (Renard *et al.*, 1995). Formation of a stabilized complex between newly synthesized βENaC and γENaC and the αENaC would permit maturation of the complex toward the apical membrane. This has been confirmed by Firsov *et al.* after a quantification of ENaC cell surface expression in *Xenopus* oocytes (Firsov *et al.*, 1996). When αENaC is expressed alone, a low cell surface expression is observed. Coexpressing αENaC with β and γENaC dramatically increases the cell surface expression. Due to a significant residual expression of β and γENaC RNAs in normally-fed animals, it is also likely that another mechanism also contributes to the retention of newly synthesized proteins within the cytoplasm. Using minipumps, Asher *et al.* have analyzed the initial time-course of the ENaC response to aldosterone (Asher *et al.*, 1996). They showed that short-circuit current is increased within 3 h of the infusion, before any transcriptional effects on βENaC or γENaC genes. Several mechanisms have been proposed to explain the fast effects of aldosterone on Na^+ channel activity. In A6 cells, aldosterone primarily increases the open probability of the Na^+ channels with a minor effect on the number of channels per patch (Kemendy *et al.*, 1992). Changes in internal pH (Harvey *et al.*, 1988), methylation (Sariban-Sohraby *et al.*, 1984; Frindt and Palmer, 1996) or changes in internal calcium (Petzel *et al.*, 1992) have been proposed to explain the short-term increased activity. Mobilization of ENaC subunits stored into cytoplasmic vesicles to the apical membrane has been proposed by Palmer *et al.* who first identified in toad urinary bladder a metabolically-dependent recruitment of pre-existing Na^+ channels from a reservoir of electrically undetectable channels (Palmer *et al.*, 1982).

8.3 Kidney

Frindt *et al.* showed that a low Na^+ diet causes a ~100-fold increase in the amiloride-sensitive whole-cell current measured in rat cortical collecting tubule cells (Frindt *et al.*, 1990). The effects of a low Na^+ diet on the activity of the Na^+ channel from rat cortical collecting tubules has been demonstrated by cell-attached patch clamp recordings. Pácha *et al.* reported an increase in the number of active Na^+ channels, but did not detect any modification in the open probability of the channels (Pácha *et al.*, 1993). Reif *et al.* noted the synergistic effects of antidiuretic hormone and desoxycorticosterone on Na^+ transport in perfused rat cortical collecting tubules (Reif *et al.*, 1986). The rapid activation of the Na^+

channels by vasopressin, observed within minutes is consistent with the existence of a pool of silent channels (Reif *et al.*, 1986). In rat kidney, a steroid treatment (i.e. chronic low Na^+ diet or high doses of dexamethasone) hardly affect ENaC RNA levels (Renard *et al.*, 1995). A 2.6-fold increase in expression of γENaC RNA was reported in cultured rabbit CCD cells (Denault *et al.*, 1996). Volk *et al.* have reported a 2-fold increase in αENaC transcription and no effect on βENaC and γENaC transcription in rat inner medullary collecting duct (Volk *et al.*, 1995). Such a small increase can hardly explain the large aldosterone effects on renal Na^+ channel activity (Frindt *et al.*, 1990). Immunolabelling experiments have been performed in kidney from normally fed animals or from Na^+-depleted animals by Renard *et al.* levels (Renard *et al.*, 1995). A high constitutive expression of ENaC proteins in distal segments of the kidney tubule was observed: steroid treatments had no effect on expression of αENaC protein in medullary rays, nor on γENaC protein in distal convoluted tubules levels (Renard *et al.*, 1995).

Chen *et al.* (Chen *et al.*, 1999) have found that coexpression into Xenopus oocytes of ENaC with the serine–threonine kinase SGK increases by 7-fold the amiloride-sensitive current. SGK had been previously identified as a glucocorticoid responsive gene in a mammary gland cell line. Transcription of rat SGK is indeed rapidly increased by aldosterone, making SGK a good candidate for participating to the fast stimulation of ENaC by aldosterone. Identification of genes rapidly transcribed after treatment with aldosterone is certainly an important step to understand the mechanisms of aldosterone action on ENaC in kidney.

8.4 Lung

ENaC controls the quantity and composition of the respiratory tract fluid and plays a key role in the transition from a fluid-filled to an air-filled lung at the time of birth (O'Brodovich, 1991; Strang, 1991). ENaCs mRNAs have been identified in the lung (Renard *et al.*, 1995), and the biophysical properties reported by Voilley *et al.* appear identical to those of the renal Na^+ channel (Voilley *et al.*, 1994). Around birth, an increase in Na^+ channel transcription and expression results in a switch of the ionic transport in lung from active Cl^- secretion to active Na^+ reabsorption (McDonald *et al.*, 1994; O'Brodovich *et al.*, 1993; Voilley *et al.*, 1994, 1997). This results in clearance of the pulmonary fluid as the lung switches to an air-conducting system. After inactivation of murine αENaC, deficient neonates develop respiratory distress and die with 40 h of birth from failure to clear their lungs of liquid (Hümmler *et al.*, 1996). Administration of glucocorticoids has previously been demonstrated to induce a Na^+ absorptive capacity in the immature fetal lung (Barker *et al.*, 1991), suggesting that this modification may be due to ENaC stimulation by glucocorticoids. Champigny *et al.* and Voilley *et al.* have analyzed the mechanisms of this stimulation in primary cultures of fetal rat lung epithelial cells, and showed that the ENaC activity is controlled by corticosteroids (Champigny *et al.*, 1994; Voilley *et al.*, 1997). Treatment with dexamethasone, or with RU28362, a synthetic pure glucocorticoid agonist, increases Na^+ channel activity via stimulation of the three ENaC subunits transcription, i.e. a distinct effect of those observed in kidney and in colon with aldosterone.

The increase in Na^+ channel activity observed in the lung around birth might be related to the raise of corticosteroids (Champigny *et al.*, 1994; O'Brodovich, 1991; Strang, 1991; Voilley *et al.*, 1997). However, Tchepichev *et al.* have observed differences between development and steroids, suggesting the existence of others triggers, such as change in P_{O_2} (Tchepichev *et al.*, 1995).

9. Involvement in human pathology

9.1 Liddle's syndrome

Liddle's syndrome is an autosomal dominant form of hypertension, characterized by hypokalemia and suppressed renin and aldosterone levels. The high blood pressure responds specifically to amiloride, indicative of an upregulation of ENaC activity (Liddle *et al.*, 1963). Genetic analysis demonstrates that the disorder can be attributed to a mutation in the βENaC gene (Shimkets *et al.*, 1994): introduction of a premature stop codon ($R564X_{\beta ENaC}$) truncates the cytoplasmic COOH terminal segment before the PPPXY sequence highlighted by Rotin *et al.* (1994). This proline-rich sequence in COOH terminal region of αENaC is conserved among ENaC. Rotin *et al* proposed that cytoplasmic interactions via SH_3 domains (for instance with α-spectrin) mediate an apical localization of the Na^+ channel, and would provide a novel mechanism for retaining proteins in specific membranes of polarized epithelial cells. Several similar mutations have been reported in βENaC and γENaC, that all remove the last 45–76 COOH-terminal amino acids of the proteins (Hansson *et al.*, 1995a; Jeunemaitre *et al.*, 1997; Shimkets *et al.*, 1994). The crucial role played by the PPPXY motif in βENaC and γENaC has been highlighted by the identification of the two missense mutations P616L (Hansson *et al.*, 1995b) and Y618H (Tamura *et al.*, 1996) in other Liddle's syndrome patients.

Heterologous expression of the mutant proteins leads to the expression of an overactive channel (for instance, a 3.7±0.3 fold increase of the channel activity is measured in *Xenopus* oocytes after expression of a mutated human βENaC, with normal αENaC and γENaC) (Jeunemaitre *et al.*, 1997). This increased activity is not explained by alteration of single channel conductance and of open probability, but rather by an increased number of channels inserted into the plasma membrane (Firsov *et al.*, 1996; Schild *et al.*, 1996; Snyder *et al.*, 1995). No Liddle's syndrome mutation have been detected so far in human αENaC, and the effects of αENaC *in vitro* mutations on Na^+ channel function are still unclear (Schild *et al.*, 1996; Snyder *et al.*, 1995).

Two distinct mechanisms can presently explain how Liddle's mutations increase ENaC activity:

(i) Since tyrosine from the PPPXY plays a crucial role, it is proposed that defective endocytosis would lead to an accumulation of ENaC proteins at the apical membrane (Snyder *et al.*, 1995). Such a mechanism has been described for the low density lipoprotein receptor, the lysosomal acid phosphatase and the β-adrenergic receptor. In these cases, a tyrosine residue in the context of a tight turn is crucial for endocytosis. The role of membrane biogenesis, and interaction with endosomal proteins remain to be elucidated in the case of ENaC.

(ii) Using a yeast two-hybrid system, Staub *et al.* have identified the rat protein NEDD-4 as a binding partner for the proline-rich regions of βENaC and γENaC (Staub *et al.*, 1996). Interaction between rat Nedd-4 and ENaC occurs via one of the three NEDD-4 WW domains. WW domains are 38 residues sequences that contains two conserved tryptophans at positions 7 and 29. Rat Nedd-4 also contains a Ca^{2+}-lipid binding domain, and a ubiquitin-ligase domain. Staub *et al.* proposed that association between NEDD4 and ENaC subunits might bring the Ca^{2+}-lipid binding and ubiquitin-ligase domains in close proximity with the channel, initiating its removal from the apical membrane (after mobilization in the endocytotic degradative pathway, ubiquitination, and degradation by the proteasome; Staub *et al.*, 1996). Indeed, NEDD-4 act as a negative regulator of the wild-type epithelial Na^+ channel, but is inactive on a Liddle form of the channel (Abriel *et al.*, 1999; Goulet *et al.*, 1998).

Shimkets *et al.* have investigated the role of clathrin-coated pit-mediated endocytosis on the function of normal and Liddle form of ENaC in oocytes (Shimkets *et al.*, 1997). Inhibition of endocytosis by coexpression of ENaC with a dominant-negative dynamin leads to a large increase in the activity of wild-type channels, demonstrating that normal turnover of this channel is through the clathrin-coated pit pathway. In contrast, coexpression of Liddle's mutations and dynamin mutants leads to no further increase in channel activity, consistent with one of the effects of Liddle's mutations being the loss of endocytosis of these channels.

Several other variants of β and γENaC have been identified in humans, especially in populations of African, African-American and West Indian origins (Persu *et al.*, 1998). While an increased Na^+ channel activity was detected in some of these patients (Baker *et al.*, 1998; Cui *et al.*, 1997), the expression of these variants in *Xenopus* oocytes was not associated with altered levels of current. This suggests that other factors might also be important to fine-tune the activity of the epithelial Na^+ channel.

9.2 Regulation by cyclic AMP dependent protein kinase — relevance to cystic fibrosis

In lung, β-adrenergic agonists increase lung fluid clearance and this effect is inhibited by amiloride (Berthiaume *et al.*, 1987). It is, however, unclear whether the stimulation acts at the level of ENaC, or at the level of a chloride conductance, which is also necessary for NaCl absorption (Jiang *et al.*, 1998). In rat, cyclic AMP stimulates ENaC activity in cortical collecting tubules (Frindt and Palmer, 1996; Hawk *et al.*, 1996) but not in colon (Bridges *et al.*, 1984), even though the same three subunits that make up the channel are present in these three tissue types (Renard *et al.*, 1995). In toad urinary bladder, cyclic AMP stimulation is lost after disruption of the cells (Lester *et al.*, 1988). In frog skin, frog colon, toad urinary bladder and A6 cells, cyclic AMP increases the number of conductive Na^+ channels without affecting the open probability (Helman *et al.*, 1983; Krattenmacher *et al.*, 1988; Li and Lindemann, 1982; Marunaka and Eaton,

1991). These observations are consistent with an indirect stimulatory mechanism, which does not require ENaC phosphorylation by cyclic AMP dependent protein kinase. In accordance, ENaC do not contain conserved consensus sites for phosphorylation by protein kinase A in their cytoplasmic domains (Renard *et al.*, 1994).

Cyclic AMP stimulates $^{22}Na^+$ uptake through ENaC in MDCK, Vero, and NIH-3T3 cells, while it has no effect on the current in the same cells, nor in oocytes injected with the same subunits (P. Barbry, unpublished data). A slightly different situation is observed when ENaC is coexpressed with CFTR, the cyclic AMP regulated Cl^- channel which is defective in cystic fibrosis (Mall *et al.*, 1996; Stutts *et al.*, 1995). In that case, cyclic AMP becomes inhibitory for ENaC function. This effect has been related to an approximately 2-fold increase in amiloride-sensitive Na^+ permeability, observed during cystic fibrosis (Boucher *et al.*, 1986). In *Xenopus* oocyte, cyclic AMP inhibits ENaC activity in the presence of wild-type CFTR, but not in the presence of the pathologic mutants ΔF508-CFTR and G551D-CFTR (Mall *et al.*, 1996). A direct physical interaction between the NH_2 terminal segment of αENaC and the first nucleotide binding domain CFTR has been reported using the yeast-two hybrid system (Kunzelmann *et al.*, 1997). This is nevertheless intriguing since hyper-Na^+ reabsorption is not observed in sweat gland, another CF-affected tissue, where ENaCs and CFTR are also co-expressed (Bijman and Frömter, 1986; Quinton, 1990). Also, some strains of transgenic mice, obtained by knock-out of the CFTR gene, do not develop excessive airway Na^+ reabsorption (Barbry and Lazdunski, 1996).

The cyclic AMP stimulation of ENaC in the absence of CFTR might be explained by exocytosis of vesicles containing functional ENaC complex to the apical membrane, that would increase the number of active Na^+ channels expressed at the cell surface. When coexpressed with functional CFTR, the latter would affect membrane recycling (Bradbury *et al.*, 1992), and would decrease the mobilization of the Na^+ channel to the cell surface. Alternatively, indirect control of the Na^+ channel activity by intracellular anions has been demonstrated in submandibulary salivary glands (Dinudom *et al.*, 1995; Komwatana *et al.*, 1996): increase in intracellular Na^+ and Cl^- activity would inhibit ENaC by a mechanism that would involve specific heterotrimeric G-proteins. The requirement for specific regulatory proteins would in that case explain why the effect of CFTR on ENaC activity is not ubiquitous. In A6 cells, Ling *et al.* have reported that inhibition of CFTR expression by antisense oligonucleotides increases the open probability without affecting the number of active Na^+ channels (Ling *et al.*, 1997).

The carboxyl termini of βENaC and γENaC can be phosphorylated in transfected cells expressing the three subnits of ENaC, after the cells had been treated with aldosterone, insulin, or activators of protein kinases A and C (Shimkets *et al.*, 1998). The enzyme responsible for this phosphorylation has not yet been identified. A good candidate is SGK (serum and glucocorticoid-regulated kinase), a member of the serine-threonine kinase family, which can stimulate *Xenopus* ENaC activity approximately 7-fold after coexpression in *Xenopus laevis* oocytes (Chen *et al.*, 1999).

10. Conclusion

Molecular biology of Na^+ transport has permitted the identification of the key molecules involved into Na^+ permeation through the apical membrane of high resistance epithelia. Distinct mechanisms of regulation by steroids have been identified; two human genetic diseases, that affect ion transport, have been associated with mutations of the genes; the relationship between the epithelial Na^+ channel and other proteins has been revealed, and their quaternary structure established. However, a lot of questions remain unsolved, such as the mechanisms of short term regulation of the activity of the channel, or the molecular mechanisms of cross-talk between ENaC and other apical transporters, such as CFTR, or between ENaC and basolateral transporters, such as the Na^+-K^+-ATPase.

Acknowledgments

My work is supported by the Centre National de la Recherche Scientifique (CNRS) and the Association Française de Lutte contre la Mucoviscidose (AFLM).

References

Abriel, H., Loffing, J., Rebhun, J.F., Pratt, J.H., Schild, L., Horisberger, J.D., Rotin, D. and Staub, O. (1999) Defective regulation of the epithelial Na^+ channel by Nedd4 in Liddle's syndrome. *J. Clin. Invest.* **103**: 667–673.

Årnason, S.S. and Skadhauge, E. (1991) Steady-state sodium absorption and chloride secretion of colon and coprodeum, and plasma levels of osmoregulatory hormones in hens in relation to sodium intake. *J. Comp. Physiol.* **B 161**: 1–14.

Asher, C. and Garty, H. (1988) Aldosterone increases the apical Na^+ permeability of toad bladder by two different mechanisms. *Proc. Natl Acad. Sci. USA* **85**: 7413–7417.

Asher, C., Eren, R., Kahn, L., Yeger, O. and Garty, H. (1992a) Expression of the amiloride-blockable Na^+ channel by RNA from control *versus* aldosterone-stimulated tissue. *J. Biol. Chem.* **267**: 16061–16065.

Asher, C., Singer, D., Eren, R., Yeger, O., Dascal, N. and Garty, H. (1992b) NaCl-dependent expression of an amiloride-blockable Na^+ channel in *Xenopus* oocytes. *Am. J. Physiol.* **262**: G244-G248.

Asher, C., Wald, H., Rossier, B.C. and Garty, H. (1996) Aldosterone-induced increase in the abundance of Na^+ channel subunits. *Am. J. Physiol.* **271**: C605–C611.

Baker, E.H., Dong, Y.B., Sagnella, G.A., *et al.* (1998) Association of hypertension with T594M mutation in beta subunit of epithelial sodium channels in black people resident in London. *Lancet* **351**: 1388–1392.

Barbry, P. and Hofman, P. (1997) Molecular biology of Na^+ absorption. *Am. J. Physiol.* **273**: G571–G585.

Barbry, P. and Lazdunski, M. (1996) Structure and regulation of the amiloride-sensitive epithelial Na^+ channel. In *Ion Channels* (ed. T. Narahashi). Plenum Press, New York, Vol. 4, pp. 115–167.

Barker, P.M., Walters, D.V., Markiewicz, M. and Strang, L.B. (1991) Development of the lung liquid reabsorptive mechanism in fetal sheep: synergism of triiodothyronine and hydrocortisone. *J. Physiol. London* **433**: 435–449.

Bastl, C.P. and Hayslett, J.P. (1992) The cellular action of aldosterone in target epithelia. *Kidney Int.* **42**: 250–264.

Belkin, K.J. and Abrams, T.W. (1993) FMRFamide produces biphasic modulation of the LFS motor neurons in the neural circuit of the siphon withdrawal reflex of *Aplysia* by activating Na^+ and K^+ currents. *J. Neurosci.* **13**: 5139–5152.

Bentley, P.J. (1968) Amiloride : a potent inhibitor of sodium transport across the toad bladder. *J. Physiol.* **195**: 317–330.

Berdiev, B.K., Karlson, K.H., Jovov, B., *et al.* (1998) Subunit stoichiometry of a core conduction element in a cloned epithelial amiloride-sensitive Na^+ channel. *Biophysical J.* **75**: 2292–2301.

Berthiaume, Y., Staub, N.C. and Matthay, M.A. (1987) β adrenergic agonists increase lung liquid clearance in anesthetized sheep. *J. Clin. Invest.* **79**: 335–343.

Bijman, J. and Frömter, E. (1986) Direct demonstration of high transepithelial chloride-conductance in normal human sweat duct which is absent in cystic fibrosis. *Pflügers Arch.* **407**: S123–S127.

Boucher, R.C., Stutts, M.J., Knowles, M.R., Cantley, L. and Gatzy, J.T. (1986) Na^+ transport in cystic fibrosis respiratory epithelia. Abnormal basal rate and response to adenylate cyclase activation. *J. Clin. Invest.* **78**: 1245–1252.

Bradbury, N.A., Jilling, T., Berta, G., Sorscher, E.J., Bridges, R.J. and Kirk, K.L. (1992) Regulation of plasma membrane recycling by CFTR. *Science* **256**: 530–532.

Bridges, R.J., Rummel, W. and Wolenberg, P. (1984) Effects of vasopressin on electrolyte tranport across isolated colon from normal and dexamethasone-treated rats. *J. Physiol.* **355**: 11–23.

Canessa, C.M., Horisberger, J.D. and Rossier, B.C. (1993) Epithelial sodium channel related to proteins involved in neurodegeneration. *Nature* **361**: 467–470.

Canessa, C.M., Schild, L., Buell, G., Thorens, B., Gautschi, I., Horisberger, J.D. and Rossier, B.C. (1994a) Amiloride-sensitive epithelial sodium channel is made of three homologous subunits. *Nature* **367**: 463–467.

Canessa, C.M., Merillat, A.M. and Rossier, B.C. (1994b) Membrane topology of the epithelial sodium channel in intact cells. *Am. J. Physiol.* **267**: C1682–C1690.

Chalfie, M. (1993) Touch receptor development and function in *Caenorhabditis elegans*. *J. Neurobiol.* **24**: 1433–1441.

Chalfie, M. and Wolinski, E. (1990) The identification and suppression of inherited neurodegeneration in *Caenorhabditis elegans*. *Nature* **345**: 410–416.

Champigny, G., Voilley, N., Lingueglia, E., Friend, V., Barbry, P. and Lazdunski, M. (1994) Regulation of expression of the lung amiloride-sensitive Na^+ channel by steroid hormones. *Embo J.* **13**: 2177–2181.

Chang, S.S., Grunder, S., Hanukoglu, A., *et al.* (1996) Mutations in subunits of the epithelial sodium channel cause salt wasting with hyperkalaemic acidosis, pseudohypoaldosteronism type 1. *Nature Genet.* **12**: 248–253.

Charney, A. and Feldman, G. (1984) Systemic acid-base disorders and intestinal electrolyte transport. *Am. J. Physiol.* **247**: G1–G12.

Chen, C.C., England, S., Akopian, A.N. and Wood, J.N. (1998) A sensory neuron-specific, proton-gated ion channel. *Proc. Natl Acad. Sci USA* **95**: 10240–10245.

Chen, S.Y., Bhargava, A., Mastroberardino, L., Meijer, O.C., Wang, J., Buse, P., Firestone, G.L., Verrey, F. and Pearce, D. (1999) Epithelial sodium channel regulated by aldosterone-induced protein sgk. *Proc. Natl Acad. Sci. USA* **96**: 2514–2519.

Cheng, C., Prince, L.S., Snyder, P.M. and Welsh, M.J. (1998) Assembly of the epithelial Na^+ channel evaluated using sucrose gradient sedimentation analysis. *J. Biol. Chem.* **273**: 22693–22700.

Chinet, T.C., Fullton, J.M., Yankaskas, J.R., Boucher, R.C. and Stutts, M.J. (1993) Sodium-permeable channels in the apical membrane of human nasal epithelial cells. *Am. J. Physiol.* **265**: C1050–C1060.

Chraïbi, A., Vallet, V., Firsov, D., Hess, S.K. and Horisberger, J.D. (1998) Protease modulation of the activity of the epithelial sodium channel expressed in Xenopus oocytes. *J. Gen. Physiol.* **111**: 127–38.

Clauss, W., Dürr, J.E., Guth, D. and Skadhauge, E. (1987) Effects on adrenal steroids on Na transport in the lower intestine coprodeum of the hen. *J. Membr. Biol.* **96**: 141–152.

Cornell, J.C. (1984) Seasonal changes in sodium transport and amionide sensitivity in the isolated intestine of the earthworm *Lumbricus terrestris*. *Comp. Biochem. Physiol.* **78**: 463–468.

Coscoy, S., Lingueglia, E., Lazdunski, M. and Barbry, P. (1998) The Phe–Met–Arg–Phe–amide-activated sodium channel is a tetramer. *J. Biol. Chem.* **273**: 8317–8322.

Coscoy, S., de Weille, J.R., Lingueglia, E. and Lazdunski, M. (1999) The pre-transmembrane 1 domain of acid-sensing ion channels participates in the ion pore. *J. Biol. Chem.* **274**: 10129–10132.

Cragoe, E.J. (1979) Structure activity relationships int the amiloride series. In *Amiloride and Epithelial Sodium Transport* (eds A.W. Cuthbert., G.M.J. Fanelli, and A. Scrabini), Urban and Schwarzenberg, Baltimore, pp. 1–20.

Cui, Y., Su, Y.R., Rutkowski, M., Reif, M., Menon, A.G. and Pun, R.Y. (1997) Loss of protein kinase C inhibition in the beta-T594M variant of the amiloride-sensitive Na^+ channel. *Proc. Natl Acad. Sci. USA* **94**: 9962–9966.

Denault, D.L., Fejes-Toth, G. and Naray-Fejes-Toth, A. (1996) Aldosterone regulation of sodium channel gamma-subunit mRNA in cortical collecting duct cells. *Am. J. Physiol.* **271**: C423–C428.

Dinudom, A., Komwatana, P., Young, J.A. and Cook, D.I. (1995) Control of the amiloride-sensitive Na^+ current in mouse salivary ducts by intracellular anions is mediated by a G protein. *J. Physiol.* **487**: 549–555.

Driscoll, M. and Chalfie, M. (1991) The *Mec-4* gene is a member of a family of *Caenorhabditis elegans* genes that can mutate to induce neuronal degeneration. *Nature* **349**: 588–593.

Drummond, H.A., Price, M.P., Welsh, M.J. and Abboud, F.M. (1998) A molecular component of the arterial baroreceptor. *Neuron* **21**: 1435–1441.

Duc, C., Farman, N., Canessa, C.M., Bonvalet, J.P. and Rossier, B.C. (1994) Cell-specific expression of epithelial sodium channel α, β and γ subunits in aldosterone-responsive epithelia from the rat : localization by *in situ* hybridization and immunocytochemistry. *J. Cell Biol.* **127**: 1907–1921.

Eigler, J. and Crabbe, J. (1969) Effect of diuretics on active Na-transport in amphibian membranes. In: *Renal Transport and Diuretics* (eds K. Thurau and H. Jahrmärker). Springer-Verlag KG, Berlin, pp. 195–208.

Farman, N., Talbot, C.R., Boucher, R., Fay, M., Canessa, C., Rossier, B. and Bonvalet, J.P. (1997) Noncoordinated expression of alpha-, beta-, and gamma-subunit mRNAs of epithelial Na^+ channel along rat respiratory tract. *Am. J. Physiol.* **272**: C131-C141.

Firsov, D., Schild, L., Gautschi, I., Merillat, A.M., Schneeberger, E. and Rossier, B.C. (1996) Cell surface expression of the epithelial Na channel and a mutant causing Liddle syndrome: a quantitative approach. *Proc. Natl Acad. Sci. USA* **93**: 15370–15375.

Firsov, D., Gautschi, I., Merillat, A.M., Rossier, B.C. and Schild, L. (1998) The heterotetrameric architecture of the epithelial sodium channel (ENaC). *Embo J.* **17**: 344–352.

Frindt, G. and Palmer, L.G. (1996) Regulation of Na channels in the rat cortical collecting tubule: effects of cAMP and methyl donors. *Am. J. Physiol.* **271**: F1086–1092.

Frindt, G., Sackin, H. and Palmer, L.G. (1990) Whole-cell currents in rat cortical collecting tubule : low Na^+ diet increases amiloride-sensitive conductance. *Am. J. Physiol.* **258**: F562–F567.

Frings, S., Purves, R.D. and Macknight, A.D.C. (1988) Single channel recordings from the apical membrane of the toad urinary bladder epithelial cell. *J. Membr. Biol.* **106**: 157–172.

Fromm, M., Schulzke, J.D. and Hegel, U. (1993) Control of electrogenic Na^+ absorption in rat late distal colon by nanomolar aldosterone added *in vitro*. *Am. J. Physiol.* **264**: E68–E73.

Fuller, C.M., Awayda, M.S., Arrate, M.P., Bradford, A.L., Morris, R.G., Canessa, C.M., Rossier, B.C. and Benos, D.J. (1995) Cloning of a bovine renal epithelial Na^+ channel subunit. *Am. J. Physiol.* **269**: C641–C654.

Fyfe, G.K. and Canessa, C.M. (1998) Subunit composition determines the single channel kinetics of the epithelial sodium channel. *J. Gen. Physiol.* **112**: 423–432.

García-Añoveros, J., Ma, C. and Chalfie, M. (1995) Regulation of *Caenorhabditis elegans* degenerin proteins by a putative extracellular domain. *Curr. Biol.* 5: 441–448.

García-Añoveros, J., Derfler, B., Neville-Golden, J., Hyman, B.T. and Corey, D.P. (1997) BNaC1 and BNaC2 constitute a new family of human neuronal sodium channels related to degenerins and epithelial sodium channels. *Proc. Natl Acad. Sci. USA* **94**: 1459–1464.

George, A.L., Staub, O., Geering, K., Rossier, B. and Kleyman, T.R. (1989) Functional expression of the amiloride-sensitive sodium channel in *Xenopus* oocytes. *Proc. Natl Acad. Sci. USA* **86**: 7295–7298.

Goldstein, O., Asher, C. and Garty, H. (1997) Cloning and induction by low NaCl intake of avian intestine Na^+ channel subunits. *Am. J. Physiol.* **272**: C270–C277.

Goulet, C.C., Volk, K.A., Adams, C.M., Prince, L.S., Stokes, J.B. and Snyder, P.M. (1998) Inhibition of the epithelial Na^+ channel by interaction of Nedd4 with a PY motif deleted in Liddle's syndrome. *J. Biol. Chem.*, **273**: 30012–30017.

Green, K.A., Falconer, S.W. and Cottrell, G.A. (1994) The neuropeptide Phe–Met–Arg–Phe–NH_2 (FMRFamide) directly gates two ion channels in an identified Helix neurone. *Pflugers Archiv. Eur. J. Physiol.* **428**: 232–40.

Gross, J.B. and Kokko, J.P. (1977) Effects of Aldosterone and potassium-sparing diuretics on electrical potential differences across the distal nephron. *J. Clin. Invest* **59**: 82–89.

Gründer, S., Firsov, D., Chang, S.S., Jaeger, N.F., Gautschi, I., Schild, L., Lifton, R.P. and Rossier, B.C. (1997) A mutation causing pseudohypoaldosteronism type 1 identifies a conserved glycine that is involved in the gating of the epithelial sodium channel. *EMBO J.* **16**: 899–907.

Guy, H.R. and Durell, S.R. (1996) Developing three-dimensional models of ion channel proteins. In *Ion Channels* (ed. T. Narahashi). Plenum Press, New York, Vol. **4**, pp. 1–40.

Hamilton, K.L. and Eaton, D.C. (1986) Single channel recordings from two types of amiloride-sensitive epithelial Na^+ channels. *Membr. Biochem.* **6**: 149–171.

Hansson, J.H., Nelson-Williams, C., Suzuki, H., *et al.* (1995a) Hypertension caused by a truncated epithelial sodium channel g subunit: genetic heterogeneity of Liddle syndrome. *Nature Genet.* **11**: 76–82.

Hansson, J.H., Schild, L., Lu, Y., Wilson, T.A., Gautschi, I., Shimkets, R., Nelson-Williams, C., Rossier, B.C. and Lifton, R.P. (1995b) A *de novo* missense mutation of the beta subunit of the epithelial sodium channel causes hypertension and Liddle syndrome, identifying a proline-rich segment critical for regulation of channel activity. *Proc. Natl Acad. Sci. USA* **92**: 11495–11499.

Harvey, B.J., Thomas, S.R. and Ehrenfeld, J. (1988) Intracellular pH controls cell membrane Na and K conductances and transport in frog skin epithelium. *J. Gen. Physiol.* **92**: 767–791.

Hawk, C.T., Li, L. and Schafer, J.A. (1996) AVP and aldosterone at physiological concentrations have synergistic effects on Na^+ transport in rat CCD. *Kidney Int. Suppl.* **57**: S35–41.

Helman, S.I., Cox, T.C. and Van Driessche, W. (1983) Hormonal control of apical membrane Na transport in epithelia. Studies with fluctuation analysis. *J. Gen. Physiol.* **82**: 201–220.

Hinton, C.F. and Eaton, D.C. (1989) Expression of amiloride-blockable sodium channels in *Xenopus* oocytes. *Am. J. Physiol.* **257**: C825–C829.

Huang, M. and Chalfie, M. (1994) Gene interactions affecting mechanosensory transduction in *Caenorhabditis elegans*. *Nature* **367**: 467–470.

Hummler, E., Barker, P., Gatzy, J., Beermann, F., Verdumo, C., Schmidt, A., Boucher, R. and Rossier, B.C. (1996) Early death due to defective neonatal lung liquid clearance in alpha-ENaC-deficient mice. *Nature Genet.* **12**: 325–328.

Ishabashi, K. and Marumo, F., (1998) Molecular cloning of a DEG/ENaC sodium channel cDNA from human testis. *Biochem. Biophys. Res. Comm.* **245**: 589–593.

Iwasa, K.H., Mizuta, K., Lim, D.J., Benos, D.J. and Tachibana, M. (1994) Amiloride-sensitive channels in marginal channels in the stria vascularis of the guinea pig cochlea. *Neurosci. Lett.* **172**: 163–166.

Jeunemaitre, X., Bassilana, F., Persu, A., Dumont, C., Champigny, G., Lazdunski, M., Corvol, P. and Barbry, P. (1997) Genotype-phenotype analysis of a newly discovered family with Liddle's syndrome. *J. Hypertens.* **15**: 1091–1100.

Jiang, X., Ingbar, D.H. and O'Grady, S.M. (1998) Adrenergic stimulation of Na^+ transport across alveolar epithelial cells involves activation of apical Cl-channels. *Am. J. Physiol.* **275**: C1610–1620.

Kaczorowski, G.J., Barros, F., Dethmers, J.K., Trumble, M.J. and Cragoe, E.J. (1985) Inhibition of Na/Ca exchange in pituitary plasma membrane vesicles by analogues of amiloride. *Biochemistry* **24**: 1394–1403.

Kemendy, A.E., Kleyman, T.R. and Eaton, D.C. (1992) Aldosterone alters the open probability of amiloride-blockable sodium channels in A6 epithelia. *Am. J. Physiol.* **263**: C825–C837.

Killick, R. and Richardson, G.P. (1997) Isolation of chicken alpha ENaC splice variants from a cochlear cDNA library. *Biochim. Biophys. Acta* **1350**: 33–37.

Kim, M., Yoo, O.J. and Choe, S. (1997) Molecular assembly of the extracellular domain of P2X2: an ATP-gated ion channel. *Biochem. Biophys. Res. Commun.* **240**: 618–622.

Kizer, N., Guo, X.L. and Hruska, K. (1997) Reconstitution of stretch-activated cation channels by expression of the alpha-subunit of the epithelial sodium channel cloned from osteoblasts. *Proc. Natl Acad. Sci. USA* **94**: 1013–1018.

Koefoed-Johnsen V, U.H. (1958) The nature of the frog skin potential. *Acta Physiol. Scand.* **42**: 298–308.

Komwatana, P., Dinudom, A., Young, J.A. and Cook, D.I. (1996) Cytosolic Na^+ controls and epithelial Na^+ channel via the Go guanine nucleotide-binding regulatory protein. *Proc. Natl Acad. Sci. USA* **93**: 8107–8111.

Kosari, F., Sheng, S., Li, J., Mak, D.O., Foskett, J.K. and Kleyman, T.R. (1998) Subunit stoichiometry of the epithelial sodium channel. *J. Biol. Chem.* **273**: 13469–13474.

Krattenmacher, R., Fischer, H., Van Driessche, W. and Clauss, W. (1988) Noise analysis of cAMP-stimulated Na current in frog colon. *Pflügers Arch.* **412**: 568–573.

Kretz, O., Barbry, P., Bock, R. and Lindemann, B. (1999) Differential expression of RNA and protein of the three pore-forming subunits of the amiloride-sensitive epithelial sodium channel in taste buds of the rat. *J. Histochem. Cytochem.* **47**: 51–64.

Kroll, B., Bautsch, W., Bremer, S., Wilke, M., Tümmler, B. and Frömter, E. (1989) Expression of Na^+ channel from mRNA respiratory epithelium in *Xenopus* oocytes. *Am. J. Physiol.* **257**: L284–L288.

Kunzelmann, K., Kiser, G.L., Schreiber, R. and Riordan, J.R. (1997) Inhibition of epithelial Na^+ currents by intracellular domains of the cystic fibrosis transmembrane conductance regulator. *FEBS Lett.* **400**: 341–344.

Lai, C.C., Hong, K., Kinnell, M., Chalfie, M. and Driscoll, M. (1996) Sequence and transmembrane topology of MEC-4: an ion channel subunit required for mechanotransduction in *Caenorhabditis elegans*. *J. Cell Biol.* **133**: 1071–1081.

Lester, D.S., Asher, C. and Garty, G. (1988) Characterization of cAMP-induced activation of epithelial sodium channels. *Am. J. Physiol.* **254**: C802–C808.

Letz, B., Ackermann, A., Canessa, C.M., Rossier, B.C. and Korbmacher, C. (1995) Amiloride-sensitive sodium channels in confluent M-1 mouse cortical collecting duct cells. *J. Membrane Biol.* **148**: 127–141.

Lewis, S.A. and Alles, W.P. (1986) Urinary kallikrein: a physiological regulator of epithelial Na^+ absorption. *Proc. Natl Acad. Sci. USA* **83**: 5345–5348.

Li, J.H.-Y. and Lindemann, B. (1982) Movement of Na and Li across the apical membrane of frog skin. In *Basic Mechanisms in the Action of Lithium* (eds H.M. Emrich, J.B. Aldenhoff and L.D. Lux). Excerpta Med., Amsterdam, pp. 28–35.

Lichtenberg, D., Robson, R.J. and Dennis, E.A. (1992) Solubilization of phospholipids by detergents. Structural and kinetic aspects. *Biochim. Biophys. Acta* **737**: 285–304.

Liddle, G.W., Bledsoe, T. and Coppage, W.S. (1963) A familial renal disorder simulating primary aldosteronism but with negligible aldosterone secretion. *Trans. Assoc. Am. Physicians* **76**: 199–213.

Lindemann, B. and Van Driessche, W. (1977) Sodium-specific membrane channels of frog skin are pores: current fluctuations reveal high turnover. *Science (Wash DC)* **195**: 292–294.

Ling, B.N., Zuckerman, J.B., Lin, C., *et al.* (1997) Expression of the cystic fibrosis phenotype in a renal amphibian epithelial cell line. *J. Biol. Chem.* **272**: 594–600.

Lingueglia, E., Voilley, N., Waldmann, R., Lazdunski, M. and Barbry, P. (1993) Expression cloning of an epithelial amiloride-sensitive Na^+ channel. A new channel type with homologies to *Caenorhabditis elegans* degenerins. *FEBS Lett.* **318**: 95–99.

Lingueglia, E., Renard, S., Waldmann, R., Voilley, N., Champigny, G., Plass, H., Lazdunski, M. and Barbry, P. (1994) Different homologous subunits of the amiloride-sensitive Na^+ channel are differently regulated by aldosterone. *J. Biol. Chem.* **269**: 13736–13739.

Lingueglia, E., Champigny, G., Lazdunski, M. and Barbry, P. (1995) Cloning of the amiloride-sensitive FMRFamide peptide-gated sodium channel. *Nature* **378**: 730–733.

Lingueglia, E., de Weille, J.R., Bassilana, F., Heurteaux, C., Sakai, H., Waldmann, R. and Lazdunski, M. (1997) A modulatory subunit of acid sensing ion channels in brain and dorsal root ganglion cells. *J. Biol. Chem.* **272**: 29778–29783.

Liu, D.T., Tibbs, G.R. and Siegelbaum, S.A. (1996a) Subunit stoichiometry of cyclic nucleotide-gated channels and effects of subunit order on channel function. *Neuron* **16**: 983–990.

Liu, J., Schrank, B. and Waterston, R. (1996b) Interaction between a putative mechanosensory membrane channel and a collagen. *Science* **273**: 361–364.

Mall, M., Hipper, A., Greger, R. and Kunzelmann, K. (1996) Wild type but not deltaF508 CFTR inhibits Na^+ conductance when coexpressed in Xenopus oocytes. *FEBS Lett.* **381**: 47–52.

Marunaka, Y. and Eaton, D.C. (1991) Effects of vasopressin, adenosine 39-59-cyclic monophosphate and cholera toxin on single amiloride-blockable Na channels in renal cells. *Am. J. Physiol.* **260**: C1071–C1084.

Matalon, S., Benos, D.J. and Jackson, R.M. (1996) Biophysical and molecular properties of amiloride-inhibitable Na^+ channels in alveolar epithelial cells. *Am. J. Physiol.* **271**: L1–L22.

Matsushita, K., MacCray, P.B., Sigmund, R.D., Welsh, M.J. and Stokes, J.B. (1996) Localization of epithelial sodium channel subunit mRNAs in adult rat lung by *in situ* hybridization. *Am. J. Physiol.* **271**: L332–L339.

McDonald, F.J., Snyder, P.M., McCray, P.B., Jr. and Welsh, M.J. (1994) Cloning, expression, and tissue distribution of a human amiloride-sensitive Na^+ channel. *Am. J. Physiol.* **266**: L728–734.

McDonald, F.J., Price, M.P., Snyder, P.M. and Welsh, M.J. (1995) Cloning and expression of the beta- and gamma-subunits of the human epithelial sodium channel. *Am. J. Physiol.* **268**: C1157–1163.

McKinnon, R. (1991) Determination of the subunit stoichiometry of a voltage-activated potassium channel. *Nature*, **350**, 232–235.

Nicke, A., Baumert, H.G., Rettinger, J., Eichele, A., Lambrecht, G., Mutschler, E. and Schmalzing, G. (1998) P2X1 and P2X3 receptors form stable trimers: a novel structural motif of ligand-gated ion channels. *EMBO J.* **17**: 3016–3028.

Noda, M., Ikeda, T., Suzuki, H., Takeshima, H., Takahashi, T., Kuno, M. and Numa, S. (1986) Expression of functional sodium channels from cloned cDNA. *Nature* **322**: 826–828.

O'Brodovich, H. (1991) Epithelial ion transport in the fetal and perinatal lung. *Am. J. Physiol.* **261**: C555–C564.

O'Brodovich, H., Canessa, C., Ueda, J., Rafii, B., Rossier, B.C. and Edelson, J. (1993) Expression of the epithelial Na^+ channel in the fetal rat lung. *Am. J. Physiol.* **265**: C491–C496.

Pácha, J., Frindt, G., Antonian, L., Silver, R.B. and Palmer, L.G. (1993) Regulation of Na channels of the rat cortical collecting tubule by aldosterone. *J. Gen. Physiol.* **102**: 25–42.

Palmer, L.G. (1992) Epithelial Na channels : function and diversity. *Annu. Rev. Physiol.* **54**: 51–66.

Palmer, L.G. and Frindt, G. (1986) Amiloride-sensitive Na channels from the apical membrane of the rat cortical collecting tubule. *Proc. Natl Acad. Sci. USA* **83**: 2767–2770.

Palmer, L.G., Li, J.H.-Y., Lindemann, B. and Edelman, I.S. (1982) Aldosterone control of the density of sodium channels in the toad urinary bladder. *J. Membrane Biol.* **64**: 91–102.

Persu, A., Barbry, P., Bassilana, F., Houot, A.M., Mengual, R., Lazdunski, M., Corvol, P. and Jeunemaitre, X. (1998) Genetic analysis of the beta subunit of the epithelial Na^+ channel in essential hypertension. *Hypertension* **32**: 129–137.

Petzel, D., Ganz, M.B., Nestler, E.J., Lewis, J.J., Goldenring, J., Akcicek, F. and Hayslett, J.P. (1992) Correlates of aldosterone-induced increases in Ca_i^{2+} and Isc suggest that Ca_i2^+ is the second messenger for stimulation of apical membrane conductance. *J. Clin. Invest.* **89**: 150–156.

Price, M.P., Snyder, P.M. and Welsh, M.J. (1996) Cloning and expression of a novel brain Na^+ channel. *J. Biol. Chem.* **271**: 7879–7882.

Puoti, A., May, A., Canessa, C.M., Horisberger, J.D., Schild, L. and Rossier, B.C. (1995) The highly selective low-conductance epithelial Na^+ channel of *Xenopus laevis* A6 kidney cells. *Am. J. Physiol.* **269**: C188–C197.

Quinton, P.M. (1990) Cystic fibrosis : a disease of electrolyte transport. *FASEB J.* **4**: 2709–2717.

Reif, M.C., Troutman, S.L. and Schafer, J.A. (1986) Sodium transport by rat cortical collecting tubule. Effects of vasopressin and desoxycorticosterone. *J. Clin. Invest.* **77**: 1291–1298.

Renard, S., Lingueglia, E., Voilley, N., Lazdunski, M. and Barbry, P. (1994) Biochemical analysis of the membrane topology of the amiloride-sensitive Na^+ channel. *J. Biol. Chem.* **269**: 12981–12986.

Renard, S., Voilley, N., Bassilana, F., Lazdunski, M. and Barbry, P. (1995) Localization and regulation by steroids of the α, β and γ subunits of the amiloride-sensitive Na^+ channel in colon, lung and kidney. *Pflüg. Arch. Eur. J. Physiol.* **430**: 299–307.

Rossier, B.C. and Palmer, L.G. (1992) Mechanism of aldosterone action on sodium and potassium transport. In *The Kidney: Physiology and Pathophysiology* (eds D.W. Seldin and G. Gubisch). Raven Press, New York, pp. 1373–1409.

Rotin, D., Bar-Sagi, D., O'Brodovich, H., Merilainen, J., Lehto, V.P., Canessa, C.M., Rossier, B.C. and Downey, G.P. (1994) An SH3 binding region in the epithelial Na^+ channel (∞ENaC) mediates its localization at the apical membrane. *EMBO J.* **13**: 4440–4450.

Sandle, G.I. and Binder, H.J. (1987) Corticosteroids and intestinal ion transport. *Gastroenterology* **93**: 188–196.

Sariban-Sohraby, S., Burg, M., Wiesmann, W.P., Chiang, P.K. and Johnson, J.P. (1984) Methylation increases sodium transport into A6 apical membrane vesicles: possible mode of aldosterone action. *Science* **225**: 745–746.

Schild, L., Lu, Y., Gautschi, I., Schneeberger, E., Lifton, R.P. and Rossier, B.C. (1996) Identification of a PY motif in the epithelial Na channel subunits as a target sequence for mutations causing channel activation found in Liddle syndrome. *EMBO J.* **15**: 2381–2387.

Schild, L., Schneeberger, E., Gautschi, I. and Firsov, D. (1997) Identification of amino acid residues in the alpha, beta, and gamma subunits of the epithelial sodium channel (ENaC) involved in amiloride block and ion permeation. *J. Gen. Physiol.* **109**: 15–26.

Schwiebert, E.M., Potter, E.D., Hwang, T.H., Woo, J.S., Ding, C., Qiu, W., Guggino, W.B., Levine, M.A. and Guggino, S.E. (1997) cGMP stimulates sodium and chloride currents in rat tracheal airway epithelia. *Am J Physiol* **272**: C911–922.

Shimkets, R.A., Warnock, D.G., Bositis, C.M., *et al.* (1994) Liddle's syndrome : heritable human hypertension caused by mutations in the β subunit of the epithelial sodium channel. *Cell* **79**: 407–414.

Shimkets, R.A., Lifton, R.P. and Canessa, C.M. (1997) The activity of the epithelial sodium channel is regulated by clathrin-mediated endocytosis. *J. Biol. Chem.* **272**: 25537–25541.

Shimkets, R.A., Lifton, R. and Canessa, C.M. (1998) In vivo phosphorylation of the epithelial sodium channel. *Proc. Natl Acad. Sci. USA* **95**: 3301–3305.

Snyder, P.M., McDonald, F.J., Stokes, J.B. and Welsh, M.J. (1994) Membrane topology of the amiloride-sensitive epithelial sodium channel. *J. Biol. Chem.* **269**: 24379–24383.

Snyder, P.M., Price, M.P., McDonald, F.J., Adams, C.M., Volk, K.A., Zeiher, B.G., Stokes, J.B. and Welsh, M.J. (1995) Mechanism by which Liddle's syndrome mutations increase activity of a human epithelial Na^+ channel. *Cell* **83**: 969–978.

Snyder, P.M., Cheng, C., Prince, L.S., Rogers, J.C. and Welsh, M.J. (1998) Electrophysiological and biochemical evidence that DEG/ENaC cation channels are composed of nine subunits. *J. Biol. Chem.* **273**: 681–684.

Staub, O., Dho, S., Henry, P., Correa, J., Ishikawa, T., McGlade, J. and Rotin, D. (1996) WW domains of Nedd4 bind to the proline-rich PY motifs in the epithelial Na^+ channel deleted in Liddle's syndrome. *EMBO J.* **15**: 2371–2380.

Strang, L.B. (1991) Fetal lung liquid : secretion and reabsorption. *Physiol. Rev.* **71**: 991–1133.

Strautnieks, S.S., Thompson, R.J., Gardiner, R.M. and Chung, E. (1996) A novel splice-site mutation in the gamma subunit of the epithelial sodium channel gene in three pseudohypoaldosteronism type 1 families. *Nature Genet.* **13**: 248–250.

Stutts, M.J., Canessa, C.M., Olsen, J.C., Hamrick, M., Cohn, J.A., Rossier, B.C. and Boucher, R.C. (1995) CFTR as a cAMP-dependent regulator of sodium channels [see comments]. *Science* **269**: 847–850.

Tamura, H., Schild, L., Enomoto, N., Matsui, N., Marumo, F. and Rossier, B.C. (1996) Liddle disease caused by a missense mutation of beta subunit of the epithelial sodium channel gene. *J. Clin. Invest.* **97**: 1780–1784.

Tanabe, T., Takeshima, H., Mikami, A., Flockerzi, V., Takahashi, H., Kangawa, K., Kojima, M., Matsuo, H., Hirose, T. and Numa, S. (1987) Primary structure of the receptor for calcium channel blockers from skeletal muscle. *Nature* **328**: 313–318.

Tchepichev, S., Ueda, J., Canessa, C., Rossier, B.C. and O'Brodovich, H. (1995) Lung epithelial Na channel subunits are differentially regulated during development and by steroids. *Am. J. Physiol.* **269**: C805–C812.

Turnamian, S.G. and Binder, H.J. (1989) Regulation of active sodium and potassium transport in the distal colon of the rat. *J. Clin. Invest.* **84**: 1924–1929.

Voilley, N., Lingueglia, E., Champigny, G., Mattéi, M.-G., Waldmann, R., Lazdunski, M. and Barbry, P. (1994) The lung amiloride-sensitive Na^+ channel: biophysical properties, pharmacology, ontogenesis, and molecular cloning. *Proc. Natl Acad. Sci. USA* **91**: 247–251.

Voilley, N., Bassilana, F., Mignon, C., Merscher, S., Mattéi, M.-G., Carle, G.F., Lazdunski, M. and Barbry, P. (1995) Cloning, chromosomal localization and physical linkage of the β and γ subunits of the human epithelial amiloride-sensitive sodium channel. *Genomics* **28**: 560–565.

Voilley, N., Galibert, A., Bassilana, F., Renard, S., Lingueglia, E., Le Néchet, S., Champigny, G., Hoffman, P., Lazdunski, M. and Barbry, P. (1997) The amiloride-sensitive Na^+ channel : from primary structure to function. *Comp. Biochem. Physiol.* **118**: 193–200.

Volk, K.A., Sigmund, R.D., Snyder, P.M., McDonald, F.J., Welsh, M.J. and Strokes, J.B. (1995) rENaC is the predominant Na^+ channel in the apical membrane of the rat renal inner medullary connecting duct. *J. Clin. Invest.* **96**: 2748–2757.

Wakabayashi, S., Shigekawa, M. and Pouyssegur, J. (1997) Molecular physiology of vertebrate Na^+/H^+ exchangers. *Physiol. Rev.* **77**: 51–74.

Waldmann, R., Champigny, G., Bassilana, F., Voilley, N. and Lazdunski, M. (1995a) Molecular cloning and functional expression of a novel amiloride-sensitive Na^+ channel. *J. Biol. Chem.*, **270**: 27411–27414.

Waldmann, R., Champigny, G. and Lazdunski, M. (1995b) Functional degenerin-containing chimeras identify residues essential for amiloride-sensitive Na^+ channel function. *J. Biol. Chem.* **270**: 11735–11737.

Waldmann, R., Bassilana, F., Voilley, N., Lazdunski, M. and M.-G. Mattéi, M.-G. (1996a) Assignment of the human amiloride sensitive Na^+ channel d isoform to chromosome 1p36.3 – 1p36.2. *Genomics* **34**: 262–263.

Waldmann, R., Champigny, G., Voilley, N., Lauritzen, I. and Lazdunski, M. (1996b) The mammalian degenerin MDEG, an amiloride-sensitive cation channel activated by mutations causing neurodegeneration in *C. elegans*. *J. Biol. Chem.* **271**: 10433–10436.

Waldmann, R., Bassilana, F., De Weille, J., Champigny, G., Heurteaux, C. and Lazdunski, M. (1997) Molecular cloning of a non-inactivating proton-gated Na^+ channel specific for sensory neurons. *J. Biol. Chem.* **272**: 20975–20978.

Waldmann, R., Champigny, G., Bassilana, F., Heurteaux, C. and Lazdunski, M. (1997) A proton-gated cation channel involved in acid sensing. *Nature* **386**: 173–177.

Weber, W.M., Blank, U. and Clauss, W. (1995) Regulation of electrogenic Na^+ transport across leech skin. *Am. J. Physiol.* **268**: R605-R613.

Willmann, J.K., Bleich, M., Rizzo, M., Schmidt-Hieber, M., Ullrich, K.J. and Greger, R. (1997) Amiloride-inhibitable Na^+ conductance in rat proximal tubule. *Pflugers Archiv. Eur. J. Physiol.* **434**: 173–178.

Yang, J., Jan, Y.N. and Jan, L.Y. (1995) Determination of the subunit stoichiometry of an inwardly rectifying potassium channel. *Neuron* **15**: 1441–1447.

Zeiske, W., Wills, N.K. and Van Driessche, W. (1982) Sodium channels and amiloride-induced noise in the mammalian colon epithelium. *Biochim. Biophys. Acta* **688**: 201–210.

9

The study of candidate genes in hypertension using tools of molecular biology and genetic epidemiology

Florent Soubrier, Alain Bonnardeaux, Toru Nabika,
Xavier Jeunemaitre, Pierre Corvol and François Cambien

1. Introduction

A considerable number of physiological and pharmacological studies have been performed to elucidate the pathophysiology of hypertension, and have highlighted the prominent role of angiotensin II, endothelin I, and nitric oxide for controlling blood pressure. The knowledge gained from these studies has permitted the design of efficient treatments of the disease. However, the basic mechanisms of hypertension have not been elucidated. Molecular genetics and genetic epidemiology have opened a new field for investigating the etiology of hypertension.

The aim of this chapter is to evaluate the contribution of classical physiology to the genetic epidemiology of hypertension. We also aim to review the data obtained with candidate genes, and the potential interest of candidate genes and loci which have not yet been studied in essential hypertension. These new candidate genes, or loci, come from the study of monogenic forms of hypertension, rat hereditary hypertension, and from the unexpected phenotype of mice with targeted gene disruption. The mechanisms by which these genes might be involved in essential hypertension will be considered.

2. The physiological and pathophysiological approach to hypertension

When hypertension was a major cause of early cardiac and renal failure, stroke and

Molecular Genetics of Hypertension, edited by A.F. Dominiczak, J.M.C. Connell and F. Soubrier.

vascular diseases, it was tempting to consider hypertension as a specific entity. The historical debate between Platt and Pickering ended with the widely accepted notion that blood pressure has a unimodal distribution (Ward, 1995).

Based on physiological investigations of hypertensives compared to control subjects, several authors have proposed basic mechanisms that could explain hypertension. Some of the theories, proposed to explain hypertension, supposed that common pathophysiological mechanisms were present in all forms of hypertension, whereas others hypothesized the existence of subgroups of hypertensives covering definite entities, with specific underlying mechanisms. Although secondary forms of hypertension and monogenic forms of hypertension represent examples of these entities, they are distinct and their phenotypes are very specific, as compared to essential hypertension.

Unifying theories of hypertension have attributed the responsibility of high blood pressure to one major system of blood pressure regulation.

According to the theory of Guyton, hypertension is due to a failure of the kidney to eliminate enough salt and water at normal pressure levels, and this abnormality precedes the increase in total peripheral resistance (Guyton, 1991).

Brenner and colleagues proposed that sodium sensitivity of blood pressure is based on a decrease in the whole kidney ultrafiltration coefficient and/or increase in tubular reabsorption (Kimura *et al.*, 1994). Therefore, sodium sensitivity and glomerular hypertension reflects adaptations necessary to overcome defective sodium excretion by the kidney. The reduced nephron number accompanying intrauterine growth retardation could explain the higher blood pressure in adult individuals born with a low body weight (Kimura *et al.*, 1994).

Laragh and his colleagues have emphasized the role of nephron heterogeneity (Sealey *et al.*, 1988). According to this theory, a minor population of nephrons is ischemic, with reduced perfusion pressure, reduced blood flow, and reduced glomerular filtration rate. These ischemic glomeruli release an abnormally high level of renin which generates an excess of angiotensin II, to which normal nephrons are exposed. The consequence is an enhanced tubular reabsorption of sodium and an increased tubulo-glomerular feedback-mediated afferent vasoconstriction. This hypothesis is based on morphological studies of kidneys from hypertensives, as well as biochemical and pharmacological studies of the renin–angiotensin system.

The concept of modulation and nonmodulation was proposed by Hollenberg and Williams to explain hypertension in a subgroup of hypertensives (Hollenberg and Williams, 1995; Lifton *et al.*, 1989). The term modulation reflects the vascular and adrenal responses to angiotensin II, which differ as a consequence of the opposite changes in the number of angiotensin II binding sites in both tissues. With a low salt diet, angiotensin II generation increases, ensuring the biosynthesis of the salt-preserving hormone aldosterone in the adrenals owing to an increased number of receptors. In arteries, desensitization to elevated angiotensin II results in a decreased number of binding sites. According to the nonmodulator concept, a group of hypertensives, with high or normal renin, are characterized by an absence of the normal shift of their renal and adrenal responses to angiotensin II in response to changes in sodium intake. Their adrenal response to angiotensin II is blunted, and their renal vascular response is increased and set to the level that would normally correspond to a low sodium intake. The identification of patients with modulation and nonmodulation is based on a physiological investigation under a low salt diet.

Some theories have emphasized the role of ion transport and ion excretion, since both an increased concentration of sodium in vascular smooth muscle cells might increase their tone and sodium retention in the kidney might be responsible for hypertension (Hilton, 1986). Functional tests have been designed in order to detect abnormalities of ion transport in accessible cells such as red or white blood cells and have been performed in hypertensive and normotensive subjects. An increased activity of Na–lithium countertransport (SLC) is present in a large proportion of hypertensives (Canessa *et al.*, 1980). Whether SLC reflects the activity of the ubiquitous Na^+–H^+ antiport is not established. Several studies suggest that an altered antiport activity might be a causal factor in hypertension (Rosskopf *et al.*, 1993). Salt sensitivity, a related phenotype characterized by an increase in blood pressure in response to salt loading, was found to be more frequent in hypertensives than in normotensives (Luft *et al.*, 1988).

The kidney has been designated most often as the culprit organ in hypertension. This is based on basic physiology of blood pressure and body fluid homeostasis but also on experimental data. Among these data, cross transplantation of kidneys from genetically hypertensive rats to normotensive rats, and vice versa, is known to relieve or induce hypertension depending on the origin of the kidney, and this is a good argument for a prominent role of the kidney (Ferrari and Bianchi, 1995).

Other theories on hypertension have focused on the role of the central nervous system, or on the role of the efferent nerves of the renal sympathetic nervous system (Ferrari and Bianchi, 1995), of insulin resistance through increased sodium reabsorption (de Fronzo, 1981; Weidman and Ferrari, 1991).

The general theories of hypertension have major drawbacks. They try to explain all forms of essential hypertension by an abnormality of a single function of the organism. Since genetic epidemiology data support the role of more than one gene, this would imply that the various genes involved would result in an abnormality of a single function. Physiological abnormalities observed in hypertension can also be attributed to common consequences of various mechanisms of hypertension, all concurring to produce renal vascular damage.

However, this classical view of hypertension has the merit of designating potential candidate genes to be studied by molecular genetics and genetic epidemiology and to supply the basis for studying genotype–phenotype relationships.

3. The genetic epidemiology approach to hypertension

Another approach to hypertension is to consider blood pressure as a quantitative trait and hypertension as the top of the distribution of this trait in the population. The classical biometric approach was followed by the development of segregation analyses, which allowed the so-called genetic and environmental components of blood pressure to be estimated by statistical methods applied to phenotypic data.

Various types of biometric studies have corroborated the role of genetic determinants of blood pressure. Such studies include analysis of family correlations, adopted children studies, comparisons between mono- and dizygotic twins. These studies concluded that approximately 30% of blood pressure variance is genetically determined (Ward, 1995). More recently, segregation analysis, considering blood pressure as a quantitative trait, was used to model the mode of transmission of the trait, and to assess the respective roles of genetic determinant and environment in blood pressure variance. Different studies suggested the existence of a major gene effect, that is, the effect of a single gene on blood pressure parameters. A study performed in French-Canadians detected a major gene effect on systolic blood pressure (SBP) in parents but the effect was not present in offspring, suggesting a difference in the penetrance (probability of developing the disease in the presence of the genetic factor) of the gene according to age (Rice *et al.*, 1990). The existence of a major gene influencing SBP, with a penetrance varying with age and sex was also found in another study (Perusse *et al.*, 1991). More recently, another study suggested the existence of a recessive gene with a frequency of 0.23 acting on diastolic blood pressure (DBP) change over a 7.2-year period of follow-up (Cheng *et al.*, 1995). The genetic effect on DBP increase was age-dependent since the genotype effect decreased with age (Cheng *et al.*, 1995). In contrast, Morton *et al.* did not find any evidence for a major gene effect on SBP and DBP (Morton *et al.*, 1980).

Hypertension is most commonly considered to be due to several genes, although the number of genes and the effect of each of these genes is unknown (discussed in Chapter 1). According to genetic epidemiology data, hypertension would result from the polymorphic variation of gene sequences at more than one locus. In order to reconcile the frequency of hypertension with the requirement for more than one predisposing gene, the existence of different genotype combinations should be postulated, from individual to individual. This explains the genetic heterogeneity of hypertension, and the difficulty in identifying hypertension genes.

4. Molecular genetics of candidate genes

The evaluation of the role of candidate genes in hypertension by genetic epidemiology methods has required the development of various molecular tools which were developed on a limited number of these genes.

The first polymorphisms to be used were single nucleotide polymorphisms (SNP) detected by restriction fragment length analysis of genomic DNA in southern blot experiments. DNA amplification by the polymerase chain reaction (PCR) elicited the isolation of a large number of SNPs by various techniques of mutation detection described below, and in several candidate genes. The low informativity of these biallelic markers (as defined by the proportion of subjects heterozygous for the marker) can be overcome by combining different SNPs in haplotypes which can be deduced from parental genotypes or estimated using statistical methods (since parental genotypes are usually not available for hypertensive patients, actual haplotypes are not determined; Soubrier *et al.*, 1990).

SNPs are currently detected by methods which use DNA amplification as a first step, allowing a direct access to the individual genomic sequence. These methods are based on several physicochemical properties of DNA, and the most commonly used are the single strand conformation analysis (SSCA; Orita *et al.*, 1989) and denaturing gradient gel electrophoresis (DGGE; Cotton, 1993). The first method (SSCA) is based on the difference of conformation that a single nucleotide change can impart to single stranded DNA in nondenaturing conditions of electrophoresis. This method allows the detection of single strand conformation polymorphism (SSCP; Orita *et al.*, 1989).

The DGGE method takes advantage of different melting conditions of the double-stranded fragments of DNA, depending on its base composition or on the presence of heteroduplex DNA, due to the presence, in the latter case, of noncomplementary nucleotides on the two strands at one or more positions (Lerman *et al.*, 1986).

The chemical cleavage method is certainly the most reliable but requires the use of toxic chemicals (Cotton, 1993). The method was adapted to fluorescently labeled DNA and automated sequencers (Verpy *et al.*, 1994).

Probably one of the most promising method for mutation detection, enabling high-throughput analysis, is the denaturing high pressure liquid chromatography (HPLC; Huber and Berti, 1996). This technique allows the detection of heteroduplexes with a high sensitivity (>95%) in large fragments of DNA (around 800 bp). In view of the large throughput which can be reached for mutation detection with denaturing HPLC, it is expected that several hundred candidate genes will be investigated for detecting SNPs in the near future.

The interest in the detection of SNPs, usually biallelic, by exhaustive investigation of the candidate genes in a series of selected patients and controls, is multiple. Firstly, SNPs may represent candidate functional variants, because they cause a change in the coding sequence (however, neutral mutations can sometimes modify gene function if they create a splicing site inside an exon), or are located within regulatory sequences of the gene. Secondly, they may represent neutral markers which could be in linkage disequilibrium with putative undiscovered functional polymorphisms of the gene, lying in remote and unexplored regulatory regions. Therefore, they are the most suitable markers to be used in linkage disequilibrium studies.

Additional methods enable the genotyping of known polymorphisms in a large series of patients. These methods also use a first step PCR amplification of genomic DNA followed by hybridization of the PCR product with allele-specific oligonucleotides (ASO). Several other methods exist, which can be used for the detection and genotyping of these markers. The highly polymorphic markers used for family studies consist of multiallelic tandem repeat of elementary motifs. The most widely and evenly distributed markers are the n(dCdA) dinucleotide repeat, or n(dGdT) on the complementary strand. For the specific purpose of investigating the role of candidate genes, specific microsatellite markers were identified on, or close to, these genes. This is usually performed by cloning large genomic fragments around the gene to be studied, either from lambda phages or cosmids, and more recently from new vectors such as bacterial artificial chromosomes (BAC; Shizuya *et al.*, 1992) and P1 artificial chromosomes (PAC; Ioannou *et al.*, 1994), which represent more adapted tools since their large inserts (around 120–150 kb) enable more microsatellites to be identified and used.

The need for specific development of highly polymorphic markers will be circumvented progressively by the availability of a dense physical and genetic map of the genome (Deloukas *et al.*, 1998). Since the accurate physical position of candidate genes can be obtained electronically for several of them, and given the availability of more than 5000 mapped microsatellite markers, the need for specifically developed markers will disappear.

5. Study design

5.1 Linkage studies

Since hypertension is a multifactorial disease with non-Mendelian inheritance, linkage analysis using parametric methods in families (LOD-score method, logarithm of the odds ratio) is less appropriate than nonparametric approaches.

The affected sib-pair method allows linkage to be studied between a marker and the disease and offers a number of advantages for the study of candidates genes in hypertension (Suarez *et al.*, 1978). It is a robust method which does not assume any defined genetic model, specifying the mode of transmission of the gene effect, its penetrance, and the gene frequency. Families can be additively studied and the method accommodates heterogeneity, even if this last factor decreases the power of the study. This method is based on the comparison of the observed concordance for the marker alleles in affected sibs and the expected concordance, under the null hypothesis of absence of linkage between the marker and the disease locus. Indeed, if the disease locus is located close to the marker locus, the allele sharing will be increased and will exceed the average identity by descent between sibs of 50%. The affected sib-pair method is more powerful and reliable when parental genotypes are determined. In the absence of parental genotypes, or if the marker is not fully informative, the expected concordance relies on the marker allele frequencies in the population studied. Highly informative markers are required for these studies, such as microsatellites (di or trinucleotide repeats).

The sib-pair method is also appropriate for quantitative traits. In this case, squared intrapair differences for quantitative phenotype values are regressed on the intrapair genotype concordance. If significantly lower intrapair differences are observed in pairs with concordant alleles than in those with discordant alleles, the marker is potentially linked to the locus determining the trait (Amos *et al.*, 1989). Using this method, Wilson *et al.* found a linkage of DBP to a locus on chromosome 1 (Wilson *et al.*, 1991).

Parametric linkage analysis by LOD-score have been also used in hypertension by Julier *et al.* (1997). They obtained similar results with this method to those obtained with the nonparametric sib-pair method in the analysis of the ACE locus on chromosome 17.

5.2 Linkage disequilibrium studies

The classical association study, or case–control study, is widely used since it offers sensitivity and apparent simplicity. The allele or genotype frequencies for a marker are compared in a series of cases, for instance hypertensive patients, and a series of control normotensive subjects, matched for all important parameters including age and ethnicity. One of the major drawbacks of association study is the possibility of spurious association due to unrecognized stratification of the populations studied. Ethnic origin is one of the main causes of problems encountered since the allele frequency can vary widely between ethnic groups. The interethnical difference for allele frequency is particularly pronounced for the angiotensinogen (AGT) polymorphism (M235T), for which the frequency of the 235T allele is far more frequent in Blacks than in Caucasians (Bloem *et al.*, 1995).

The hypothesis underlying this kind of study is the presence of a linkage disequilibrium between the marker allele and the disease allele. The presence of a linkage disequilibrium supposes that the distance between the marker and the disease locus is very small (a few hundred kilobase pairs, and probably less), depending on the number of generations since the occurrence of the founder effect. In addition, two markers, even at short distance, are not necessarily in linkage disequilibrium, and this implies that several markers should be studied in order to investigate a locus through an association study. Increased informativity is another reason for studying several markers. Association studies are only applicable to candidate genes, or a series of selected genes inside a genome region. A genome-screen strategy cannot be envisioned unless it is performed on young isolated populations (in which linkage disequilibrium is conserved over large regions) or if significant progress is made in DNA technology, for example, the development of DNA-chip technology (Hacia *et al.*, 1996) and the identification and mapping of thousands of SNPs (Wang *et al.*, 1998).

More recently, association studies were designed in which fictitious controls were constituted by the parental pool of nontransmitted alleles (Falk and Rubinstein, 1987). In the transmission disequilibrium test, the preferential transmission of an allele from parents, heterozygous for the marker, is tested (Spielman *et al.*, 1993). These methods avoid, in theory, the occurrence of spurious association due to incorrect match between cases and controls, that may, for example, result from unrecognized population stratification.

The interpretation of candidate gene studies should be careful since the candidate is not necessarily the gene responsible for the disease linkage and the use of markers surrounding the candidate genes may help interpret the data.

Linkage or association data with candidate genes can be advantageously supported by biological phenotypes, specific for the gene, called intermediate phenotypes. These phenotypes can reflect gene expression (in the wide sense), can be associated with gene polymorphisms, and be related to high blood pressure through well-characterized biochemical mechanisms. These genotype–phenotype relationships allow defined mechanisms to be proposed in order to support the hypothesis of the involvement of the gene. The case of AGT is probably the best characterized example (see Chapter 10).

6. The study of candidate genes

6.1 The renin–angiotensin system genes

The most important genes coding for the major proteins of the renin–angiotensin systems were studied to establish their role in hypertension. The angiotensinogen gene is reviewed in detail in another chapter (Chapter 10).

The renin gene. Since renin is the limiting enzyme of the renin–angiotensin system, the renin gene was one of the first to be investigated, in rats as well as in humans. Several restriction fragment length polymorphisms (RFLPs) were initially identified on the renin gene and were used in association studies (Morris and Griffiths, 1988; Soubrier *et al.*, 1990). These association studies belong to the first generation, since they include a limited number of patients and controls. However, they consistently provided negative results, which can be interpreted as a lack of power or as true negative results.

Two family studies were performed with renin gene markers. The first study included large pedigrees with hypertensive and normotensive subjects. A LOD-score was calculated with renin gene markers, and was negative, as expected from the family structure and the type of the disease (Naftilan *et al.*, 1989). An affected sib-pair study was performed in French families, which used a combination of RFLPs, grouped in haplotypes (Jeunemaitre *et al.*, 1992b). The haplotype frequencies were estimated by a maximum likelihood method, since parental genotypes were not known. No excess of concordance for estimated renin haplotypes was observed in affected hypertensive sibs.

A linkage of hypertension with the renin gene cannot be definitively ruled out since the renin gene has not been investigated with highly informative markers in large studies. An extensive search for polymorphisms in the coding region in hypertensive patients was negative by SSCA (X. Jeunemaitre, personal communication), suggesting that, if a common variant was to be found, it should lie in regulatory regions of the gene, which were not well defined until recently. A potent enhancer located ~6 kb upstream the transcription start of the renin gene was identified by transfection experiments, and is responsible for transcription activation of the renin gene in chorionic cells, and probably in other cell types (Germain *et al.*, 1998).

The angiotensin I converting enzyme gene. The ACE gene is interesting for several reasons. Firstly, ACE has a key role in angiotensin II generation and bradykinin degradation, and may thus markedly affect vascular wall reactivity and morphology, as well as kidney function. Its concentration is not rate-limiting in the circulation, but this is not known for tissue ACE. Secondly, the polymorphism of the ACE gene strongly modulates plasma ACE level. Therefore, any relationship found with the ACE gene marker would be readily attributable to the biological phenotype associated with this polymorphism. Thirdly, the ACE gene lies within the genomic region which is linked with hypertension in the SHRSP strain of rat. The homologous rat ACE gene lies on chromosome 10 within a conserved linkage group, located on chromosome 17 in humans. More recently, the mapping of the BP locus was refined and data obtained with congenic strains are compatible with the existence of two genes for hypertension located on chromosome 10 (Kreutz *et al.*, 1995).

Various types of studies related to hypertension were performed with ACE gene markers. Several association studies used the deletion/insertion (D/I) polymorphism as the marker genotype. Most of these studies were negative (Harrap *et al.*, 1993). One study was positive and an interesting but speculative hypothesis was proposed to explain an excess of II genotypes in older hypertensive patients

(Morris *et al.*, 1994). This hypothesis is based on the previously found association of the DD genotype with myocardial infarction and suggests that this association might cause an increased risk of death in subjects with the DD genotype and severe hypertension. This selection would result in a relative increase with age of hypertensives bearing the II genotype.

Another type of study was performed with the D/I polymorphism in normal nuclear families and no relationship with blood pressure was found (Tiret *et al.*, 1992). Using a highly informative marker on the GH gene, which is located in close proximity to the ACE gene, Jeunemaitre *et al.* did not find any evidence for linkage in hypertensive families from Utah (Jeunemaitre *et al.*, 1992a).

A linkage with the ACE locus with DBP or mean blood pressure was reported by two independent studies (Fornage *et al.*, 1998; O'Donnell *et al.*, 1998). In both studies, several hundred of families were collected, without selection for a particular level of blood pressure but representing large samples of population. The statistical methodology used to quantify, by a family approach, the effect of the ACE locus on blood pressure is slightly different in the two studies. The study by O'Donnell *et al.* (1998) used a classical approach, with the SIBPAL program, to test for linkage by relating the quantitative or qualitative trait difference to genotype resemblance in sib-pairs. Fornage *et al.* (1998) used a different methodology aimed at quantifying the part of the variance of the quantitative trait (blood pressure) which is determined by the marker locus. The markers used at the ACE locus were the same in the two studies, a highly polymorphic and complex tandem repeat (di- and tetranucleotide repeat) marker located within the growth hormone (*GH*) gene, for which no recombination is observed with the ACE gene in humans. The D/I polymorphism of the ACE gene was also used by O'Donnell *et al.* in the linkage and the association analyses.

The study of O'Donnell shows a linkage of the ACE locus markers in the whole panel of families (O'Donnell *et al.*, 1998). After subdivision according to sex, there was a marginally significant linkage with DBP in male (P=0.02 and P=0.04, for ACE and *hGH* respectively) but not in female sibling pairs. Surprisingly, the nominal P value for linkage was more significant with the less informative marker, in this case the D/I polymorphism, both in male-only and in sex-pooled analyses. A borderline P value (P=0.047) was obtained for linkage of the ACE locus to hypertension taken as a dichotomous status, and only in men.

Fornage *et al.* similarly found that the ACE locus is linked to DBP and mean blood pressure in adolescents, at mean age of 15 years (Fornage *et al.*, 1998). The analysis on the whole group gave results not very different for the ACE marker (P=0.04) from what was obtained with the AGT (P=0.06) or the angiotensin II-type I-receptor marker (P=0.10). The analyses of siblings having a family history of hypertension, and the analyses of male sibships led to more significant results, the variance explained by the ACE gene reaching 30% for DBP and mean blood pressure, with P values below 0.005.

Thus, results obtained with the ACE locus itself are controversial. However, the ACE gene is located in close proximity, but is distinct from another locus, located about 15 cM centromeric on chromosome 17, which is linked to monogenic forms of hypertension and to essential hypertension (see below).

Angiotensin type I receptor. Since it is a cellular component of the RAS, less accessible to quantitative measurements, few data are available concerning the biology of the AT1R in hypertension, that might support its implication as a candidate gene. Comparisons of angiotensin II effects in normotensives to those observed in hypertensives is an indirect way to analyze the receptor itself. Increased pressor response to angiotensin II infusion was observed in hypertensives (Meier *et al.*, 1981) and the angiotensin II type I receptor might be involved in the pathogenesis of hypertension in a subgroup of hypertensives (Lifton *et al.*, 1989).

There is little genetic epidemiology data available for the AT1R gene. We initially performed an extensive search for polymorphisms in a panel of 60 hypertensive subjects, and found no substitutive polymorphisms of the coding region (Bonnardeaux *et al.*, 1994). However, some neutral substitutions were found in the coding region, together with polymorphisms of the 3′ untranslated region. We found that the 1166C allele of the A1166C polymorphism was more frequent in hypertensives than in normotensives, and more precisely, this association was restricted to severe hypertensives (*Table 1*). No difference in allele frequency was found between the normotensive and the hypertensive group for the two other polymorphisms studied. An informative dinucleotide microsatellite marker was also used in a sib-pair study involving 267 pairs from 138 pedigrees, and no significant linkage was found.

This discrepancy between positive results found in association studies and negative results for the linkage study in hypertensive sibs can be explained in two different ways. Either the association found with the A1166C polymorphism is spurious, or the sib-pair study is not sensitive enough to detect the linkage, due to its lower power compared to an association study. The difference in the power of linkage versus linkage disequilibrium studies has been recently been emphasized (Risch and Merikangas, 1996). In the case of AT1R, a TDT and HRR analysis was performed with the AT1R microsatellite in the French panel of families studied initially (see above) and results were negative (A. Bonnardeaux, unpublished data). Another study performed in Australian subjects found a positive association with the A1166C polymorphism of AT1R (Wang *et al.*, 1997), but it was not the case for another study performed in German hypertensives (Schmidt *et al.*, 1997).

The AT1R polymorphism was also investigated in relation with the pulse wave velocity (PWV) in a group of mild and moderate hypertensives and a group of

Table 1. Genotype and allele frequencies of polymorphism A1166C in different subsets of hypertensives as compared to normotensive controls

Genotypes	Age of onset <40 years	Severe hypertension (two drugs or DBP >105 mmHg)	Normotensive controls
AA	35.2	38.3	51.3
AC	51.9	44.2	40.6
CC	13.0	17.5	8.1
Significance of the comparison with controls	$P<0.02$	$P<0.01$	

Modified from Bonnardeaux, A. *et al.* (1994) Angiotensin II type I receptor gene polymorphisms in human essential hypertension, *Hypertension*, vol. 24, pp. 63–69, with permission from Lippincott, Williams & Wilkins.

normotensive subjects. In hypertensives, an increased PWV was observed in subjects bearing the 1166CC genotype as compared to those bearing the A1166C and AA1166 genotype (Benetos *et al.*, 1996). These results thus suggest that this polymorphism is associated with increased arterial wall stiffness in hypertensives. The A1166C polymorphism was analyzed jointly with the ACE D/I polymorphism to assess their effect on the risk of myocardial infarction (MI). An interaction was found between the two genes, since the increased risk of MI associated with the DD genotype was restricted to the subjects bearing the 1166C allele, either in the heterozygous, or in the homozygous state (Tiret *et al.*, 1994).

As a consequence of the absence of biological intermediate phenotype associated with this polymorphism, it is difficult to propose and to test hypotheses for the involvement of the AT1R gene in hypertension. Since no alterations of the coding regions of the gene has been detected in hypertensives, polymorphisms of the AT1R gene should be located in regulatory regions of the gene, resulting in modification of the receptor number more likely than in receptor affinity. These alterations would result in an increased signal in vascular smooth muscle cells, and aldosterone-secreting cells in adrenals, leading to vasoconstriction and sodium retention. This hypothesis needs to be tested by measuring gene expression and receptor numbers in hypertensives according to the A1166C genotype, which is not easily feasible.

6.2 The endothelial nitric oxide (NO) synthase gene

The identification of NO as the endothelium-derived relaxing factor (EDRF), and the molecular cloning of three isoforms of NO synthase (NOS) were important steps for genetic studies of this vasodilator system. The endothelial isoform of NO synthase (eNOS), also called NOS III, represents a major candidate gene since this isoform is responsible for the constitutive generation of NO from arginine by endothelial cells, which in turn activates guanylate cyclase in vascular smooth muscle cells. Several lines of evidence suggest that blood pressure level results from an equilibrium between a vasoconstrictor tone (mainly the renin–angiotensin and endothelin systems), and a vasodilator tone, due to nitric oxide, prostacyclin and other less characterized factors, such as endothelium-derived hyperpolarizing factor (EDHF) which has been recently proposed to be potassium (Edwards *et al.*, 1998). The role of eNOS in basal vascular tone was established by both pharmacological and genetic experiments. Chronic and acute administration of NOS inhibitors increase blood pressure in rats, but these experiments do not permit the identification of the NOS isoform responsible for the vasodilator tone (Ruilope *et al.*, 1994). The involvement of the eNOS gene was demonstrated by basal blood pressure increase in mice homozygous for the knock-out of the eNOS gene, as compared to wild-type mice (Huang *et al.*, 1995a,b).

Some data suggest that the NO/cGMP vasodilator pathway might be impaired in hypertensive subjects, in particular the decrease in endothelium-dependent vasodilation by acetylcholine, an inducer of NO release by endothelial cells (Linder *et al.*, 1990; Panza *et al.*, 1990). In these studies, the NO formation step appeared to be implicated since the arterial response to the NO donor nitroprusside was normal.

A search for polymorphisms of the coding region, within major functional regions of the eNOS gene, was performed by SSCA in a series of hypertensives (Bonnardeaux *et al.*, 1994). No frequent polymorphism of the coding region was revealed by this approach, but neutral markers were identified in introns. They were subsequently used in a case–control study, which did not reveal any difference in allele frequency between hypertensives and normotensives.

We, and others, determined the structure of the human eNOS gene and identified a highly informative dinucleotide repeat inside intron 13 of the gene (Marsden *et al.*, 1993; Nadaud *et al.*, 1994). We used this marker in a sib-pair study. The 269 pairs gave a significant power to the study. No excess of concordance in the affected sibs was observed despite the high heterozygosity of the marker and the relatively large number of hypertensive sib-pairs analyzed.

A similar study was performed by Hunt *et al.* (1996) in 194 hypertensive sib-pairs from Utah. They did not observe any significant excess of allele sharing in these siblings.

These two studies do not support a prominent role of the eNOS gene in essential hypertension. An investigation of intermediate phenotypes related to the endothelial NO/cGMP pathway would be of interest. If a relation with the various polymorphisms of the eNOS gene was found, this would constitute an argument to design more specific studies aimed at finding subtle regulatory abnormalities of the gene in a subgroup of hypertensive patients.

6.3 The endothelin system genes

Endothelin-1 (ET-1) is a 21-amino acid peptide, which is cleaved from a larger precursor, 38 amino acids in length, by the endothelin converting enzyme (ECE). ET-1, which is the main endothelin secreted by the endothelium, binds to the ET_A receptor on vascular smooth muscle cells and provokes vasoconstriction. The genes coding for ET-1, the ET_A receptor and ECE are therefore potential candidate genes for hypertension. Other genes of the endothelin system such as the endothelin II and III and the type B receptor subtype, also represent possible candidate genes.

There is some pharmacological evidence suggesting that endothelin-I contributes to basal vascular tone, since an ECE antagonist is able to decrease the basal level of blood pressure (Haynes and Webb, 1994). A mixed ET_A and ET_B antagonist is also able to decrease blood pressure in humans (Schiffrin, 1998). Using *in situ* hybridization with an endothelin antisense probe in small arteries from a gluteal fat biopsy, Schiffrin *et al.* described an enhanced ET-1 mRNA in moderate and severe hypertensives, as compared to normotensive controls (Schiffrin *et al.*, 1997).

Surprisingly, few studies are available on the relationships between these genes and human hypertension. Polymorphisms of the ET-1 gene and the ET_A receptor gene were described and nonsignificant or marginally significant associations with hypertension were observed (Stevens and Brown, 1995). In a study of normotensive subjects, no relation was found between a *Taq* I polymorphism of the ET-1 gene and SBP or DBP (Berge and Berg, 1992). Other polymorphisms of the ET-1 gene are available on the web site CANVAS (http://ifr69.vjf.inserm.fr/~canvas/).

Additional studies are therefore required to evaluate the roles of these candidate genes. It is worth noting that the ECE gene is located on chromosome 1p36, on which a locus for systolic BP was found by sib-pair analysis with a protein marker (Wilson *et al.*, 1991).

6.4 Genes of the lipid and glucose metabolism

Several genes of the lipid and glucose metabolism were studied, mostly in case–control studies. The interest in these genes comes from the proposed common physiopathology for obesity, insulin resistance, atherosclerosis and hypertension. A common syndrome, supposedly linking these diseases, is called syndrome X, and it is therefore tempting to find the common genetic factor that might contribute to this syndrome. These studies were performed in particular groups of hypertensives, such as obese or diabetic hypertensives.

In obese hypertensives, Zee *et al.* found an increased frequency of one allele of the ApaLI RFLP of the low density lipoprotein receptor gene (Zee *et al.*, 1992). The same group also found a significant difference of allele frequency for an insulin receptor gene polymorphism in hypertensives (Ying *et al.*, 1991). A rare mutation of the glucagon receptor gene, the Gly40Ser substitution, was found more frequently in noninsulin-dependent diabetes mellitus (NIDDM), and the 40ser allele was shown, by *in vitro* studies, to have a decreased affinity for glucagon (Hansen *et al.*, 1996). This mutation was studied in hypertensives by Chambers and Morris who found the 40ser allele more frequently in hypertensives (3.8%) than in normotensives (1%) (Chambers and Morris, 1996; Morris *et al.*, 1997).

The *Xba* I RFLP of the glycogen synthase gene was studied in a group of hypertensives and a group of normotensives (Schalin-Jantti *et al.*, 1996). The A2 allele was found to be more frequent in hypertensives with a family history of NIDDM than in normotensive subjects without a family history of NIDDM. This allele was also associated with a decreased rate of insulin-stimulated glucose storage in hypertensives.

The lipoprotein lipase locus was studied in normotensive sibs of Chinese origin, using three different highly polymorphic markers (Wu *et al.*, 1996). The sib-pair linkage approach was applied to SBP and DBP, analyzed as quantitative traits. The basis of the analysis is to compare the variation of the trait between siblings as a function of the markers shared identical-by-descent (IBD). Variables were adjusted for age, sex and body mass index (BMI) by linear regression analysis. The significance of the regression of the squared variable difference versus the number of alleles shared IBD among sib-pairs was tested by Student's *t* statistics. Significant regression was found for SBP with three markers centered by the lipoprotein lipase gene.

7. Candidate genes from the signal transduction pathway

The possible implication of the polymorphism of the G protein as a genetic mechanism of predisposition to hypertension has been highlighted by the identification of a variant of the β-3 subunit of G protein (GNB3) by Siffert *et al.* (1998). An increase in sodium–proton exchange is repeatedly observed in cells from essential hypertensive subjects. The hypothesis that this increased ion transport activity could be due to enhanced intracellular transduction, was tested by these authors, who observed an increased cell activation restricted to receptors coupled to pertussis toxin (PTX)-sensitive G proteins in lymphoblasts from hypertensive subjects (Siffert *et al.*, 1998). A splice variant (GNB3-s) of the GNB3 gene, consisting of an in frame deletion of 123 bp in exon 9, was found in subjects bearing at least one T allele of a C825T polymorphism of the GNB3 gene. It is not known whether the C825T polymorphism is in complete linkage disequilibrium with another functional polymorphism responsible for the splice variant, or if the T allele is causative. The TC and CC genotypes were found more frequently in hypertensives versus normotensives (53.1 vs. 44.0%, $P=0.008$).

Three independent studies were performed after this first publication. In a population-based sample of German subjects, the T allele of the C825T polymorphism of the GNB3 gene was marginally associated with DBP, but not with SBP (Schunkert *et al.*, 1998). In a selected series of hypertensives with strong family history of hypertension, a higher frequency of the T allele was found in hypertensives (0.43 vs. 0.25; $P=0.00002$; Benjafield *et al.*, 1998). In contrast, a study performed in Japanese hypertensives was negative, since no difference in the allelic frequency of the C825T polymorphism was found between hypertensives and normotensives (Kato *et al.*, 1998).

Several other genes of the signal transduction pathway have yet to be tested for their association and linkage with hypertension. For example, the EP2 receptor of PGE2 is a strong candidate, since mice homozygous for the knock-out of this gene exhibit salt-sensitive systolic hypertension (Kennedy *et al.*, 1999).

8. Candidate regions found in the rat model

To date, 11 chromosomal regions representing putative quantitative trait loci (QTLs) for blood pressure were identified by linkage analysis of an F_2 population of rats derived from crosses between normotensive and hypertensive strains of rats. Two were found on chromosome 1, two on chromosome 10, and one each on chromosomes 2, 3, 7, 9, 10, 13 (Rapp and Deng, 1995). Although some candidate genes were found in some of these regions, no definite mutation identified in the hypertensive strain gene can definitively account for the blood pressure difference, except for studies performed with the 11β hydroxylase gene (see Chapter 4).

Linkage analyses provide statistical estimates of QTL localization only in the 15–20 cM range. Identifying the genes responsible for a quantitative trait is a difficult task. Finding the genes will require the construction of congenic strains to narrow the region, followed by exon trapping and cDNA capture strategies, such as those used to map quantitative traits in obese or tubby mice (Kleyn *et al.*, 1996).

Loci genetically linked to hypertension were found by various strategies, mostly by positional cloning. However, the SA gene was found by an alternative approach, the differential genetic screening (Iwai and Inagami, 1991). Indeed, the SA gene was found to be expressed at a higher level in the SHR strain of rat than in the WKY (Wistar–Kyoto) strain. Subsequent linkage analyses, in different strains of hypertensive rats using polymorphic markers of the SA gene, confirmed that the SA locus was linked to hypertension (Harris *et al.*, 1993; Nara *et al.*, 1993; Samani *et al.*, 1993).

We conducted a study on the human counterpart of the SA locus in hypertensives. This study consisted of performing an association and a linkage study with several genetic markers within the human locus, homologous to the rat SA locus (Nabika *et al.*, 1995). We cloned the human SA cDNA, which displays a high nucleotide similarity with the rodent gene (85%). We detected biallelic polymorphisms in introns of the human SA gene by SSCA. Polymorphic dinucleotide repeats were identified in a yeast artificial chromosome (YAC) which contained the SA gene. These markers were used to map the gene on human chromosome 16, by genetic linkage in the CEPH panel of families. Using the same panel of families and the same case and control group used for the eNOS gene study (Bonnardeaux *et al.*, 1995), both association and linkage studies were negative with these markers. Additional markers of the IGF2 gene, located on human chromosome 11 in humans but in a region close to the SA gene on rat chromosome 1, were also tested and gave negative results (Nabika, *et al.*, 1995).

These data are difficult to interpret since it is not known whether the SA gene itself is responsible for the linkage in the rat model (Rapp, 1998). Therefore, adjacent regions of the rat genome might be involved instead, making the study of the region rather complex in humans since linkage groups are not conserved extensively in this region. In addition, the SA gene may be at play in rat hypertension but not in human hypertension, either because the gene product has a different physiological role (which is unknown in the case of the SA gene), or because there is no common functional variant in humans.

As previously mentioned, a rat chromosome 10 locus responsible for high blood pressure and containing the ACE gene was identified in the stroke-prone spontaneously hypertensive rat (SHRSP) and WKY cross (Kreutz *et al.*, 1995a,b). The region of linkage is rather wide and two loci for blood pressure are probably present in this region. A study was performed on a large number of families, using several highly informative microsatellite markers of the homologous region of the human genome on chromosome 17 (Julier *et al.*, 1997). This study pointed out two regions in which the most significant nominal values for linkage were found. One was around the growth hormone and ACE locus, and the other one, about 10–15 cM towards the centromere, was around the anion exchanger 1 (band 3) gene. The latter region coincides with one of the loci which were mapped for a monogenic form of hypertension, the pseudo-hypoaldosteronism type II (PHAII), or Gordon's syndrome (Mansfield *et al.*, 1997). At the present time, since no gene has been identified yet, it is not known whether it is the same gene which is responsible for PHAII and for predisposition to essential hypertension in this region, and if it corresponds to the same gene responsible for hypertension in the rat model (Soubrier, 1998).

The other rat hypertension loci have not been extensively tested as yet in humans. The identification of the rat genes with the use of congenic strategies and physical mapping will enable easier investigation of these genes in human hypertension.

9. Genes responsible for monogenic hypertension or hypotension

These genes are major candidates since mutations on these genes are able to raise blood pressure at high levels. Mutations such as those found in monogenic hypertension, resulting in major functional effects, are not expected to be found in essential hypertension. These mutations are responsible for the very specific clinical and biological pictures of these syndromes, which do not compare with essential hypertension. However, mutations in the regulatory regions or even mutations of the coding region, but with a weaker effect, could be expected.

9.1 The epithelial sodium channel (ENaC)

The ENaC is composed of three subunits, alpha, beta and gamma, each including two transmembrane domains and a proline rich region in the 3′ region. Premature stop codons of the 3′ region deleting the proline rich region, or mutations of the proline rich region provoke an increased number of ENaC molecules to be present at the cell surface. These mutations were shown to be responsible for Liddle's syndrome.

Physical mapping and linkage studies showed that the beta and gamma subunits are located in close proximity to the SA gene, thus raising the hypothesis that one of these genes is involved in rat hereditary hypertension, instead of the SA gene. This hypothesis is not supported by experimental data at the present time (Huang *et al.*, 1995). As a consequence of the vicinity of the two genes, our study of the SA locus in human hypertension has also investigated the locus of the beta and gamma subunits of the ENaC gene. Indeed, microsatellite markers located 2 cM from the ENaC beta and gamma subunit genes were used in our analysis of the SA locus. Therefore, in our panel of families, a linkage of hypertension to this locus, although not definitively eliminated, is not likely.

Another approach consists of looking for mutations in the ENaC subunit genes in large series of severely hypertensive patients with hypokalemia, under the hypothesis that minor mutations of the gene might be responsible for attenuated forms of Liddle's syndrome. In a series of 525 subjects of Caucasian and Afro-Caribbean origin, Persu *et al.* analyzed the sequence of the last exon of the ENaC β subunit gene (Persu *et al.*, 1998). They found several polymorphisms, including the T594M variant, which was also found more frequently in black hypertensives than in black normotensives in another independent study (Baker *et al.*, 1998). The functional effects of these variants were tested by site-directed mutagenesis of the ENac β-subunit gene, followed by expression in *Xenopus* oocytes and functional analysis of the channel containing the mutated β-subunit (Persu *et al.*, 1998). None of the variants found exhibited any significant functional effects, ruling out the hypothesis that these variants could affect the function of the ENac in subjects bearing these alleles. Even a more extensive investigation of the gene in 101 hypertensives with low-renin hypertension was negative in this study.

Interestingly, this gene was found to be responsible for pseudohypoaldosteronism type I through mutations invalidating the function of the gene, instead of the mutation responsible for a gain of function as observed in the Liddle syndrome (Chang *et al.*, 1996).

9.2 The CYP11B2 gene

The CYP11B2 gene (aldosterone synthase) was shown to be responsible for glucocorticoid-remediable aldosteronism (GRA), due to the appearance of a chimeric gene secondary to an unequal crossing-over (Lifton *et al.*, 1992; see Chapter 8). The clinical and biological picture of the GRA is easily distinguishable from common essential hypertension, but the CYP11B2 is a candidate gene for hypertension, due to its role in aldosterone biosynthesis. We have tested a possible linkage between the gene and hypertension in a group of hypertensive families, using a highly polymorphic microsatellite, which was identified in close proximity to the CYP11B2 and the CYP11B1 gene (Brand *et al.*, 1998a). We did not find any support for linkage with hypertension in a group of 292 hypertensive sibling pairs using this microsatellite marker. We also studied the C-344T polymorphism of the gene in an association study, comparing the frequency of the alleles in a series of 380 hypertensive subjects and of 293 normotensive subjects. We observed that the T allele was more frequent in hypertensives (0.56 vs. 0.49; $P=0.01$; *Table 2*). Therefore, this association study is compatible with a modest implication of this gene in hypertension. Further studies are required to confirm this association and to elucidate its biological basis, since the C-344T polymorphism corresponds to a SF-1 binding site, but was shown to be nonfunctional (Clyne *et al.*, 1997).

9.3 The 11 beta hydroxysteroid dehydrogenase (11HSD) gene

The function of this enzyme is to catalyse the metabolism of cortisol into cortisone, which is unable to bind the mineralocorticoid receptor (MR), and thus ensuring a protection of the MR by preventing the binding of cortisol to the MR. Two isoforms of this enzyme exist and both were cloned. The 11HSD1 enzyme is mainly expressed in the liver, uses $NADP^+$ or NADPH as cofactor and catalyzes 11β-dehydrogenation and 11-oxoreduction. The 11HSD2 isoform is expressed in the kidney and in the placenta, uses NAD^+ as a cofactor and catalyzes the 11β-dehydrogenation of cortisol. These two isoforms do not only differ by their site of expression but also by their biochemical characteristics (see Chapter 7). Only the kidney isoform is involved in a Mendelian form of hypertension, the apparent mineralocorticoid excess (Chapter 7). This enzymatic mechanism of selectivity of the MR, although not the only one, is predominant *in vivo*, since the loss of function of the two copies of the gene is responsible for a severe form of genetic hypertension.

It was proposed that hypertension is associated with an abnormal excretion of cortisol metabolites characterized by a higher ratio of cortisol metabolites

Table 2. Comparison of allele and genotype frequencies in controls and hypertensive cases for the C-344T polymorphism of the CYP11B2 gene

	TT	CT	CC	χ^2_2	T	C	χ^2_1
Controls	71 (24.2%)	144 (49.2%)	78 (26.6%)		286 (48.8%)	300 (51.2%)	
Cases 1	116 (30.5%)	194 (51.1%)	70 (18.4%)	7.537 $P = 0.023$	426 (56.1%)	334 (43.9%)	6.687 $P = 0.010$
Cases 2	98 (28.8%)	176 (51.8%)	66 (19.4%)	5.052 $P = 0.080$	372 (54.7%)	308 (45.3%)	4.157 $P = 0.041$
Cases 3	88 (31.2%)	149 (52.8%)	45 (16.0%)	10.55 $P = 0.005$	325 (57.6%)	239 (42.4%)	8.625 $P = 0.003$
Cases 4	83 (29.0%)	149 (52.1%)	54 (18.9%)	5.300 $P = 0.071$	315 (55.1%)	257 (44.9%)	4.303 $P = 0.038$
Cases 5	78 (27.2%)	149 (51.9%)	60 (20.9%)	2.700 $P = 0.259$	305 (53.1%)	269 (46.9%)	2.006 $P = 0.157$
Cases 6	43 (29.7%)	73 (50.3%)	29 (20.0%)	2.865 $P = 0.239$	159 (54.8%)	131 (45.2%)	2.579 $P = 0.108$

Cases 1: the most hypertensive person of the family.
Cases 2: (DBP ⩾95 mmHg or AHD ⩾1) and (onset <60 years).
Cases 3: (onset <45 years).
Cases 4: (DBP ⩾100 mmHg or AHD ⩾2).
Cases 5: (DBP ⩾95 mmHg or AHD ⩾1) and (BMI ⩽27 kg m^{-2}).
Cases 6: (DBP ⩾100 mmHg or AHD ⩾2) and (onset <45 years) and (BMI ⩽27 kg m^{-2}).

Reprinted from Brand, E. *et al.* (1998) Structural analysis and evaluation of the aldosterone synthase gene in human hypertension, *Hypertension*, vol. 32, pp. 198–204, with permission from Lippincott, Williams & Wilkins.

compared to cortisone metabolites, potentially reflecting a defect of this selectivity in hypertension (Soro *et al.*, 1995;Walker *et al.*, 1991). We performed a linkage and an association study with the 11HSD2 gene by identifying a microsatellite marker and an SNP on the 11HSD2 gene (Brand *et al.*, 1998b). The linkage study in hypertensive sib-pairs was negative, but this result must be interpreted with caution in view of the moderate informativity of the microsatellite marker. Similarly, an association study performed with a neutral marker was also negative, but the informativity of this SNP was rather low. Indeed, an intensive search for variants by SSCA in the coding and intronic sequence of the 11HSD2 gene was negative in hypertensive patients.

9.4 Pseudohypoaldosteronism type II or Gordon's syndrome

Gordon's syndrome is probably the form of monogenic hypertension which most closely mimics essential hypertension.The blood pressure level is not very high and the biological phenotype is rather attenuated, composed of fluctuating hyperkalemia with hyperchloremic acidosis, and a pronounced beneficial effect of salt restriction or treatment with thiazides (Gordon *et al.*, 1970).This mild phenotype might leave a large proportion of patients with Gordon's syndrome undiagnosed, if hyperkalemia, the most evident sign, is not recognized. Two loci have been identified for this disease, clearly showing the heterogeneity of this syndrome (Mansfield *et al.*, 1997). One is located on chromosome 1 and the other one on chromosome 17. The chromosome 17 locus was also found independently to be linked to this syndrome in another pedigree (O'Shaughnessy *et al.*, 1998). This chromosome 17 locus was also shown to be linked to essential hypertension, as mentioned above (Julier *et al.*, 1997).

9.5 Sodium transporters

Two sodium transporters, the thiazide-sensitive Na–Cl cotransporter (TSC), the Na–K–2Cl cotransporter (NKCC2), and the ATP sensitive K channel (ROMK) were recently shown to be responsible for two autosomal recessive syndromes of inherited hypokalemia with alkalosis. Mutations affecting both alleles of the TSC were found in patients with the Gitelman's syndrome, which is associated with hypocalciuria and hypomagnesemia (Simon *et al.*, 1996c). Frameshift and nonconserved mutations of the NKCC2 gene and of the ATP-sensitive K^+ channel gene were found in patients with the Bartter's syndrome, which presents with hypercalciuria and volume depletion (Simon *et al.*, 1996a,b).

Heterozygosity for mutations of one of these two channels could be protective against hypertension, and cause mild spontaneous or diuretic induced hypokalemia. In order to be responsible for hypertension, a gain of function must be hypothesized, which would result in increased sodium reabsorption. Until now, there were no studies to search for this kind of mutations in patients with hypertension.

10. Conclusion

The genes and genotypes predisposing to hypertension will be difficult to recognize. This is due to the low genetic determination of the disease, its genetic heterogeneity, and expected low genotypic risk ratios for the disease, and also to the influence of environmental factors. Simulations have shown that the number of families with the disease required to establish a linkage in this kind of disease is huge (more than 1000) and at the present time, no linkage study in hypertension has reached the required number of families (Risch and Merikangas, 1996).

Theoretically, projects can be designed with the appropriate power. They should include the collection of thousands of families with severely hypertensive siblings. Additional groups of patients and well matched control subjects should allow large association studies to be performed. Intrafamilial association and linkage tests should be possible, by collecting young hypertensives with available parental genotypes. These different panels would allow positive results to be confirmed independently by different statistical tests.

The availability of high throughput technologies for detecting mutations and for genotyping will probably favor the investigation of hundreds of candidate genes for association with SNPs. Two elements are very important for interpreting the large amount of data which will be generated. Firstly, the easy access to results and the possibility of replicating rapidly the results in independent studies. To achieve this goal the initiative of creating a web site for cardiovascular genetics where polymorphisms and results of different studies are easily accessible is very beneficial (http: //ifr69.vjf.inserm.fr/~canvas/). Secondly, the functional analysis of the variants tested will be essential. For this purpose, a wide variety of experiments will be required, from *in vitro* studies to *in vivo* knock-out and knock-in experiments. This is the only way to establish biological significance for an association between a variant and a disease.

References

Amos, C.I., Elston, R.C., Wilson, A.F. and Bailey-Wilson, J.E. (1989) A more powerful robust sib-pair test of linkage for quantitative traits. *Genet. Epidemiol.* **6**: 435–449.

Baker, E.H., Dong, Y.B., Sagnella, G.A., *et al.* (1998) Association of hypertension with T594M mutation in beta subunit of epithelial sodium channels in black people resident in London. *Lancet* **351**: 1388–1392.

Benetos, A., Gautier, S., Ricard, S., *et al.* (1996) Influence of angiotensin converting enzyme and angiotensin II type 1 receptor gene polymorphisms on aortic stiffness in normotensive and hypertensive patients. *Circulation* **94**: 698–703.

Benjafield, A.V., Jeyasingam, C.L., Nyholt, D.R., Griffiths, L.R. and Morris, B.J. (1998) G-protein beta3 subunit gene (GNB3) variant in causation of essential hypertension. *Hypertension* **32**: 1094–1097.

Berge, K.E. and Berg, K. (1992) No effect of a Taq I polymorphism in DNA at the endothelin I (EDN1) locus on normal blood pressure level or variability. *Clin. Genet.* **41**: 90–95.

Bloem, L.J., Manatunga, A.K., Tewksbury, D.A. and Pratt, J.H. (1995) The serum angiotensinogen concentration and variants of the angiotensinogen gene in white and blacks children. *J. Clin. Invest.* **95**: 948–953.

Bonnardeaux, A., Davies, E., Jeunemaitre, X., *et al.* (1994) Angiotensin II type I receptor gene polymorphisms in human essential hypertension. *Hypertension* **24**: 63–69.

Bonnardeaux, A., Nadaud, S., Charru, A., Jeunemaitre, X., Corvol, P. and Soubrier, F. (1995) Lack of evidence for linkage of the endothelial cell nitric oxide synthase gene to essential hypertension. *Circulation* **91**: 96–102.

Brand, E., Chatelain, N., Mulatero, P., Fery, I., Curnow, K., Jeunemaitre, X., Corvol, P., Pascoe, L. and Soubrier, F. (1998a) Structural analysis and evaluation of the aldosterone synthase gene in human hypertension. *Hypertension* **32**: 198–204.

Brand, E., Kato, N., Chatelain, N., *et al.* (1998b) Structural analysis and evaluation of the 11b-hydroxysteroid dehydrogenase type 2 (11β-HSD2) gene in human essential hypertension. *J. Hypertension* **16**: 1627–1633.

Canessa, M., Adragna, N., Solomon, H., Connolly, T.M., Tosteson, B.S. and Tosteson, D.C. (1980) Increased sodium–lithium countertransport in red cells of patients with essential hypertension. *N. Engl. J. Med.* **302**: 772–776.

Chambers, S.M. and Morris, B.J. (1996) Glucagon receptor gene mutation in essential hypertension. *Nature Genet.* **12**: 122.

Chang, S.S., Grunder, S., Hanukoglu, A., *et al.* (1996) Mutations in subunits of the epithelial sodium channel cause salt wasting with hyperkalaemic acidosis, pseudohypoaldosteronism type 1. *Nature Genet.* **12**: 248–253.

Cheng, L.S.C., Carmelli, D., Hunt, S.C. and Williams, R.R. (1995) Evidence for a major gene influencing 7-year increases in diastolic blood pressure with age. *Am. J. Hum. Genet.* **57**: 1169–1177.

Clyne, C.D., Zhang, Y., Slutsker, L., Mathis, J.M., White, P.C. and Rainey, W.E. (1997) Angiotensin II and potassium regulate human CYP11B2 transcription through common *cis*-elements. *Mol. Endocrinol.* **11**: 638–649.

Cotton, R.G.H. (1993) Current methods for mutation detection. *Mutation Res.* **285**: 128–144.

de Fronzo, R.A. (1981) The effect of insulin on sodium metabolism. *Diabetelogia* **21**: 165–171.

Deloukas, P., Schuler, G.D., Gyapay, G., *et al.* (1998) A physical map of 30,000 human genes. *Science* **282**: 744–746.

Edwards, G., Dora, K.A., Gardener, M.J., Garland, C.J. and Weston, A.H. (1998) K^+ is an endothelium-derived hyperpolarizing factor in rat arteries. *Nature* **396**: 269–272.

Falk, C.T. and Rubinstein, P. (1987) Haplotype relative risks: an easy reliable way to construct a proper control sample for risk calculations. *Am. J. Hum. Genet.* **51**: 227–233.

Ferrari, P. and Bianchi, G. (1995) Lessons from experimental genetic hypertension. In: *Hypertension Pathophysiology, Diagnosis and Management* (ed. J.H. Laragh and B.M. Brenner). Raven Press, New York, pp. 1261–1287.

Fornage, M., Amos, C.I., Kardia, S., Sing, C.F., Turner, S.T. and Boerwinckle, E. (1998) Variation in the region of the angiotensin-converting enzyme gene influences interindividual differences in blood pressure levels in young Caucasian males. *Circulation* **97**: 1773–1779.

Germain, S., Bonnet, F., Philippe, J., Fuchs, S., Corvol, P and Pinet, F. (1998) A novel distal enhancer confers chorionic expression on the human renin gene. *J. Biol. Chem.* **25**: 25292–25300.

Gordon, R.D., Geddes, R.A., Pawsey, C.G. and O'Halloran, M.W. (1970) Hypertension and severe hyperkalaemia associated with suppression of renin and aldosterone and completely reversed by dietary sodium restriction. *Australas. Ann. Med.* **19**: 287–294.

Guyton, A.C. (1991) Blood pressure control. Special role of the kidneys and body fluids. *Science* **252**: 1813–1816.

Hacia, J.G., Brody, L.C., Chee, M.S., Fodor, S.P.A. and Collins, F. (1996) Detection of heterozygous mutations in BRCA1 using high density oligonucleotides arrays and two-colour fluorescence analysis. *Nature Genet.* **14**: 441–447.

Hansen, L.H., Abrahamsen, N., Hager, J., Jelinek, L., Kindsvogel, W., Froguel, P. and Nishimura, E. (1996) The Gly40Ser mutation in the human glucagon receptor gene associated with NIDDM results in a receptor with reduced sensitivity to glucagon. *Diabetes* **45**: 725–730.

Harrap, S.B., Davidson, H.R., Connor, J.M., Soubrier, F., Corvol, P., Fraser, R., Foy, C.J.W. and Watt, G.C.M. (1993) The angiotensin I-converting enzyme gene and predisposition to high blood pressure. *Hypertension* **21**: 455–460.

Harris, E.L., Dene, H. and Rapp, J.P. (1993) SA gene and blood pressure cosegregation using Dahl salt-sensitive rats. *Am. J. Hypertension* **6**: 330–334.

Haynes, W.G. and Webb, D.J. (1994) Contribution of endogenous generation of endothelin-1 to basal vascular tone. *Lancet* **344**: 852–854.

Hilton, P.J. (1986) Cellular sodium transport in essential hypertension. *N. Engl. J. Med.* **314**: 222–229.

Hollenberg, N.K. and Williams, G.H. (1995) Abnormal renal function, sodium-volume homeostasis, and renin system behavior in normal-renin essential hypertension. In: *Hypertension Pathophysiology, Diagnosis and Management* (ed. J.H. Laragh and B.M. Brenner). Raven Press, New York, pp. 1837–1856.

Huang, P., Huang, Z., Mashimo, H., Bloch, K.D., Moskowitz, M.A., Bevan, J.A. and Fishman, M.C. (1995a) Hypertension in mice lacking the gene for endothelial nitric oxide synthase. *Nature* **377**: 239–242.

Huang, H., Pravenec, M., Wang, J.-M., Kren, V., St Lezin, E., Szpirer, C., Szpirer, J. and Kurtz, T. (1995b) Mapping and sequence analysis of the gene encoding the beta subunit of the epithelial sodium channel in experimental models of hypertension. *J. Hypertension* **13**: 1247–1251.

Huber, C.G. and Berti, N. (1996) Detection of partial denaturation in AT-rich DNA fragments reversed-phase chromatography. *Anal. Chem.* **68**: 2959–2965.

Hunt, S.C., Williams, C.S., Sharma, A.M., Inoue, I., Williams, R.R. and Lalouel, J. (1996) Lack of linkage between the endothelial nitric oxide synthase gene and hypertension. *J. Hypertension* **10**: 27–30.

Ioannou, P.A., Amemiya, C.T., Garnes, J., Kroisel, P.M., Shizuya, H., Chen, C., Batzer, M.A. and de Jong, P.J. (1994) A new bacteriophage P1-derived vector for the propagation of large human DNA fragments. *Nature Genet.* **6**: 84–89.

Iwai, N. and Inagami, T. (1991) Isolation of preferentially expressed genes in the kidneys of hypertensive rats. *Hypertension* **17**: 161–169.

Jeunemaitre, X., Lifton, R.P., Hunt, S.C., Williams, R.R. and Lalouel, J.M. (1992a) Absence of linkage between the angiotensin converting enzyme locus and human essential hypertension. *Nature Genet.* **1**: 72–75.

Jeunemaitre, X., Rigat, B., Charru, A., Houot, A.M., Soubrier, F. and Corvol, P. (1992b) Sib-pair linkage analysis of renin gene haplotypes in human essential hypertension. *Hum. Genet.* **88**: 301–306.

Julier, C., Delépine, M., Keavney, B., *et al.* (1997) Genetic susceptibility for human familial essential hypertension in a region of homology with blood pressure linkage on rat chromosome 10. *Hum. Mol. Genet.* **6**: 2077–2085.

Kato, N., Sugiyama, T., Morita, H., Kurihara, H., Yamori, Y. and Yazaki, Y. (1998) G protein beta3 subunit variant and essential hypertension in Japanese. *Hypertension* **32**: 935–938.

Kennedy, C.R., Zhang, Y., Brandon, S., *et al.* (1999) Salt-sensitive hypertension and reduced fertility in mice lacking the prostaglandin EP2 receptor. *Nature Med.* 5: 217–220.

Kimura, G., Frem, G.J. and Brenner, B.M. (1994) Mechanisms of salt sensitivity in hypertension. *Curr. Opin. Nephrol. Hypertension* **3**: 1–12.

Kleyn, P.W., Fan, W., Kovats, S.G., *et al.* (1996) Identification and characterization of the mouse obesity gene tubby: a member of a novel gene family. *Cell* **85**: 281–290.

Kreutz, R., Hübner, N., James, M.R., Bihoreau, M.T., Gauguier, D., Lathrop, G.M., Ganten, D. and Lindpaintner, K. (1995a) Dissection of a quantitative trait locus for genetic hypertension on rat chromosome 10. *Proc. Natl Acad. Sci. USA* **92**: 8778–8782.

Lerman, L.S., Silverstein, K. and Grinfeld, E. (1986) Searching for gene defects by denaturing gradient gel electrophoresis. *Cold Spring Harbor Symp. Quant. Biol. LI*: 285–297.

Lifton, R.P., Hopkins, P.N., Williams, R.R., Hollenberg, N.K., Williams, G.H. and Dluhy, R.G. (1989) Evidence for heritability of non-modulating essential hypertension. *Hypertension* **13**: 884–889.

Lifton, R.P., Dluhy, R.G., Powers, M., Rich, G.M., Cook, S., Ulick, S. and Lalouel, J.M. (1992) A chimaeric 11b-hydroxylase/aldosterone synthase gene causes glucocorticoid-remediable aldosteronism and human hypertension. *Nature* **355**: 262–265.

Linder, L., Kiowski, W., Buhler, F.R. and Luscher, T.F. (1990) Indirect evidence for release of endothelium-derived relaxing factor in human forearm circulation *in vivo*. *Circulation* **81**: 1762–1767.

Luft, F.C., Miller, J.Z., Cohen, S.J., Fineberg, N.S. and Weinberger, M.H. (1988) Heritable aspects of salt sensitivity. *Am. J. Cardiol.* **61**: 1H–6H.

Mansfield, T.A., Simon, D.B., Farfel, Z., *et al.* (1997) Multilocus linkage of familial hyperkalaemia and hypertension, pseudohypoaldosteronism type II, to chromosomes 1q31–42 and 17p11-q21. *Nature Genet.* **16**: 202–205.

Marsden, P.A., Heng, H.H., Scherer, S.W., Stewart, R.J., Hall, A.V., Shi, X.M., Tsui, L.C. and Schappert, K.T. (1993) Structure and chromosomal localization of the human constitutive endothelial nitric oxide synthase gene. *J. Biol. Chem.* **268**: 17478–17488.

Meier, A., Weidmann, P., Grimm, M., Keusch, G., Glück, Z., Minder, I. and Ziegler, W.H. (1981) Pressor factors and cardiovascular pressor responsiveness in borderline hypertension. *Hypertension* **3**: 367–372.

Morris, B.J. and Griffiths, L.R. (1988) Frequency in hypertensives of alleles for a RFLP associated with the renin gene. *Biochem. Biophys. Res. Commun.* **150**: 219–224.

Morris, B.J., Zee, R.Y. and Schrader, A.P. (1994) Different frequencies of angiotensin-converting enzyme genotypes in older hypertensive individuals. *J. Clin. Invest.* **94**: 1085–1089.

Morris, B.J., Jeyasingam, C.L., Zhang, W., Curtain, R.P and Griffiths, L.R. (1997) Influence of family history on frequency of glucagon receptor Gly40Ser mutation in hypertensive subjects. *Hypertension* **30**: 1640–1641.

Morton, N.E., Gulbrandsen, C.L., Rao, D.C., Rhoads, C.G. and Kagan, A. (1980) Determinants of blood pressure in Japanese-American families. *Human Genet.* **53**: 261–266.

Nabika, T., Bonnardeaux, A., James, M., Julier, C., Jeunemaitre, X., Charru, A., Corvol, P., Lathrop, M. and Soubrier, F. (1995) Evaluation of the SA gene locus in human hypertension. *Hypertension* **25**: 6–13.

Nadaud, S., Bonnardeaux, A., Lathrop, M. and Soubrier, F. (1994) Gene structure, polymorphism and mapping of the human endothelial nitric oxide synthase gene. *Biochem. Biophys. Res. Commun.* **198**: 1027–1033.

Naftilan, A., Williams, R., Burt, D., Paul, M., Pratt, R.E., Hobart, P., Chirgwin, J. and Dzau, V. (1989) A lack of genetic linkage of renin gene restriction fragment length polymorphisms with human hypertension. *Hypertension* **14**: 219–224.

Nara, Y., Nabika, T., Ikeda, K., Sawamura, M., Mano, M., Endo, J. and Yamori, Y. (1993) Basal high blood pressure cosegregates with the loci on chromosome 1 in the F2 generation from crosses between normotensive Wistar Kyoto rats and stroke-prone spontaneously hypertensive rats. *Biochem. Biophys. Res. Commun.* **194**: 1344–1351.

O'Donnell, C.J., Lindpaintner, K., Larson, M.G., Rao, V.S., Ordovas, J.M., Schaefer, E.J., Myers, R.H. and Levy, D. (1998) Evidence from association and genetic linkage of the angiotensin converting enzyme locus with hypertension and blood pressure in men but not in women in the Framingham heart study. *Circulation* **97**: 1766–1772.

Orita, M., Iwahana, H., Kanazawa, H., Hayashi, K. and Sekiya, T. (1989) Detection of polymorphisms of human DNA by gel electrophoresis as single-strand conformation polymorphisms. *Proc. Natl Acad. Sci. USA* **86**: 2766–2770.

O'Shaughnessy, K.M., Fu, B., Johnson, A. and Gordon, R.D. (1998) Linkage of Gordon's syndrome to the long arm of chromosome 17 in a region recently linked to familial essential hypertension. *J. Hum. Hypertension* **12**: 675–678.

Panza, J.A., Quyyumi, A.A., Brush, J.E. and Epstein, S.E. (1990) Abnormal endothelium-dependent vascular relaxation in patients with essential hypertension. *N. Engl. J. Med* **323**: 22–27.

Persu, A., Barbry, P., Bassilana, F., Houot, A.M., Mengual, R., Lazdunski, M., Corvol, P. and Jeunemaitre, X. (1998) Genetic analysis of the beta subunit of the epithelial Na^+ channel in essential hypertension. *Hypertension* **32**: 129–37.

Perusse, L., Moll, P.P. and Sing, C.F. (1991) Evidence that a single gene with gender- and age dependent effects influences systolic blood pressure determination in a population-based sample. *Am. J. Hum. Genet.* **49**: 94–105.

Rapp, J. (1998) The Sa gene: what does it mean? *Hypertension* **32**: 647–648.

Rapp, J.P. and Deng, A.Y. (1995) Detection and positional cloning of blood pressure quantitative trait loci: is it possible? Identifying the genes for genetic hypertension. *Hypertension* **25**: 1121–1128.

Rice, T., Bouchard, C., Borecki, I.B. and Rao, D.C. (1990) Commingling and segregation analysis of blood pressure in a French-Canadian population. *Am. J. Hum. Genet.* **46**: 37–44.

Risch, N. and Merikangas, K. (1996) The future of genetic studies of complex human diseases. *Science* **273**: 1516–1517.

Rosskopf, D., Düsing, R. and Siffert, W. (1993) Membrane sodium-proton exchange and primary hypertension. *Hypertension* **21**: 607–617.

Ruilope, L.M., Lahera, V., Rodicio, J.L. and Romero, J.C. (1994) Participation of nitric oxide in the regulation of renal function: possible role in the genesis of arterial hypertension. *J. Hypertension* **12**: 625–631.

Samani, N.J., Lodwick, D., Vincent, M., *et al.* (1993) A gene differentially expressed in the kidney of the spontaneously hypertensive rat cosegregates with increased blood pressure. *J. Clin. Invest.* **92**: 1099–1103.

Schalin-Jantti, C., Nikula-Ijas, P., Huang, X., *et al.* (1996) Polymorphism of the glycogen synthase gene in hypertensive and normotensive subjects. *Hypertension* **27**: 67–71.

Schiffrin, E.L. (1998) Endothelin and endothelin antagonists in hypertension. *J. Hypertension* **16**: 1891–1895.

Schiffrin, E.L., Deng, L.Y., Sventek, P. and Day, R. (1997) Enhanced expression of endothelin-1 gene in resistance arteries in severe human essential hypertension. *J Hypertension* **15**: 57–63.

Schmidt, S., Beige, J., Walla-Friedel, M., Michel, M.C., Sharma, A.M. and Ritz, E. (1997) A polymorphism in the gene for the angiotensin II type 1 receptor is not associated with hypertension. *J. Hypertension* **15**: 1385–1388.

Schunkert, H., Hense, H.W., Doring, A., Riegger, G.A. and Siffert, W. (1998) Association between a polymorphism in the G protein beta3 subunit gene and lower renin and elevated diastolic blood pressure levels. *Hypertension* **32**: 510–513.

Sealey, J.E., Blumenfeld, J.D., Bell, G.M., Pecker, M.S., Sommers, S.C. and Laragh, J.H. (1988) Nephron heterogeneity with unsuppressible renin secretion: a cause of essential hypertension. *J. Hypertension* **6**: 763–777.

Shizuya, H., Birren, B., Kim, U.J., Mancino, V., Slepak, T., Tachiiri, Y. and Simon, M. (1992) Cloning and stable maintenance of 300-kilobase pair fragments of human DNA in E. Coli using an F-factor-based vector. *Proc. Natl Acad. Sci. USA* **89**: 8794–8797.

Siffert, W., Rosskopf, D., Siffert, G., *et al.* (1998) Association of a human G-protein beta3 subunit variant with hypertension. *Nature Genet.* **18**: 45–48.

Simon, D.B., Karet, F.E., Hamdan, J.M., Di Pietro, A., Sanjad, S.A. and Lifton, R. (1996a) Bartter's syndrome, hypokalaemic alkalosis with hypercalciuria, is caused by mutations in the Na–K–2Cl cotransporter NKCC2. *Nature Genet.* **13**: 183–188.

Simon, D.B., Karet, F.E., Rodriguez-Soriano, J., Hamdan, J.H., DiPietro, A., Trachtman, H., Sanjad, S.A. and Lifton, R.P. (1996b) Genetic heterogeneity of Bartter's syndrome revealed by mutations in the K+ channel, ROMK. *Nature Genet.* **14**: 152–156.

Simon, D.B., Nelson-Williams, C., Johnson Bia, M., *et al.* (1996c) Gitelman's variant of Bartter's syndrome, inherited hypokalaemic alkalosis, is caused by mutations in the thiazide-sensitive Na–Cl cotransporter. *Nature Genet.* **12**: 24–30.

Soro, A., Ingram, M.C., Tonolo, G., Glorioso, N. and Fraser, R. (1995) Evidence of coexisting changes in 11b-hydroxysteroid dehydrogenase and 5b-reductase activity in subjects with untreated essential hypertension. *Hypertension* **25**: 67–70.

Soubrier, F. (1998) Blood pressure gene at the angiotensin I-converting enzyme locus: chronicle of a gene foretold. *Circulation* **97**: 1763–1765.

Soubrier, F., Jeunemaitre, X., Rigat, B., Houot, A.-M., Cambien, F. and Corvol, P. (1990) Similar frequencies of renin gene restriction fragment length polymorphisms in hypertensive and normotensive subjects. *Hypertension* **16**: 712–717.

Spielman, R.S., McGinnis, R.E. and Ewens, W.J. (1993) Transmission test for linkage disequilibrium: the insulin gene region and insulin dependent diabetes mellitus (IDDM). *Am. J. Human Genet.* **52**: 506–516.

Stevens, P.A. and Brown, M.J. (1995) Genetic variability of the ET-1 and the ETA receptor genes in essential hypertension. *J. Cardiovasc. Pharmacol.* **26 (suppl. 3)**: S9–S12.

Suarez, B.K., Rice, J. and Reich, T. (1978) The generalized sib pair IBD distribution: its use in the detection of linkage. *Ann. Hum. Genet.* **42**: 87–94.

Tiret, L., Rigat, B., Visvikis, S., Breda, C., Corvol, P., Cambien, F. and Soubrier, F. (1992) Evidence from combined segregation and linkage analysis, that a variant of the angiotensin I-converting enzyme (ACE) gene controls plasma ACE levels. *Am. J. Hum. Genet.* **51**: 197–210.

Tiret, L., Bonnardeaux, A., Poirier, O., *et al.* (1994) Synergistic effects on angiotensin-converting enzyme and angiotensin-II type 1 receptor gene polymorphisms on risk of myocardial infarction. *Lancet* **344**: 910–913.

Verpy, E., Biasotto, M., Meo, T. and Tosi, M. (1994) Efficient detection of point mutations on color-coded strands of target DNA. *Proc. Natl Acad. Sci. USA* **91**: 1873–1877.

Walker, B.R., Stewart, P.M., Padfield, P.L. and Edwards, C.R.W. (1991) Increased vascular sensitivity to glucocorticoids in essential hypertension: 11b-hydroxysteroid dehydrogenase deficiency revisited. *J. Hypertension* **9**: 1082–1083.

Wang, D.G., Fan, J.B., Siao, C.J., *et al.* (1998) Large-scale identification, mapping, and genotyping of single- nucleotide polymorphisms in the human genome. *Science* **280**: 1077–1082.

Wang, W.Y., Zee, R.Y. and Morris, B.J. (1997) Association of angiotensin II type 1 receptor gene polymorphism with essential hypertension. *Clin. Genet.* **51**: 31–34.

Ward, R. (1995) Familial aggregation and genetic epidemiology of blood pressure. In: *Hypertension Pathophysiology, Diagnosis and Management* (ed. J.H. Laragh and B.M. Brenner). Raven Press, New York, pp. 67–88.

Weidman, P and Ferrari, P. (1991) Hypertension in the diabetic: central role of sodium. *Diabetes Care* **14**: 220–232.

Wilson, A.F., Elston, R.C., Tran, L.D. and Siervogel, R.M. (1991) Use of the robust sib-pair method to screen for single-locus, multiple-locus, and pleiotropic effects: application to traits related to hypertension. *Am. J. Hum. Genet.* **48**: 862–872.

Wu, D.-A., Bu, X., Warden, C.H., *et al.* (1996) Quantitative trait locus mapping of human blood pressure to a genetic region at or near the lipoprotein lipase gene locus on chromosome 8p22. *J. Clin. Invest.* **97**: 2111–2118.

Ying, L.-H., Zee, R.Y., Griffiths, L.R. and Morris, B.J. (1991) Association of a RFLP for the insulin receptor gene but not insulin with essential hypertension. *Biochem. Biophys. Res. Comm.* **181**: 486–492.

Zee, R.Y.L., Griffiths, L.R. and Morris, B.J. (1992) Marked association of a RFLP for the low density lipoprotein receptor gene with obesity in essential hypertension. *Biochem. Biophys. Res. Comm.* **189**: 965–971.

10

Angiotensinogen and hypertension

Xavier Jeunemaitre, Anne-Paule Gimenez-Roqueplo, Jérôme Célérier, Florent Soubrier and Pierre Corvol

1. Introduction

Research on the molecular genetics of human hypertension aims to identify the loci involved in the regulation of blood pressure, to detect gene variants, to associate them with intermediate phenotypes, and ultimately to estimate their quantitative effects on blood pressure level and their interaction with the main environmental factors. The main strategy that has been used so far relies on the study of candidate genes, genes that might contribute to the variance of blood pressure because of their well-known effect on the cardiovascular system. The genes of the renin–angiotensin system have been extensively studied because of the well-documented role of this system on salt and water homeostasis and vascular tone. This review will discuss briefly the structure of the angiotensinogen (AGT) gene, its tissue expression and regulation, the main characteristics of the protein, and will analyse in detail the different genetic studies which have been performed related to human hypertension and to some other cardiovascular diseases.

2. The angiotensinogen gene

2.1 Localization

A single human AGT gene has been localized to chromosome 1q42–3 by *in situ* hybridization (Gaillard-Sanchez *et al.*, 1990; Isa *et al.*, 1990) whereas the renin gene has been assigned to the 1q32 region. The issue of the genetic proximity of these two genes has been explored using a GT microsatellite repeat at the renin locus (76% heterozygosity), a GT microsatellite at the AGT locus (Kotelevtsev *et al.*, 1991), and other markers selected from the Consortium map of chromosome 1 (Dracopoli *et al.*, 1991). Analysis of these markers on pedigrees selected by the Centre d'Etudes du Polymorphisme Humain (CEPH) demonstrated a weak linkage between these two loci (LOD score = 4.89) for a recombination estimate

Molecular Genetics of Hypertension, edited by A.F. Dominiczak, J.M.C. Connell and F. Soubrier.

θ=0.30 (Jeunemaitre and Lifton, 1993).The AGT and renin loci do not belong to a syntenic region. In the mouse, the AGT gene is located on chromosome 8 (Clouston *et al.*, 1988) whereas REN 1 and 2 genes are located on chromosome 1 (Chirgwin *et al.*, 1984; Pravenec *et al.*, 1991). In the rat, the AGT and the renin genes are located on chromosomes 19 and 13.

2.2 Structure

The human AGT cDNA is 1455 nucleotides in length, with coding for 485 amino acids (Kageyama *et al.*, 1984). The gene contains five exons and four introns which span 13 kb (*Figure 1*). The first exon (37 bp) codes for the 5′ untranslated region of the mRNA. There are two potential AGT sites and the second exon codes for a signal peptide of 24 or 33 residues and the first 252 amino acids (59%) of the mature protein. The mature angiotensinogen consists of 452 amino acid residues, with the first 10 amino acids corresponding to angiotensin I (Ang I) and the other larger portion to des(Ang I)angiotensinogen. Exons 3 and 4 code for 90 and 48 amino acids, respectively. Exon 5 contains a short coding sequence (62 amino acids), followed by a long 3′ untranslated sequence with two different polyadenylation signals, accounting for the existence of two different mRNA species differing in length by 200 nucleotides (Gaillard *et al.*, 1989). A similar gene structure is shared by the genes of the serpin superfamily, as far as exon number, size and splicing sites are concerned (Doolittle, 1983; Tanaka *et al.*, 1984).

2.3 Promoter region

Angiotensinogen is secreted through the constitutive pathway and most of its regulation occurs at the transcriptional level. Glucocorticoids, estrogens, Ang II and thyroid hormones are the main hormonal stimuli of the AGT expression. Brasier *et al.* (1989) proposed the concept that induction of the AGT gene by these diverse physiological stimuli is mediated through changes in the nuclear abundance of sequence-specific nuclear factors.

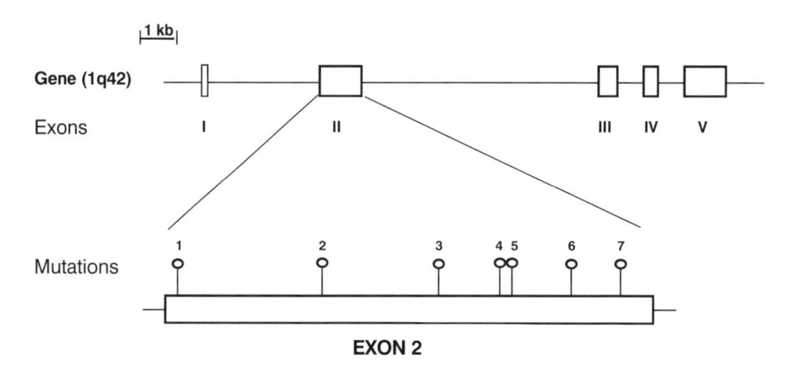

Figure 1. Angiotensinogen gene: genomic structure and missense mutations. The amino acid substitutions found in the second exon of the gene are represented by 1=L10P, 2=T104M, 3=T174M, 4= L209I, 5=L211R, 6=M235T, 7=Y248C.

The *in vitro* transcriptional regulation of rat angiotensinogen expression in the liver by glucocorticoids and cytokines has been recently described (Brasier *et al.*, 1989; Brasier and Li, 1996). In mouse, recent studies have indicated that the proximal promoter region of the mouse angiotensinogen gene (–96 to +22) was sufficient for expression of angiotensinogen in mouse fibroblast lines during their differentiation into adipocytes (Tamura *et al.*, 1994). Two factors, one liver-specific (AGF2) and one ubiquitous nuclear factor (AGF3), bind to the proximal promoter element (–96 to +52) and to the core promoter element (–6 to +22), respectively. These two factors seem to act synergistically (Tamura *et al.*, 1994).

Ang II could stimulate angiotensinogen synthesis (Eggena *et al.*, 1993) through the activation of the nuclear-kappa B transcription factor (Brasier and Li *et al.*, 1996). Ang II could also increase AGT mRNA levels by acting on the stability of the AGT mRNA (Klett *et al.*, 1988). A differentiation-specific element (DSE), located at –1000 bp in the 5′-flanking region of the rat angiotensinogen gene, could act as a binding-site for a pou-homeodomain class of transcription factors and thus play an important role in the developmental switch for the expression of angiotensinogen during the differentiation of fibroblasts to adipocytes (McGehee *et al.*, 1993).

Less work has been performed on the human AGT promoter. Three putative glucocorticoid responsive-elements (GRE) are located at –126 to –133 bp, –202 to –217 bp and –670 to –675 bp to the initial transcription site. Potential estrogen (–324 to –338 bp) and thyroid receptor responsive elements (–370 to –384 bp) are present in the 5′ flanking sequence. Other sequences potentially involved in the transcriptional regulation of the human AGT gene include a cAMP (cyclic adenosine 3′,5′-monophosphate) responsive element (–833 to –839 bp), an acute phase responsive element (APRE) (–270 to –278 bp) and a heat shock element (–561 to –573 bp). The core promoter region of the human AGT gene has been analysed by the Fukamizu's group in HepG2 cells (Yanai *et al.*, 1996). Electromobility shift assays demonstrated that an ubiquitously expressed nuclear factor could bind to a region between positions –25 to –1, denoted AGCE1. This region is probably more complex with different nuclear factors binding to its 5′ or 3′ side (Inoue *et al.*, 1997), which may be important in the general rate of reaction of transcription initiation but also in determining the pattern of the AGT gene expression. In addition, a human angiotensinogen enhancer factor has been identified in the 3′-downstream region of the gene, which may play an important role in the activation of the angiotensinogen promoter (Nibu *et al.*, 1994). All these studies point out the important role of these regulatory regions where several natural variants of the human AGT gene have been detected (*Figure 2*).

3. Role of angiotensinogen

3.1 Angiotensinogen as the substrate for renin

Human angiotensinogen is a glycoprotein which is cleaved in its *N*-terminal part by renin to generate the inactive decapeptide angiotensin I (Ang I). The K_m of renin for angiotensinogen is about $1.25 \pm 0.1\ \mu mol\ l^{-1}$, more than ten fold lower than that for

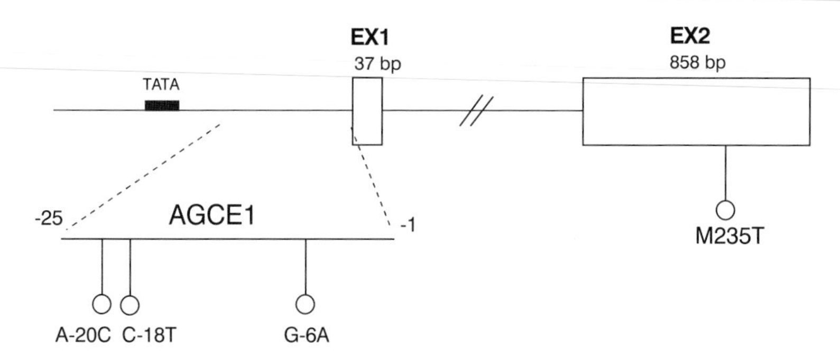

Figure 2. Polymorphisms of the promoter region of the human angiotensinogen gene. The three polymorphisms are in strong-linkage disequilibrium with M235T. The G–6A substitution has been found to be associated with an increased angiotensinogen expression. Adapted from Jeunemaitre (1992); Yanai (1996) and Inoue (1997).

the homologous synthetic tetradecapeptide renin substrate (20.7 ± 7 μmol l^{-1} (Cumin *et al.*, 1987; Tewksbury, 1990), stressing the importance of the entire molecule in the kinetics of the reaction. Its molecular mass varies between 55 and 65 kDa due to various degrees of glycosylation. We recently produced recombinant human angiotensinogen in CHO cells and showed that the four putative *N*-linked glycosylation sites (Asn–X–Ser–/Thr) of the protein can be occupied, with preferential glycosylation of the Asn^{14} and the Asn^{271}, as well as the importance of the first N-terminal site (Asn^{14}) in the kinetics of the renin hydrolysis (Gimenez-Roqueplo *et al.*, 1998a). In this work, we were able to produce a fully deglycosylated *N*–4 angiotensinogen which had a higher catalytic efficiency for human renin than the wild-type protein (k_{cat}/K_m =5.0 vs. 1.6 $\mu M^{-1}s^{-1}$).

Among the cascade of enzymatic events of the circulating renin–angiotensin system, the first step – that is the reaction of renin on angiotensinogen – is the rate-limiting one. Because plasma angiotensinogen concentration is in the micromolar range whereas plasma renin concentration is in the picomolar concentration, angiotensinogen is usually considered as a 'reservoir' for the action of renin (Tewksbury, 1990). Short-term regulation of the renin–angiotensin system does not seem to depend on changes in plasma angiotensinogen concentration. For example, rapid adaptation to changes in sodium intake is mediated by abrupt modification of renin release (which can increase up to 10-fold) by the renal juxtaglomerular (JG) cells (occurring within seconds or minutes), but changes in plasma angiotensinogen concentration are rather slow (occuring over hours or days). However, since plasma angiotensinogen concentration is about 1 μmol l^{-1} close to the K_m of its reaction with renin, variations of its concentrations can actually influence the rate of Ang I generation (Gould and Green, 1971).

3.2 Other functions?

The classical biochemical aspects of human angiotensinogen have already been reviewed in detail (Lynch and Peach, 1991; Tewksbury, 1990). We will only focus here on recent results giving some insights on its structure compared to other serpins and on its polymerization during pregnancy under a high molecular weight form of angiotensinogen.

Angiotensinogen belongs to the serpin (*ser*ine *pro*tease *in*hibitor) superfamily which comprises many genes such as α_1-antitrypsin, α_1-antichymotrypsin and antithrombin III. The sequence homology between angiotensinogen and these proteins (around 20% of identical amino acids) suggests that these genes evolved from a primitive common ancestor through a series of gene duplications, insertions and deletions (Doolittle, 1983). Based on a computer-predicted secondary structure, the distribution of α-helical segments and β-sheet sections is relatively similar in angiotensinogen and α_1-antitrypsin (Carrell and Boswell, 1986). Angiotensinogen is thought to have no serine protease inhibitory activity since its *C*-terminal part seems to have lost the property to be converted from a stressed to a relaxed form; this conformational change being a requirement for the inhibitory function of the serpin's reactive center (Huber and Carrell, 1989). However, several arguments favor the hypothesis that human angiotensinogen has a serpin-like structure with an accessible loop cleavable by some proteases. Stein *et al.* (1989) were able to show that it was efficiently cleaved by the neutrophil elastase, mainly between residues 410 and 411. Tewksbury (1990) and Mast *et al.* (1991) demonstrated that *Staphylococcus aureus* V8 protease cleaves human angiotensinogen between Glu^{408} and Ser^{409}. More recently we were able to produce *in vitro* a cleaved form of human angiotensinogen, the main cleavage site being located in the same region (J. Célérier, unpublished data). The generated *C*-terminal peptide was required for a correct folding of the protein and computer modelling predicts that it is buried in the core of the molecule. Thus, the question of the presence or the absence of a serine protease inhibitory activity *in vivo* is not yet resolved.

During pregnancy, there is a parallel rise of plasma angiotensinogen and estrogens levels (Skinner *et al.*, 1972) and the appearance of a high molecular weight form of angiotensinogen in plasma (15–30%) and in amniotic fluid (50–80%) (Tewksbury and Dart, 1982). This high molecular weight form of angiotensinogen has been recently shown to contain a covalent complex between angiotensinogen and two other proteins (Oxvig *et al.*, 1995). A major part is a 2:2 complex of angiotensinogen and the proform of eosinophil major basic protein (proMBP) from placental origin, the covalent links corresponding to disulfide bridges between the molecules (*Figure 3*). The amino acid sequence of human angiotensinogen contains four cysteine residues at positions 18, 138, 232 and 308. Only Cys^{18} and Cys^{138} are conserved in the other mammalian species cloned so far. We demonstrated the presence of a disulfide bridge between Cys^{18} and Cys^{138} (Gimenez-Roqueplo *et al.*, 1998b) in accordance with the results obtained by the analysis of a chimeric protein composed of the *N*-terminus of angiotensinogen and α_1-antitrypsin (Streatfeild-James *et al.*, 1998). We carried out site-directed mutagenesis of single cysteine residues and combination of them in eight mutants of human angiotensinogen. Analysis of these mutants demonstrated that Cys^{232} is involved in the *in vitro* formation of the complex AGT–proMBP and a possible interaction between M^{235} and Cys^{232} which could interplay with the formation of this complex (*Figure 4*) (Gimenez-Roqueplo *et al.*, 1998b).

(a)

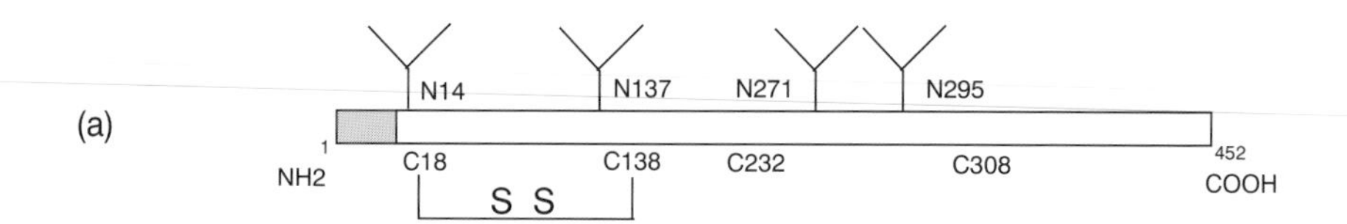

(b)

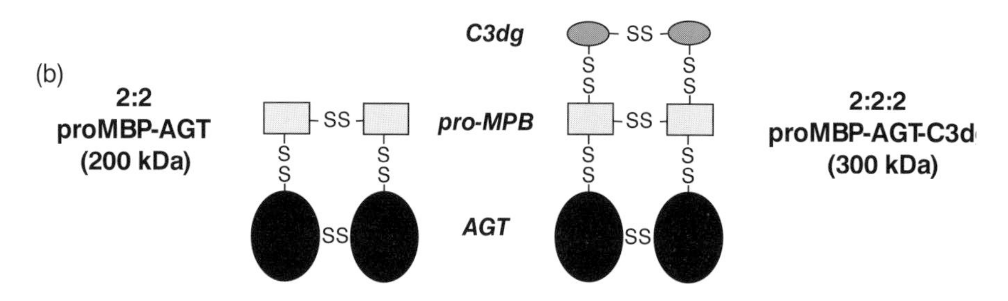

Figure 3. The two forms of angiotensinogen. (a) LMW AGT: low molecular weight angiotensinogen; (b) HMW AGT: high molecular weight angiotensinogen.

4. Angiotensinogen expression

4.1 Liver

Plasma angiotensinogen levels reflects mainly its synthesis by the liver where it is constitutively secreted (Takahashi *et al.*, 1997). In a great variety of *in vivo* and *in vitro* models, investigators have repeatedly demonstrated that administration of glucocorticoids (especially dexamethasone), estrogens, Ang II and thyroid hormones stimulate the synthesis and release of angiotensinogen (see reviews by Deschepper and Hong-Brown, 1993; Lynch and Peach, 1991; Robertson, 1993; Tewksbury, 1990). The steroid effects are neutralized by anti-glucocorticoids and

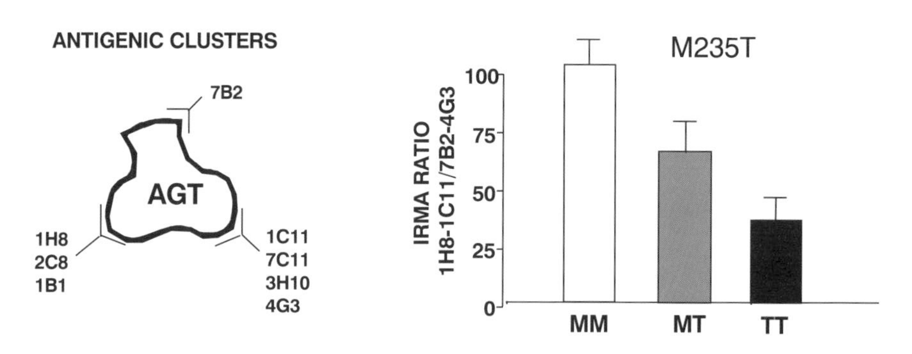

Figure 4. Epitope mapping of human angiotensinogen by monoclonal antibodies and immunological recognition of the M235T polymorphism. Adapted from results in Cohen *et al.* (1996).

anti-estrogens; thyroidectomy decreases plasma angiotensinogen levels and T3 administration restores them.

A dose-dependent increase of plasma angiotensinogen occurs with administration of oral contraceptive pills containing synthetic estrogens (Derkx *et al.*, 1986) as well as a mean 3–5 mmHg increase in blood pressure and subtle changes in renal blood (Hollenberg *et al.*, 1976). However, no direct relation has been observed between the increase in plasma angiotensinogen and blood pressure level. This stimulatory effect is not observed with percutaneous estrogen administration (DeLignieres *et al.*, 1986), probably due to the absence of accumulation of estradiol in the hepatic cells. The estrogen-induced increased expression of angiotensinogen is tissue-dependent: although the estrogen effect on liver is important, no stimulating effect has been observed in adipose tissue and contradictory results have been reported in the brain (Bunnemann *et al.*, 1993). In kidney, Campbell *et al.* (1991) showed that estrogen-induced increased plasma angiotensinogen levels were associated with an increased renal level of angiotensin peptides despite reduced plasma levels of renin and angiotensin peptides.

4.2 Other tissue localizations

Angiotensinogen is synthesized in many different tissues. Even though this expression is less abundant than in the liver, angiotensinogen appears as an important component of the extravascular local renin angiotensin system. Since the K_m of renin for angiotensinogen is relatively high, Ang I production rate of such sytems could be controlled by local angiotensinogen concentration rather than by variations in renin concentration. We will summarize only these tissue expressions which have already been reviewed (Lynch and Peach, 1991; Phillips *et al.*, 1993; Tewksbury, 1990).

The brain, large arteries, heart, kidney and adipose tissues are established sites of angiotensinogen synthesis. In the brain, *in situ* hybridization and immunohistochemistry studies have shown that glial cells are the most important source (Bunnemann, *et al.*, 1993; Lynch *et al.*, 1987; Stornetta *et al.*, 1988). Astrocytes, together with endothelial cells, are one of the main components of the blood–brain barrier; interestingly this barrier is impaired in AGT-KO mice, providing evidence that astrocytes with angiotensins are required for its functional maintenance (Kakinuma *et al.*, 1998).

Angiotensinogen is also expressed in vascular walls both in the adventitia and the medial smooth muscle cell layer (Campbell and Habener., 1986; Cassis *et al.*, 1988; Naftilan *et al.*, 1991; Ohkubo *et al.*, 1986). Increased expression of angiotensinogen mRNA has been found in the medial layer of the injured aorta, suggesting a role of angiotensinogen in the myointimal proliferation following vascular injury (Rakugi *et al.*, 1993). Because the main angiotensinogen mRNA in the rat aorta seems to be located in the brown adipose tissue (Campbell and Habener, 1987), the presence of highly vascularized adipose tissue surrounding these vessels raises the possibility of a local renin system in which adipose cells synthesize angiotensinogen. Ang II could then be generated through circulating renin and adventitial and (or) endothelial angiotensin converting enzyme (ACE).

There is, indeed, evidence for a local vascular generation of angiotensin from angiotensinogen in isolated perfused rat blood vessels (Hilgers *et al.*, 1993).

In the heart, angiotensinogen mRNA seems mostly expressed in the atria (Campbell and Habener, 1986; Lindpaintner *et al.*, 1990). The presence in the heart of mRNA of different components of the RAS, including Ang II receptors, supports the hypothesis of a physiological role of the RAS in cardiac function. Increased ACE activity and angiotensinogen expression have also been demonstrated following experimental left ventricular hypertrophy (Schunkert *et al.*, 1990), suggesting that the cardiac renin system may have an important role in modulating growth and hypertrophy of the heart.

Angiotensinogen mRNA is detectable in the kidney (Campbell and Habener, 1986; Fried and Simpson, 1986), essentially in proximal tubular cells (Ingelfinger *et al.*, 1990). Activation of the renin–angiotensin system is age-related and measurement of intrarenal angiotensin peptides in rat kidneys showed 3–6-fold higher Ang I and Ang II concentrations in newborns compared to adults (Tamura, K. *et al.*, 1996). Alteration of renal hemodynamics caused by experimental heart failure induces a renal-specific increase in angiotensinogen mRNA, suggesting its contribution to the activation of the intrarenal renin (Schunkert *et al.*, 1992). Renal angiotensinogen level, which depends also on circulating levels (Misumi *et al.*, 1989), could be an important rate-limiting factor for the local generation Ang I, at least during converting enzyme inhibition, and therefore a determinant in the generation of renal angiotensin peptides and in the fine tuning of renal vascular resistances (Campbell *et al.*, 1991a,b).

White adipose tissue is a rich source of angiotensinogen which is expressed during the differentiation of pre-adipocytes into adipocytes (Saye *et al.*, 1990). The recruitment of new fat cells could be enhanced by Ang II, via the release of prostaglandins which are adipogenic (Darimont *et al.*, 1994). In rats, angiotensinogen expression in white adipocytes is influenced by nutrition with reduction during fasting conditions and an increase with refeeding (Frederich *et al.*, 1992). An age-related decline in angiotensinogen protein and mRNA expression in retroperitoneal depots (Harp *et al.*, 1995) suggest its role in adipose tissue growth.

Finally, angiotensinogen mRNA has been detected in several other human tissues such as the adrenal gland where it is expressed in both cortex and medulla (Racz *et al.*, 1992), and where it could contribute, together with the locally produced renin, to a paracrine generation of Ang II and to the regulation of aldosterone secretion.

5. The AGT gene in experimental models of hypertension

5.1 Rodents strains

The analysis of the effect of the AGT gene on blood pressure level in strains selected for high blood pressure mainly led to negative results. The AGT locus does not cosegregate with blood pressure in F_2 rats derived from a cross between stroke-prone spontaneously hypertensive (SHR-SP) and Wistar–Kyoto (WKY; Hübner *et al.*, 1994). A missense mutation (Ile to Val at position 154 of the mature protein) in the SHR-SP angiotensinogen was found but did not show a linkage with the increase in plasma Ang II observed in these (Hübner *et al.*, 1995). However, a few studies suggested a role for AGT. Lodwick *et al.* (1995) showed a cosegregation of the rat AGT locus with a specific and mild pulse pressure increase in F_2 rats derived from a cross of the SHR with normotensive WKY rats. This locus accounted for approximately 20% of the genetic variance of this phenotype. The AGT coding sequence and plasma angiotensinogen were similar in hypertensive and normotensive rats but there was a difference in AGT mRNA levels between the two strains, especially in SHR aorta. More recently, Tamura, K. *et al.* (1996) found differences in tissue-specific regulation of angiotensinogen expression in SHR compared with WKY rats.

On the whole, the findings from SHR or SHR-SP and humans appear to differ, as they also differ in a number of identified high blood pressure loci in hypertensive rats which were not found linked to hypertension in humans. This is, for example, the case for the SA gene (Iwai and Inagami, 1992; Nabika *et al.*, 1995). This emphasizes that disease-relevant genetic loci in man and rats cannot be assumed to coincide.

5.2 Transgenic animals with AGT overexpression

Another possibility for exploring the role of angiotensinogen in blood pressure regulation is to create transgenic animals overexpressing a candidate gene (Thompson *et al.*, 1996). Because of the species specificity of the renin–angiotensinogen reaction, mice or rats expressing either the *human* renin or the *human* angiotensinogen gene do not develop hypertension. Kimura *et al.* (1992) generated transgenic mice expressing the entire rat AGT gene including 1.6 kb of 5′ flanking sequences. These animals developed hypertension, especially males, and angiotensinogen was overexpressed in liver and brain. In this study, specific cerebral expression seemed to be a prerequisite for the development of a hypertensive phenotype. Other transgenic animals have been generated by Ohkubo *et al.* (1990), who introduced in mice either the *rat* renin gene, the *rat* AGT gene, or both, under the control of the mouse metallothionine I promoter. A similar chimeric RAS was constructed through cross-mating separate lines of transgenic mice carrying either the human renin or angiotensinogen genes (Fukamizu *et al.*, 1993). Chronic overproduction of Ang II in the double transgenic mice resulted in a resetting of the baroreflex control of heart rate to a higher blood pressure without significantly changing the gain or the sensitivity of the reflex (Merrill *et al.*, 1996). These experiments documented *in vivo* the species-specificity of the renin–angiotensinogen reaction which had already been established *in vitro*. Although these transgenic experiments clearly demonstrated that overexpression of the renin and angiotensinogen genes led to increased blood pressure, their relevance to the pathogenesis of human hypertension is questionable.

An interesting model of hypertension in pregnant mice has been recently reported: Takimoto *et al.* (Takimoto *et al.*, 1996) observed that transgenic females expressing human angiotensinogen developed a transient elevation of blood pressure in late pregnancy, when mated with transgenic males expressing human renin. Blood pressure returned to normal after delivery, renal histology showed glomerulopathy lesions, and the placenta displayed gross alterations, all abnormalities also found in human pregnancy-induced hypertension. Interestingly, this result was not observed with the reverse combination, that is, female renin transgenics with male angiotensinogen transgenics. In this later situation, the pups were hypertensive but no hypertension developed in the mothers, pointing out the role of placental renin in the development of maternal hypertension.

5.3 Transgenic mice with AGT inactivation and duplication

The technology of homologous recombination allows the generation of animals with a specific gene inactivation or duplication. This approach has been used by Smithies and Kim (1994) to evaluate the hypothesis that genetically determined elevated angiotensinogen levels might predispose to hypertension. These authors used targeted gene disruption and duplication to generate mice having one, two, three or four copies of the angiotensinogen gene. Plasma angiotensinogen also increased progressively, according to the AGT gene copies number. Interestingly, blood pressure paralleled the AGT copy number, being 122, 129 and 138 mmHg for the 1 copy, 2 copies (wild type) and 3 copies (Kim *et al.*, 1995). Renal blood flow was increased in the 1 copy animals, showing the importance of the renin substrate concentration in the determination of renal vascular resistances. These results directly demonstrate that small elevations of plasma angiotensinogen, in the order of magnitude of those associated with the M235T polymorphism in humans, might play a role in renal vascular resistance and blood pressure. They also showed that the renin–angiotensin system plays an important role in the maturation of the kidney. The analysis of preweaning and some adult AGT–/– mice showed marked renal abnormalities, consisting of severe renal vascular hypertrophy, cortical thinning, severe focal fibrosis, tubular atrophy and chronic inflammation (Niimura *et al.*, 1995). These abnormalities can be rescued by complementation with both human renin and AGT genes, providing an interesting model to test the significance of variants of the human renin–angiotensin system by a combined transgenic and gene targeting approach (Davisson *et al.*, 1997).

6. The AGT gene and human essential hypertension

6.1 Linkage studies

The potential role of the AGT gene in human essential hypertension has been tested using a highly polymorphic dinucleotide GT repeat (80% heterozygosity) displaying 10 alleles and located in the 3′ region of the angiotensinogen gene (Kotelevtsev *et al.*, 1991). Evidence for linkage was suggested in two large series of hypertensive sibships which yielded a total of 379 sib pairs (Salt Lake City, Utah, U.S.A. and Paris, France) (Jeunemaitre *et al.*, 1992). When hypertensive sib pairs were stratified according to hypertension severity (characterized by a diastolic blood pressure greater than 100 mmHg or taking two or more antihypertensive medications), a 17% excess of allele sharing was found in both the Utah and Paris groups (*Table 1*). It is interesting to note that significant linkage was obtained only in male pairs, while no excess of shared angiotensinogen alleles was observed in female–female comparisons, suggesting the influence of an epistatic hormonal phenomenon. From this study, it was estimated that mutations at the AGT locus might be a predisposing factor in at least 3–6% of hypertensive individuals younger than 60 years of age.

Subsequently, Caulfield *et al.* (1994) showed a strong linkage between the AGT locus (25% excess of concordance) and essential hypertension in a study involving 63 Caucasian families recruited in London. This linkage was also observed in the subgroup of patients with diastolic pressure above 100 mmHg but was not different among male–female pairs. A surprisingly strong association between alleles of the AGT-GT repeat and essential hypertension may have favored the strength of linkage using the Affected Pedigree Member Method.

More recently, a large European study evaluated the contribution of the AGT locus using the same polymorphic marker in 350 European families involving 630 affected sib pairs (Brand *et al.*, 1998). No evidence for linkage was found either in the whole panel or in family subsets selected for severity or early onset of disease, even though a tendency was observed in male–male pairs. Similarly, the analysis of 310 hypertensive pairs from a rural area of Central China revealed no evidence for linkage either with the microsatellite marker or the M235T and T174M polymorphisms (Niu *et al.*, 1998). Conversely, in Mexican Americans, a marginally significant linkage was found in 180 affected individuals belonging to 46 large families living in San Antonio (Atwood *et al.*, 1997). Caulfield *et al.* also found linkage between the AGT locus and high blood pressure in 63 African Carribean families (Caulfield *et al.*, 1995). Overall, these conflicting results could reflect

Table 1. Angiotensinogen and essential hypertension: linkage studies

Populations	Sibships	Excess of concordance	Significance	References
Caucasians				
1. Utah	244 sib pairs	3.8%	ns	Jeunemaitre, 1992
	50 more severe	17.1%	$P<0.01$	
Paris	135 sibpairs	7.7%	$P<0.05$	
	60 more severe	15.3%	$P<0.02$	
2. London	63 multiplex	25.9%	$P<0.001$	Caulfield, 1994
	31 more severe	25.1%	$P<0.001$	
3. Europe	630 sibpairs	<1%	ns	Brand, 1998
	295 more severe	<1%	ns	
Chinese				
Anhui province	310 sibpairs	<1%	ns	Niu, 1998
	147 top 5th percentile		ns	
Afro-Caribeans				
St Vincent	63 sibpairs	$T=3.07$	$P=0.001$	Caulfield, 1995
Mexican-Americans				
San-Antonio	36 families, 180 hypertensives		$P<0.02$	Atwood, 1997

racial/ethnic differences with regard to the pathogenesis of hypertension, the relatively modest role of the angiotensinogen locus on blood pressure level and the difficulty in finding linkage in a such complex disorder.

6.2 Case–control studies

In our initial search to detect mutations that may affect this locus, several missense mutations were found, mainly located in the exon 2 of the AGT gene (*Figure 2*). Most of them were rare and detected in few hypertensive probands. Two polymorphisms T174M (Thr→Met at position 174 of the mature angiotensinogen) and M235T (Met→Thr at position 235) were in complete linkage disequilibrium. The M235T polymorphism which showed association with hypertension and plasma angiotensinogen was then extensively studied in many case–control studies.

We will summarize the main findings associated with this polymorphism. However, some words of caution are needed before interpretation of the corresponding results. Firstly, the frequency of the AGT T235 allele varies strongly according to ethnic groups. It is more frequent in the Asian (approximately 0.75) than in the Caucasian population (0.40) and is by far the predominant allele in the African population (0.90–0.95) (*Table 2*). As a consequence, false positive results may arise from population admixture but also false negative results may be obtained in populations in which this allele is predominant, due to the limited statistical power of the corresponding association studies. Secondly, as for other susceptibility genes involved in multifactorial diseases, the observation of positive or negative results is strongly constrained by the sample size of the study. For example, at least 400 cases and 400 controls are needed to replicate our initial findings with 80% power to detect a difference in the T235 allele frequency from 0.38 to 0.46 between hypertensives and controls (Jeunemaitre *et al.*, 1992). However, most of the results reported so far concern relatively small numbers of patients. Finally, different genetic backgrounds, different choice criteria (such as blood pressure level, age of onset of hypertension, weight) or choice of different subsets of hypertensives might influence the results and explain some divergence. The results will be presented according to the different ethnic groups even though the homogeneity of each group itself is questionable.

Table 2. Average frequency of the AGT 235T allele according to ethnic origin

Ethnic origin	Frequency (T allele)
Caucasians	0.40
Indian-Americans	0.65
Japanese	0.70
African-Americans	0.85
Afro-Caribbeans	0.85
Africans	0.95

The M235T polymorphism and essential hypertension

1. Caucasians

There are divergent results concerning the association between the M235T polymorphism and essential hypertension in Caucasians. In our original study (Jeunemaitre *et al.*, 1992), a positive association was detected. In both Utah and Paris groups, the T235 allele was found more frequently in hypertensive probands (0.46) than in controls (0.38), especially in the more severe index cases (allele frequency 0.50). In a subsequent study on 136 mild to moderate hypertensives from Paris, we found that the T235 allele was more frequent in these unselected hypertensive patients than in controls, although the significance was reached only for patients with a family history of hypertension (Jeunemaitre *et al.*, 1993). Schmidt *et al.* (1995) found also a higher frequency of the T235 allele in subjects with a family history and early onset of hypertension. Recently, three large studies found a positive association between the M235T polymorphism and essential hypertension. The T235 allele frequency was 0.47 in a series of 477 probands of hypertensive families and 0.38 (P=0.004) in the 364 Caucasian controls (Jeunemaitre *et al.*, 1997). In another case–control study involving 779 hypertensive subjects and 532 Caucasian controls, Tiret *et al.* (1998) found a significant increase in T235 allele frequency which was restricted to men (0.46 vs. 0.40, P=0.01). A similar trend was observed with women whose hypertension was diagnosed before 45 years. In a cross-sectional sample of 634 middle-aged subjects from the Monica Augsburg cohort, Schunkert *et al.* (1997) found that individuals carrying at least one copy of the T235 allele had higher systolic and diastolic blood pressures and were more likely to use an antihypertensive drug.

Conversely, Caulfield *et al.* did not find an association between hypertension and the M235T polymorphism but compared only 64 controls and 63 hypertensive probands of British families (Caulfield *et al.*, 1994). Similar results were obtained by Barley *et al.* (1994) on 64 cases and 74 controls. These two studies performed in UK individuals showed similar frequencies in hypertensives, but quite different frequencies in normotensives (0.50 and 0.40, respectively). Bennett *et al.* (1993) failed also to find an association using only 92 individuals, offsprings of two hypertensive parents and ascertained through self-referral. On a population-based selection of 104 hypertensives with mild hypertension and 195 matched normotensives, Fornage *et al.* (1995) did not find any association but a surprising difference in T235 allele frequency between normotensive men (0.47) and women (0.36). More recently, Hingorami *et al.* (1996) did not find any difference in M235T frequency between 223 hypertensive subjects and 187 normotensive individuals originated from the East Anglia region of the U.K. A more powerful study performed in Finland by Kiema *et al.* (1996) showed negative results in 508 mild hypertensives and 523 population-based controls. More recent case–control studies dealing with a limited number of subjects were also negative (Fardella *et al.*, 1998; Fernandez-Llama *et al.*, 1998; Vasku *et al.*, 1998).

The association of the T235 allele with hypertension was replicated in a sample of the Framingham study (T frequency 0.51 vs. 0.39 in cases and controls, respectively), as well as in white subjects of the ARIC study when the effects of body mass index, triglycerides and the presence of coronary heart disease were controlled (Borecki *et al.*, 1997). A positive association was also found in 229 hypertensive subjects of Bedouin descent originated from Arab Emirates (Frossard *et al.*, 1998).

It is difficult to form conclusions about the role of the M235T polymorphism in the Caucasian population. A meta-analysis was performed on all relevant articles published between 1992 and 1996, involving 5493 patients (Kunz *et al.*, 1997). The AGT T235 allele was significantly associated with hypertension, in subjects with positive family history of hypertension (odds ratio: 1.42, $P<0.0001$) and those with more severe hypertension (odds ratio: 1.34, $P<0.0001$). It is probably crucial here to emphasize the importance of sample size when one wants to test such an association in a complex disease like hypertension. The contribution of the T235 allele may be also dependent of interacting factors such as sex, body mass index and estrogen status.

2. Japanese

A reduced heterogeneity of the pathophysiology of hypertension due to a more homogeneous genetic and environmental background of the population could explain the more homogeneous results which have been reported in the Japanese population. An increased frequency of the T235 allele was consistently observed in the Japanese population ranging from 0.65 (Nishimura *et al.*, 1995) to 0.75 (Hata *et al.*, 1994). A significant increase in hypertensive subjects has been reported by several groups having selected subjects from multiple centers such as Yamanashi and Nagoya (Hata *et al.*, 1994), Osaka (Kamitani *et al.*, 1994; Sato *et al.*, 1997) and 11 rural communities (Nishimura *et al.*, 1995). A multiple regression analysis in 347 subjects, selected in the Department of Medicine of the Shiga University, showed that the T235 allele was a significant predictor of both diastolic and systolic blood pressure in the population less than 50 years, whereas body mass index was the only predictor of blood pressure in older subjects (Iwai *et al.*, 1995).

3. Africans and African Americans

The T235 allele is by far the most frequent allele in Africans (around 0.90) and in African Americans (around 0.80; Rutledge *et al.*, 1994), frequencies which are consistent with an approximate 25% admixture of genes of European origin in these populations (Rotimi *et al.*, 1996). The distribution of genotypes for M235T revealed similar frequencies for T235 in the 63 hypertensive index cases (0.88), the 150 hypertensive individuals (0.86) and the 93 normotensive controls (0.84) recruited from the African Carribean population studied by Caulfield *et al.* (1995). A weak association between blood pressure and angiotensinogen has been found in a community survey in Jamaica, but the effect was more marked in conjunction with plasma ACE level (Forrester *et al.*, 1996).

The M235T polymorphism has been evaluated in other ethnic backgrounds (Fardella *et al.*, 1998). A positive association was found between systolic blood pressure and the T235 allele in 497 adult native Canadians from an isolated community in Northern Ontario (Hegele *et al.*, 1997). As already mentioned, the interpretation of these studies is difficult due to the high prevalence of the T235 allele in these populations.

The T174M polymorphism and essential hypertension. The T174M polymorphism has a frequency of about 0.10 to 0.15 in all ethnic groups studied so far. It is in complete linkage disequilibrium with M235T. It does not induce any epitopic change in the immunological recognition of the protein (Cohen *et al.*, 1996). In our initial work, this polymorphism was found to be associated with severe hypertension (Jeunemaitre *et al.*, 1992). At that time, two possibilities were suggested (i) a true association of the 174M allele with hypertension, (ii) an association which would only be the consequence of an hitch-hiking effect due its linkage disequilibrium with 235T. Since that work, most of the other case–control studies failed to replicate an association between T174M and hypertension. In an extensive work involving 477 probands of hypertensive families and 364 Caucasian controls as well as a series of 92 hypertensives and 122 controls from Japan, allele frequencies averaging 0.10 were observed in both populations independently of the hypertension status (Jeunemaitre *et al.*, 1997). In another large study, Tiret *et al.* found also very similar 174M frequencies in 802 hypertensive subjects and 658 Caucasian controls (Tiret *et al.*, 1998).

However, a few independent studies suggested a possible association between this polymorphism and hypertension. In a genetic isolate of 741 individuals, the Hutterian Brethren in North America, there was a significant association between resting blood pressure and the 174M allele (Hegele *et al.*, 1994). This effect was restricted to systolic blood pressure and was only observed in men. The same group also found an association between the ApoB codon 4154, the angiotensinogen T174M polymorphism and blood pressure (Hegele *et al.*, 1996) as well as an association with markers of the chromosome 1q region and the waist to hip circumference in men (Hegele *et al.*, 1995). Interaction with weight was also observed by Tiret *et al.* (1995). A higher frequency of the 174M allele was only observed in one limited study in Japanese individuals (Morise *et al.*, 1995).

Polymorphisms of the promoter region. We found a G to A substitution located at position –6 upstream of the initial transcription site which is of the same frequency and in almost complete linkage disequilibrium with the T235 allele in the Caucasian, Japanese and Afro Carribean populations (Inoue *et al.*, 1997; Jeunemaitre *et al.*, 1997). In these studies, the haplotype combining both G-6A and M235T polymorphisms was the one associated with hypertension. *In vitro* experiments suggest that the G-6A nucleotide substitution affects the basal transcription rate of the angiotensinogen. In an extensive work using different cell lines and multiple lengths of human AGT promoters, Lalouel and colleagues found 20–40% differences in promoter activities containing either the G or A nucleotide (Inoue *et al.*, 1997). In addition, site-directed mutagenesis confirmed the significance of the –6 region in transcriptional activity and binding of at least two distinct nuclear transcription factors. If the increase of AGT expression induced by the G-6A substitution *in vitro* could be extrapolated to the *in vivo* situation, we might be able to explain the association of the M235T polymorphism with increased plasma angiotensinogen concentrations.

Other polymorphisms of the promoter region of the AGT gene have been found in strong linkage disequilibrium with M235T (*Figure 2*). Among them, the A-20C and C-18T variants are located in the core promoter region of the gene comprised between positions –25 to –1 and denoted AGCE1, which has been demonstrated to bind a ubiquitously expressed nuclear factor (Yanai *et al.*, 1996). Because of its low frequency in several populations (1–3%, personal results), it is difficult to assess the association of the C-18T polymorphism with hypertension. However, Sato *et al.* found a significant increase of this allele when comparing 180 hypertensive Japanese (3.5%) and 194 controls (1%, $P<0.01$) (Sato *et al.*, 1997). The A-20C could be more interesting for two reasons: (i) its higher allele frequency (around 20%), (ii) its potential impact *in vitro*, since the A to C change creates a sequence consensus to an estrogen receptor element. Analysing 186 hypertensive Japanese subjects, Ishigami *et al.* showed a weak but significant correlation between A-20C, plasma angiotensinogen concentration and essential hypertension ($P=0.05$) (Ishigami *et al.*, 1997).

The M235T polymorphism as a marker of a causative mutation? There is no evidence that T235 directly affects the function, the secretion or the metabolism of angiotensinogen. The comparison of plasma samples from 235 TT and MM homozygotes did not show any difference in the K_m for renin (personal unpublished results). Similarly, analysis of the recombinant M235 and T235 proteins showed no difference in protein secretion and in K_m, K_{cat} and catalytic efficiencies (Inoue *et al.*, 1997). However, the substitution of a methionine by a threonine residue has some structural implications. The 235 amino acid is located in a region with little conservation but directly adjacent to the s3A sheet which is constituted by very well conserved residues among the members of the serpin family (Huber and Carrell, 1989). By comparison with the three-dimensional structure of α_1-antitrypsin, the corresponding s3A sheet and the 235 amino acid of the angiotensinogen molecule should be exposed to the surface. Indeed, using multiple monoclonal antibodies and two different immunometric assays, Cohen *et al.* (1996) showed an epitopic change, allowing the discrimination of homozygous and heterozygous plasma for this allele (*Figure 4*). This immunological tool might prove to be interesting in genotyping large populations from plasma. Even though there is no evidence that this amino acid change can induce a functional modification of angiotensinogen, it may play a role on the formation of high-molecular weight angiotensinogen (see below).

In the absence of a causal link between the M235T amino acid substitution and angiotensinogen function, the other hypothesis is that this variant could serve as a marker for unknown molecular variant(s), directly mediating predisposition. In this case, it is logical to assume that only a subset of the genes carrying T235 also harbor the functional mutation(s). To address this question, we identified and genotyped a series of 10 diallelic AGT polymorphisms and characterized their haplotype distribution in a large series of hypertensive cases and normotensive controls of Caucasian and Japanese (Jeunemaitre *et al.*, 1997). Partition of the T235 allele with diallelic markers did not unambiguously pinpoint a subset of T235 haplotypes responsible of the association with hypertension. From this haplotype analysis of angiotensinogen in different populations (Jeunemaitre *et al.*, 1997), the high prevalence of the T235 allele in the majority of human populations and its presence in primate monkeys (Inoue *et al.*, 1997), it appears that T235 is the ancestral allele and that M235 is the neomorph. If this allele is associated with an increased activity of the renin–angiotensin–aldosterone system, one might speculate that it would have conferred an advantage for retaining salt and for controlling intravascular volume and blood pressure, at a time when the access to salt for humans was limited.

6.3 A comprehensive pathway between the M235T/G–6A polymorphism and hypertension

Relation with plasma angiotensinogen. In rodents, the relation between plasma angiotensinogen and blood pressure relies on several lines of evidence. A blood pressure decrease was observed after injection of angiotensinogen antibodies in rats, which was dependent on sodium balance (Gardes *et al.*, 1982). Conversely, a blood pressure increase was observed by Ménard *et al.* (1991) after injection of pure angiotensinogen in salt-depleted rats. In mice carrying one to four copies of the murine angiotensinogen gene, there was an 8-mmHg blood pressure increase per copy which was roughly proportional to the increase of plasma angiotensinogen concentration (Kim *et al.*, 1995). Finally, antisense strategies for reducing hypertension which have been recently reviewed (Phillips, 1997) confirm the importance of the *in vivo* angiotensinogen concentration on the state of activation of the renin system and on blood pressure level. Hypertension and liver angiotensinogen expression was reduced by using AGT antisense oligodeoxynucleotides (AS-ODNs) delivered into the portal vein of SHR rats (Tomita *et al.*, 1995) or peripherally when coupled to carrier molecules targeted to the liver (Makino *et al.*, 1998). A long-lasting decrease in blood pressure was also observed when AGT AS-ODNs were delivered intra-cerebroventricularly, supporting the concept of centrally mediated regulation of hypertension (Gyurko *et al.*, 1993; Peng *et al.*, 1998).

In humans, a relation between plasma angiotensinogen and blood pressure level has been suggested from several studies. A high correlation between the renin substrate and blood pressure ($r=0.39$, $P<0.0001$) was first reported in a large study involving 574 subjects (Walker *et al.*, 1979). Higher levels of plasma angiotensinogen have also been observed in hypertensives and offspring of hypertensive parents compared to normotensives (Fasola *et al.*, 1968). In a study aimed to evaluate the genetic determinants of blood pressure by comparing offspring of parents having either high or low blood pressure ('four-corners' approach), Watt *et al.* (1992) found a significant increase in plasma angiotensinogen in offspring with high blood pressure and high parental blood pressure. Finally, two cases of hypertension associated with hepatic cell tumors producing large amounts of angiotensinogen have been reported (Kew *et al.*, 1989; Ueno *et al.*, 1984). All these findings suggest that increase in plasma angiotensinogen level in humans might contribute to elevated blood pressure.

In our initial study involving French and Utah families, a positive association was found between the M235T polymorphism and plasma angiotensinogen concentration, with about a 20% increase in plasma angiotensinogen in subjects carrying the T235 allele (Jeunemaitre *et al.*, 1992). This association was observed both in males and females, appeared to be codominant and was confirmed in another sample of patients with essential hypertension (Jeunemaitre *et al.*, 1992). Bloem *et al.* (1995) reported also that plasma angiotensinogen levels were around

13% higher in normotensive white children bearing the T235 allele. In addition, the mean plasma angiotensinogen concentration in African American children was 19% higher than in Caucasians. A more recent evaluation of this African American group by Bloem and colleagues (Bloem *et al.*, 1997) showed an association between a haplotype containing the T235 allele and another AGT gene polymorphism located in the promoter region of the gene. In a large sample of the Monica Augsburg cohort, a mild codominant and significant increase of plasma angiotensinogen concentration was also observed according to the M235T polymorphism (Schunkert *et al.*, 1997). Because of the intra- and interassay variability of the plasma angiotensinogen measurement and the mild association with the M235T polymorphism, a large number of individuals are probably required to detect this relation (*Table 3*).

The significance of elevated plasma angiotensinogen levels was recently extended to the short negative feedback of Ang II on renin release. In an analysis of 228 men belonging to the Monica cohort, Danser *et al.* (1998) showed a codominant negative relationship between the M235T genotype and plasma renin concentration. Thus, the rate of Ang II formation in the circulation will not be dependent of the M235T genotype under most circumstances. However, if renin levels are elevated, especially in conditions of salt deprivation, increased angiotensinogen levels may result in accelerated Ang II formation and thus contribute to the developement of hypertension and other cardiovascular disorders. Chronically increased plasma angiotensinogen could also facilitate hypertension in predisposed individuals who would have in addition an abnormal short feedback loop between Ang II and renin release.

The M235T polymorphism and the non-modulation phenotype. Williams and Hollenberg (1991) have shown in a series of studies that nonmodulation is a trait characterized by a failure to modulate renal, vascular and adrenal zona glomerulosa responsiveness to Ang II during a high Na^+ intake. Patients with this trait may represent up to 40% of essential hypertensives. These patients have a reduced renal vascular response to Ang II during a high salt diet. Nonmodulators could, therefore, be a discrete subset of the essential hypertension population. This trait seems to be genetically inherited (Williams *et al.*, 1992). Using the same protocol, the same group recently showed that there was a blunted renal vascular

Table 3. Relation between plasma angiotensinogen concentration and the angiotensinogen M235T polymorphism in Caucasians

	MM	MT	TT	Significance
Utah HTN Probands[a]	1422 (67)	1479 (109)	1641 (33)	$P<0.005$
Paris HTN Probands[a]	1085 (32)	1318 (55)	1514 (29)	$P<0.001$
Paris HTN[b]	990 (45)	1046 (63)	1130 (28)	$P<0.03$
Children[c]	1438 (50)	1577 (72)	1646 (26)	$P<0.005$
Monica Cohort[d]	940 (216)	1010 (291)	1080 (103)	$P<0.01$

The angiotensinogen concentration is expressed in ng Ang I ml^{-1}. The number of subjects is given in parenthesis. HTN = hypertensive subjects. Adapted from [a]Jeunemaitre (1992), [b]Jeunemaitre (1993), [c]Bloem (1995), [d]Schunkert (1997).

response to Ang II in homozygous patients carrying the T235 genotype compared to the heterozygous or homozygous M235 genotype (Hopkins *et al.*, 1996; *Figure* 5). In addition, obesity was found to interact significantly with genotype and enhanced the blunting effect on renal vascular response. These results that are in accordance with the increased renal blood flow in mice bearing one AGT copy (Kim *et al.*, 1995), suggest that patients carrying the T235 genotype produce more intra-renal Ang II thereby contributing to a disturbed renal physiology.

6.4 Other rare missense mutations

In addition to the T174M and M235T diallelic polymorphisms, other rare missense mutations have been described at the AGT locus. Most of them have been described in few Caucasian hypertensive probands (L359M, V388M; Jeunemaitre *et al.*, 1992) and in rare African American subjects (L209I, L211R; Hixson and Powers, 1995), without further analysis of their potential pathophysiological consequences. Two mutations located in exon 2 of the AGT gene have been thoroughly analyzed.

The mutation occurring at the site of cleavage of angiotensinogen (Leu_{10}–Val_{11}) by renin was initially found in a patient with pre-eclampsia (Inoue *et al.*, 1995). The kinetic consequences of this mutation were especially interesting to study since it constitutes the core of the renin–angiotensinogen reaction which is strongly species specific: human renin cleaves a Leu–Val bond in human angiotensinogen whereas rat renin hydrolyzes a Leu–Leu bond in rat angiotensinogen (Bohlender *et al.*, 1996; Bouhnik *et al.*, 1981). When the L10F mutation was expressed *in vitro*, it resulted in a 2-fold increase of the catalytic efficiency of renin acting on the corresponding human angiotensinogen produced by transient expression (Inoue *et al.*, 1995). In addition, ACE exhibited more than a 2-fold increased catalytic efficiency for the mutant F10–Ang I decapeptide with phenylalanine at residue 10 when compared to the natural L10–Ang I. It was therefore logical to suspect that these kinetic differences may significantly affect the production of Ang II and alter the function of the systemic or the local renin system in this form of hypertension induced by pregnancy.

We were able to test this hypothesis in a patient with hypertension, heterozygous for the L10F mutation (X. Jeunemaitre *et al.*, unpublished data). The use of Ang I antibodies specific to either the L10–Ang I or the F10–Ang I,

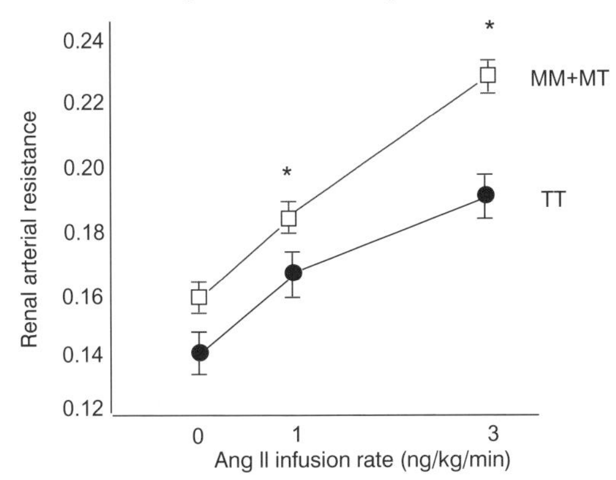

Figure 5. Renal arterial resistance according to the AGT M235T genotype. Compared with the others, patients bearing the 235 TT genotype showed a blunted response following Ang II infusion. Adapted from results in Hopkins *et al.* (1996).

allowed to confirm *in vivo* the results obtained by Inoue *et al. in vitro* (Inoue *et al.*, 1995). A pharmacological blockade of the Ang II generation by an ACE inhibitor lead to a 3-fold increase in renin secretion. Following this increase in renin, we could estimate that the generation of F10–Ang I was 4–6-fold greater than that of L10–Ang I. Taking into account the greater catalytic efficiency of ACE for this mutant form, the heterozygous L10F behaved as if one his AGT allele was much more effective in the Ang II generation than the other.

Another interesting mutation is the one found to replace a tyrosine residue by a cysteine at position 248 (Y248C) of angiotensinogen. We identified one pedigree where several members of the family were heterozygous for this mutation (Gimenez-Roqueplo *et al.*, 1996). Compared to other members of the family, heterozygous C248 individuals had a 40% decrease in plasma angiotensinogen concentration and a similar reduction of the Ang I production rate. This decrease was possibly due to an altered secretion of angiotensinogen, since the C248 mutation resulted *in vitro* in an abnormally glycosylated and secreted protein. The use of angiotensinogen monoclonal antibodies revealed up to a 7.5-fold difference in epitopic recognition of the wild-type and mutant proteins, but similar kinetic constants of Ang I production with human renin were observed.

7. Angiotensinogen, obesity and blood pressure

Angiotensinogen is abundantly expressed in human adipose tissue and together with other markers of adipocyte function, its production appears at the time of differentiation from fibroblast to pre-adipocyte cell lines (Saye *et al.*, 1990). Ang II could be one major mechanism of stimulation of the development of adipose tissue (Ailhaud, 1997). Angiotensinogen expression in white adipocytes is influenced by nutrition, with a 15–30% reduction during fasting conditions and a 2-fold increase in re-fed rats (Frederich *et al.*, 1992). An increased number of adipocytes could lead to a parallel increase in angiotensinogen production resulting into vasoconstriction and increased peripheral resistance. The nutritional status could play a role in influencing blood supply to adipose tissue and thus affect vascular resistance and blood pressure in obese individuals.

This hypothesis has not been confirmed but several observations are of interest. Plasma angiotensinogen levels have been shown to correlate strongly with blood pressure during weight loss (Eggena *et al.*, 1991). In white and black children and adolescents, a strong relationship was observed between body mass index and plasma angiotensinogen concentration (Bloem *et al.*, 1995). A further analysis of this population showed the combined influence of gender, weight, sex hormones and a haplotype (–1074t and 235T variants) of the AGT gene (Pratt *et al.*, 1998). In a study involving 1557 participants from four black African populations, Cooper *et al.* (1998) found that body mass index was highly correlated to plasma angiotensinogen levels. Controlling for age, significant correlations were also observed between plasma angiotensinogen and systolic blood pressure for two of the four populations. A high correlation between plasma angiotensinogen and body mass index (r=0.29, P<0.001) and leptin levels (r=0.40, P<0.001) was also observed recently in 91 healthy male students (Schorr *et al.*, 1998). A study of 67 obese Japanese women showed that those who were hypertensive exhibited a higher intra-abdominal/subcutaneous area fat index than the normotensives (Kanai *et al.*, 1990). In the same study, a strong and significant correlation was found between this fat index and systolic and diastolic blood pressure (r=0.62 and 0.63, respectively), independently of age and body mass index. Overall, these results suggest that angiotensinogen expression and regulation in adipose tissue may interact with blood pressure levels.

Insulin resistance may represent a potential link by means of increased angiotensinogen secretion from adipose tissue, especially in obese subjects (Aubert *et al.*, 1998). It is interesting to note that insulin causes an increase in AGT transcription and production in vascular smooth vessels cells and thus could participate in vascular hypertrophy through the trophic effects of Ang II (Kamide *et al.*, 1998). Polymorphisms of both the AGT and the ACE genes have been associated with insulin resistance (Cambien *et al.*, 1998; Huang *et al.*, 1998; Sheu *et al.*, 1998).

8. The angiotensinogen gene and pre-eclampsia

Pre-eclampsia complicates about 5% of all pregnancies and is a leading cause of maternal, fetal and neonatal mortality and morbidity. Early reports have suggested a genetic component of this complex condition (Chesley and Cooper, 1986; Sutherland *et al.*, 1981) with a mode of transmission and maternal–fetal genotype interactions which remain controversial (Arngrimsson *et al.*, 1995). In addition to the high circulating concentration of angiotensinogen in plasma and in the amniotic fluid, angiotensinogen is expressed in and around human spiral arteries during the first trimester, and may play a role in pregnancy-induced remodelling of these vessels (Morgan *et al.*, 1997). Given the role of the renin–angiotensin system in the regulation of blood volume, the considerable physiological changes in plasma volume and vascular resistances during pregnancy and the relation found between the AGT gene and essential hypertension, it was logical to test this locus in pre-eclampsia.

Analysis of the allelic inheritance of the AGT dinucleotide repeat in 52 Icelandic and Scottish families, Arngrimsson *et al.* (1993) showed a significant linkage between the AGT locus and pre-eclampsia. Ward *et al.* (1993) found a significant association between the angiotensinogen M235T variant and pre-eclampsia in both Caucasian and Japanese samples. In this study, 20% of Caucasian women homozygous for the 235T allele suffered of pre-eclampsia compared with less than 1% of Caucasian women homozygous for the 235M allele. The association in Japanese women was confirmed in a larger case–control study involving 139 women with pregnancy-induced hypertension and 278 matched controls (Kobashi *et al.*, 1995). However, other studies limited in their sample size found no indication of association or linkage between pre-eclampsia and the AGT gene in Caucasian women (Guo *et al.*, 1997; Morgan *et al.*, 1995; Wilton *et al.*, 1995). Interestingly, nulligravid women homozygous in T235 have a reduced plasma volume during the follicular phase of the menstrual cycle compared with those bearing the MT or MM genotype (Bernstein *et al.*, 1998).

Some of the contradictory results of the case–control studies could be explained by the difficulty in distinguishing between pre-eclampsia and other hypertensive disorders of pregnancy, which are likely to be caused by different mechanisms. The angiotensinogen gene could predispose to only one subset of pregnancy-induced hypertension. Two molecular mechanisms could explain this relationship. Firstly, in heterozygous women, angiotensinogen expression of the T235 allele in decidual spiral arteries seems higher than in the M235 allele (Morgan *et al.*, 1997). Secondly, we recently demonstrated that the presence of a Met or Thr at position 235 could interact with the Cys at position 232 of the mature protein, and thus favor or inhibit the formation of a disulfide bond with proMBP (Gimenez-Roqueplo *et al.*, 1998a). Because hydrolysis of the angiotensinogen–proMBP complex is about seven times slower than the monomeric form of angiotensinogen and the proportion of the latter was found to be higher with the T235 recombinant angiotensinogen *in vitro*, women bearing this allele could therefore possess a more active renin substrate in the amniotic fluid and placenta. Both mechanisms acting together, one linked to an increased angiotensinogen expression due to the G-6A variant, the other due an increased proportion of active monomeric form of angiotensinogen, could increase the local formation of Ang II and facilitate hyperplasia of the spiral arteries and thus a reduction of the uteroplacental blood flow.

9. Conclusions and perspectives

Molecular genetic studies suggest that angiotensinogen is a susceptibility locus for human essential hypertension and pregnancy-induced hypertension. The M235T and the G-6A polymorphisms are associated with an increased plasma angiotensinogen level which could result in turn in a small increase in the formation rate of Ang II, especially in tissues where these proteins are rate limiting for Ang II generation. This genetically chronic overstimulation of the renin system would then favor kidney sodium reabsorption, vascular hypertrophy and(or) increase of the sympathetic nervous system activity thus predisposing to hypertension and its complications. However, the effect of these polymorphisms appears relatively weak in the whole population, explaining the positive and negative findings. It is likely that this effect is modulated by a variety of interacting genes, pathophysiological conditions (gender, estrogen status, obesity) and environmental factors which remained to be determined. It is interesting to note that some positive but limited reports have suggested a relationship between genetic polymorphisms and responses to antihypertensive therapy (O'Byrne and Caulfield, 1998). Defining the subset of individuals where the AGT gene may play a more important role, like salt-sensivity (Hunt *et al.*, 1998) will be one of the important tasks of the near future.

Acknowledgments

This work was supported by grants from INSERM, Collège de France, Bristol-Myers Squibb, Hoffman La Roche, Assistance Publique-Hôpitaux de Paris, Association Claude Bernard and Association Naturalia and Biologia.

References

Ailhaud, G. (1997) Molecular mechanisms of adipocyte differentiation. *J. Endocrinol.* **155**: 201–202.

Arngrimsson, R., Purandare, S., Connor, M., Walker, J.J., Björnsson, S., Soubrier, F., Kotelevtsev, Y., Geirsson, R. T. and Björnsson, H. (1993) Angiotensinogen: a candidate gene involved in preeclampsia? *Nature Genet.* **4**: 114–115.

Arngrimsson, R., Bjornsson, H. and Geirsson, R. (1995) Analysis of different inheritance patterns in preeclampsia/eclampsia syndrome. In: *Hypertension in Pregnancy* (ed. P.C. Rubin). In: *Handbook of Hypertension*, Vol. 2 (eds W.H. Birkenhäger and J.L. Reid). Elsevier, Amsterdam, pp. 27–38.

Atwood, L.D., Kammerer, C.M., Samollow, P.B., Hixson, J.E., Shade, R.E. and MacCluer, J.W. (1997) Linkage of essential hypertension to the angiotensinogen locus in Mexican Americans. *Hypertension* **30**: 326–330.

Aubert, J., Safonova, I., Negrel, R. and Ailhaud, G. (1998) Insulin down-regulates angiotensinogen gene expression and angiotensinogen secretion in cultured adipose cells. *Biochem. Biophys. Res. Commun.* **250**: 77–82.

Barley, J., Blackwood, A., Sagnella, G., Markandu, N., MacGregor, G. and Carter, N. (1994) Angiotensinogen Met235→Thr polymorphism in a London normotensive and hypertensive black and white population. *J. Hum. Hypertens.* **8**: 639–640.

Bennet, C.L., Schrader, A.P. and Morris, B.J. (1993) Cross-sectional analysis of M235T variant of angiotensinogen gene in severe familial hypertension. *Biochem. Biophys. Res. Comm.* **197**: 833–839.

Bernstein, I.M., Ziegler, W., Stirewalt, W.S., Brumsted, J. and Ward, K. (1998) Angiotensinogen genotype and plasma volume in nulligravid women. *Obstet. Gynecol.* **92**: 171–173.

Bloem, L.J., Manatunga, A.K., Tewksbury, D.A. and Pratt, J.H. (1995) The serum angiotensinogen concentration and variants of the angiotensinogen gene in white and black children. *J. Clin. Invest.* **95**: 948–953.

Bloem, L.J., Foroud, T.M., Ambrosius, W.T., Hanna, M.P., Tewksbury, D.A. and Pratt, J.H. (1997) Association of the angiotensinogen gene to serum angiotensinogen in blacks and whites. *Hypertension* **29**: 1078–1082.

Bohlender, J., Menard, J., Wagner, J., Luft, F.C. and Ganten, D. (1996) Human renin-dependent hypertension in rats transgenic for human angiotensinogen. *Hypertension* **27**: 535–540.

Borecki, I.B., Province, M.A., Ludwig, E.H., Ellison, R.C., Folsom, A.R., Heiss, G., Lalouel, J.M., Higgins, M. and Rao, D.C. (1997) Associations of candidate loci angiotensinogen and angiotensin-converting enzyme with severe hypertension: The NHLBI Family Heart Study. *Ann Epidemiol* **7**: 13–21.

Bouhnik, J., Clauser, E., Strosberg, D., Frenoy, J.P., Menard, J. and Corvol, P. (1981) Rat angiotensinogen and des(angiotensin I)angiotensinogen: purification, characterization, and partial sequencing. *Biochemistry* **20**: 7010–7015.

Brand, E., Chatelain, N., Keavney, B., *et al.* (1998) Evaluation of the angiotensinogen locus in human essential hypertension: a European study. *Hypertension* **31**: 725–729.

Brasier, A.R. and Li, J. (1996) Mechanisms for inducible control of angiotensinogen gene transcription. *Hypertension* **27**: 465–475.

Brasier, A.R., Tate, J.E., Ron, D. and Habener, J.F. (1989) Multiple cis-acting DNA regulatory elements mediate hepatic angiotensinogen gene expression. *Mol. Endocrinol.* **3**: 1022–1034.

Bunnemann, B., Fuxe, K. and Ganten, D. (1993) The renin angiotensin system in the brain: an update 1993. *Regul. Pep.* **46**: 487–509.

Cambien, F., Leger, J., Mallet, C., Levy-Marchal, C., Collin, D. and Czernichow, P. (1998) Angiotensin I-converting enzyme gene polymorphism modulates the consequences of in utero growth retardation on plasma insulin in young adults. *Diabetes* **47**: 470–475.

Campbell, D.J. and Habener, J.F. (1986) Angiotensinogen gene is expressed and differentially regulated in multiple tissues of the rat. *J. Clin. Invest.* **78**: 31–39.

Campbell, D.J. and Habener, J.F. (1987) Cellular localization of angiotensinogen gene expression in brown adipose tissue and mesentery: quantification of messenger ribonucleic acid abundance using hybridization *in situ*. *Endocrinology* **121**: 1616–1626.

Campbell, D.J., Kladis, A., Skinner, S.L. and Whitworth, J.A. (1991a) Characterization of angiotensin peptides in plasma of anephric man. *J. Hypertension* **9**: 265–274.

Campbell, D.J., Lawrence, A.C., Tourie, A., Kladis, A. and Valentijn, A.J. (1991b) Differential regulation of angiotensin peptide levels in plasma and kidney of the rat. *Hypertension* **18**: 763–773.

Carrell, R.W. and Boswell, D.R. (1986) Serpins: the superfamily of plasma serine proteinase inhibitors. *Proteinase inhibitors* (ed. B.A. Salvesen). Elsevier Science, Amsterdam, pp. 403–420.

Cassis, L.A., Lynch, K.R. and Peach, M.J. (1988) Localization of angiotensinogen messenger RNA in rat aorta. *Circ. Res.* **62**: 1259–1262.

Caulfield, M., Lavender, P., Farrall, M., Munroe, P., Lawson, M., Turner, P. and Clark, A.J. (1994) Linkage of the angiotensinogen gene to essential hypertension. *N. Engl. J. Med.* **330**: 1629–1633.

Caulfield, M., Lavender, P., Newell-Price, J., *et al.* (1995) Linkage of the angiotensinogen gene locus to human essential hypertension in african caribbeans. *J. Clin. Invest.* **96**: 687–692.

Chesley, L.C. and Cooper, D.W. (1986) Genetics of hypertension in pregnancy: possible single gene control of pre-eclampsia and eclampsia in the descendants of eclamptic women. *Br. J. Obst. Gynaecol.* **93**: 898–903.

Chirgwin, J.M., Schaefer, I.M., Diaz, J.A. and Lalley, P.A. (1984) Mouse kidney renin gene is on chromosome 1. *Somatic Cell. Mol. Genet.* **10**: 633.

Clouston, W.M., Evans, B.A., Haralambidis, J. and Richards, R.I. (1988) Molecular cloning of the mouse angiotensinogen gene. *Genomics* **2**: 240–248.

Cohen, P., Badouaille, G., Gimenez-Roqueplo, A.-P., Mani, J. C., Guyene, T.-T., Jeunemaitre, X., Ménard, J., Corvol, P., Pau, B. and Simon, D. (1996) Selective recognition of M235T angiotensinogen variants and their determination in human plasma by monoclonal antibody-based immunoanalysis. *J. Clin. Endocrinol. Metab.* **81**: 3505–3512.

Cooper, R., Forrester, T., Ogunbiyi, O. and Muffinda, J. (1998) Angiotensinogen levels and obesity in four black populations. *J. Hypertension* **16**: 571–575.

Cumin, F., Nisato, D., Gagnol, J.P. and Corvol, P. (1987) A potent radiolabeled human renin inhibitor, [3H]SR42128: enzymatic, kinetic, and binding studies to renin and other aspartic proteases. *Biochemistry* **26**: 7615–7621.

Danser A.H.J., Derkx F.H.M., Hense H-W, Jeunemaitre X., Riegger G.A.J., Schunkert H. (1998) Angiotensinogen (M235T) and angiotensin-converting enzyme (I/D) polymorphisms in association with plasma renin and prorenin levels. *J. Hypertension* **16**: 1879–1883.

Darimont, C., Vassaux, G., Ailhaud, G. and Negrel, R. (1994) Differenciation of preadipose cells: paracrine role of prostacyclin upon stimulation of adipose cells by angiotensin-II. *Endocrinology* **135**: 2030–2036.

Davisson, R.L., Kim, H.S., Krege, J.H., Lager, D.J., Smithies, O. and Sigmund, C.D. (1997) Complementation of reduced survival, hypotension, and renal abnormalities in angiotensinogen-deficient mice by the human renin and human angiotensinogen genes. *J. Clin. Invest.* **99**: 1258–1264.

DeLignieres, B., Basdevant, A., Thomas, G., *et al.* (1986) Biological effects of estradiol–17β in postmenopausal women: oral versus percutaneous administration. *J. Clin. Endocrinol. Metab.* **62**: 536–541.

Derkx, F.H.M., Stuenkel, C., Schalekamp, M.P.A., Visser, W., Huisveld, I.H. and Schalekamp, M.A.D.H. (1986) Immunoreactive renin, prorenin and enzymatically active renin in plasma during pregnancy and in women taking oral contraceptives. *J. Clin. Endocrinol. Metab.* **63**: 1008–1015.

Deschepper, C.F. and Hong-Brown, L.Q. (1993) Hormonal regulation of the angiotensinogen gene in liver and other tissues. In: *Cellular and Molecular Biology of the Renin Angiotensin System* (eds M.K., Raizada, M.I. Philips and C. Sumners). CRC Press, Boca Raton, Fl, pp. 149–166.

Doolittle, R.F. (1983) Angiotensinogen is related to the antitrypsin–antithrombin–ovalbumin family. *Science* **222**: 417–419.

Dracopoli, N.C., O'Connell, P., Elsner, T., Lalouel, J.-M., White, R. and Buetow, K.H. (1991) The CEPH consortium linkage map of human chromosome 1. *Genomics* **9**: 686–700.

Eggena, P., Sowers, J., Maxwell, M.H., Barrett, J.D. and Golub, M.S. (1991) Hormonal correlates of weight loss associated with blood pressure reduction. *Clin. Exp. Hypertension* **13**: 1447–1456.

Eggena, P., Shu, J.H., Clegg, K. and Barrett, J.D. (1993) Nuclear angiotensin receptors induce transcription of renin and angiotensinogen mRNA. *Hypertension* **22**: 496–501.

Fardella, C.E., Claverie, X., Vignolo, P., Montero, J. and Villarroel, L. (1998) T235 variant of the angiotensinogen gene and blood pressure in the Chilean population [in process citation]. *J. Hypertension* **16**: 829–833.

Fasola, A.F., Martz, B.L. and Helmer, O.M. (1968) Plasma renin activity during supine exercise in offspring of hypertensive patients. *J. Appl. Physiol.* 25: 410–415.

Fernandez-Llama, P., Poch, E., Oriola, J., Botey, A., Rivera, F. and Revert, L. (1998) Angiotensinogen gene M235T and T174M polymorphisms in essential hypertension: relation with target organ damage [in process citation]. *Am. J. Hypertens.* **11**: 439–444.

Fornage, M., Turner, S.T., Sing, C.F. and Boerwinkle, E. (1995) Variation at the M235T locus of the angiotensinogen gene and essential hypertension: a population-based case-control study from Rochester, Minnesota. *Hum. Genet.* **96**: 295–300.

Forrester, T., McFarlane-Anderson, N., Bennet, F., *et al.* (1996) Angiotensinogen and blood pressure among blacks: findings from a community survey in Jamaica. *J. Hypertension* **14**: 315–321.

Frederich, R.C., Jr., Kahn, B.B., Peach, M.J. and Flier, J.S. (1992) Tissue-specific nutritional regulation of angiotensinogen in adipose tissue. *Hypertension* **19**: 339–344.

Fried, T.A. and Simpson, E.A. (1986) Intrarenal localization of angiotensinogen mRNA by RNA-DNA dot-blot hybridization. *Am. J. Physiol.* **250**: F374–F377.

Frossard, P.M., Hill, S.H., Elshahat, Y.I., Obineche, E.N., Bokhari, A.M., Lestringant, G.G., John, A. and Abdulle, A.M. (1998) Associations of angiotensinogen gene mutations with hypertension and myocardial infarction in a gulf population. *Clin. Genet.* **54**: 285–293.

Fukamizu, A., Sugimura, K., Takimoto, E., *et al.* (1993) Chimeric renin-angiotensin system demonstrates sustained increase in blood pressure of transgenic mice carrying both human renin and human angiotensinogen genes. *J. Biol. Chem.* **268**: 11617–11621.

Gaillard, I., Clauser, E. and Corvol, P. (1989) Structure of human angiotensinogen gene. *DNA* **8**: 87–99.

Gaillard-Sanchez, I., Mattei, M.G., Clauser, E. and Corvol, P. (1990) Assignment by in situ hybridization of the angiotensinogen gene to chromosome band 1q4, the same region as the human renin gene. *Hum. Genet.* **84**: 341–343.

Gardes, J., Bouhnik, J., Clauser, E., Corvol, P. and Ménard, J. (1982) Role of angiotensinogen in blood pressure homeostasis. *Hypertension* **4**: 185–189.

Gimenez-Roqueplo, A.P., Leconte, I., Cohen, P., *et al.* (1996) The natural mutation Y248C of human angiotensinogen leads to abnormal glycosylation and altered immunological recognition of the protein. *J. Biol. Chem.* **271**: 9838–9844.

Gimenez-Roqueplo, A.P., Celerier, J., Lucarelli, G., Corvol, P. and Jeunemaitre, X. (1998a) Role of N-glycosylation in human angiotensinogen. *J. Biol. Chem.* **273**: 21232–21238.

Gimenez-Roqueplo, A.P., Celerier, J., Schmid, G., Corvol, P. and Jeunemaitre, X. (1998b) Role of cysteine residues in human angiotensinogen. Cys232 is required for angiotensinogen-pro major basic protein complex formation. *J. Biol. Chem.* **273**: 34480–34487.

Gould, A.B. and Green, D. (1971) Kinetics of the human renin and human substrate reaction. *Cardiovasc. Res.* 5: 86–89.

Guo, G., Wilton, A.N., Fu, Y., Qiu, H., Brennecke, S.P. and Cooper, D.W. (1997) Angiotensinogen gene variation in a population case-control study of preeclampsia/eclampsia in Australians and Chinese. *Electrophoresis* **18**: 1646–1649.

Gyurko, R., Wielbo, D. and Phillips, M.I. (1993) Antisense inhibition of AT1 receptor mRNA and angiotensinogen mRNA in the brain of spontaneously hypertensive rats reduces hypertension of neurogenic origin. *Regul. Pept.* **49**: 167–174.

Harp, J.B. and Di Girolamo, M. (1995) Components of the renin–angiotensin system in adipose tissue: changes with maturation and adipose mass enlargement. *J. Gerontol. A. Biol. Sci. Med. Sci.* **50**: B270–B276.

Hata, A., Namikawa, C., Sasaki, M., Sato, K., Nakamura, T., Tamura, K. and Lalouel, J.M. (1994) Angiotensinogen as a risk factor for essential hypertension in Japan. *J. Clin. Invest* **93**: 1285–1287.

Hegele, R.A., Brunt, J.H. and Connelly, P.W. (1994) A polymorphism of the angiotensinogen gene associated with variation in blood pressure in a genetic isolate. *Circulation* **90**: 2207–2212.

Hegele, R. A., Brunt, J. H. and Connelly, P. W. (1995) Genetic variation on chromosome 1 associated with variation in body fat distribution in men. *Circulation* **92**: 1089–1093.

Hegele, R.A., Brunt, J.H. and Connelly, P.W. (1996) Genetic and biochemical factors associated with variation in blood pressure in a genetic isolate. *Hypertension* **27**: 308–312.

Hegele, R.A., Harris, S.B., Hanley, A.J., Sun, F., Connelly, P.W. and Zinman, B. (1997) Angiotensinogen gene variation associated with variation in blood pressure in aboriginal Canadians. *Hypertension* **29**: 1073–1077.

Hilgers, K.F., Hilgenfeldt, U., Veelken, R., Muley, T., Ganten, D., Luft, F.C. and Mann, J.F.E. (1993) Angiotensinogen is cleaved to angiotensin in isolated rat blood vessels. *Hypertension* **21**: 1030–1034.

Hingorami, A.D., Sharma, P., Jia, H., Hooper, R. and Brown, M.J. (1996) Blood pressure and the M235T polymorphism of the angiotensinogen gene. *Hypertension* **28**: 907–911.

Hixson, J.E. and Powers, P.K. (1995) Detection and characterization of new mutations in the human angiotensinogen gene (AGT) *Hum. Genet.* **96**: 110–112.

Hollenberg, N.K., Williams, G.H., Burger, B., Chenitz, W., Hossmand, I. and Adams, D.F. (1976) Renal blood flow and its response to angiotensin II. An interaction between oral contraceptive agents, sodium intake and the renin–angiotensin system in healthy young women. *Circ. Res.* **38**: 35–40.

Hopkins, P.N., Lifton, R.P., Hollenberg, N.K., *et al.* (1996) Blunted renal vascular response to angiotensin II is associated with a common variant of the angiotensinogen gene and obesity. *J. Hypertension* **14**: 199–207.

Huang, X.H., Rantalaiho, V., Wirta, O., Pasternack, A., Koivula, T., Hiltunen, T., Nikkari, T. and Lehtimaki, T. (1998) Relationship of the angiotensin-converting enzyme gene polymorphism to glucose intolerance, insulin resistance, and hypertension in NIDDM. *Hum Genet* **102**: 372–378.

Huber, R.A. and Carrell, R.W. (1989) Implications of the three-dimensional structure of a1-antitrypsin for structure and function of serpins. *Biochemistry* **28**: 8966–8971.

Hübner, N., Krentz, R., Takahashi, S., Ganten, D. and Lindpaintner, K. (1994) Unlike human hypertension, blood pressure in a hereditary hypertensive rat strain shows no linkage to the angiotensinogen locus. *Hypertension* **23**: 797–801.

Hübner, N., Krentz, R., Takahashi, S., Ganten, D. and Lindpaintner, K. (1995) Altered angiotensinogen aminoacid sequence and plasma angiotensin II levels in genetically hypertensive rats. *Hypertension* **26**: 279–284.

Hunt, S.C., Cook, N.R., Oberman, A., Cutler, J.A., Hennekens, C.H., Allender, P.S., Walker, W.G., Whelton, P.K. and Williams, R.R. (1998) Angiotensinogen genotype, sodium reduction, weight loss, and prevention of hypertension. *Hypertension* **32**: 393–401.

Ingelfinger, J.R., Schunkert, H., Ellison, K.E., Pivor, M., Zuo, W.M., Pratt, R. and Dzau, V.J. (1990) Intrarenal angiotensinogen: localization and regulation. *Pediatr. Nephrol.* **4**: 424–428.

Inoue, I., Rohrwasser, A., Helin, C., *et al.* (1995) A mutation of angiotensinogen in a patient with preeclampsia leads to altered kinetics of the renin–angiotensin system. *J. Biol. Chem.* **270**: 11430–11436.

Inoue, I., Nakajima, T., Williams, C.S., *et al.* (1997) A nucleotide substitution in the promoter of human angiotensinogen is associated with essential hypertension and affects basal transcription *in vitro*. *J. Clin. Invest.* **99**: 1786–1797.

Isa, M.N., Boyd, E., Morrison, N., Harrap, S., Clauser, E. and Connor, J.M. (1990) Assignment of the human angiotensinogen gene to chromosome 1q42–q43 by nonisotopic in situ hybridization [corrected] [published erratum appears in *Genomics* 1991 Aug;**10**(4): 1110]. *Genomics* **8**: 598–600.

Ishigami, T., Umemura, S., Tamura, K., *et al.* (1997) Essential hypertension and 5′ upstream core promoter region of human angiotensinogen gene [in process citation]. *Hypertension* **30**: 1325–1330.

Iwai, N. and Inagami, T. (1992) Identification of a candidate gene responsible for the high blood pressure of spontaneously hypertensive rats. *J. Hypertension* **10**: 1155–1157.

Iwai, N., Shimoike, H., Ohmichi, N. and Kinoshita, M. (1995) Angiotensinogen gene and blood pressure in the Japanese population. *Hypertension* **25**: 688–693.

Jeunemaitre, X.J. and Lifton, R.P. (1993) Genes of the renin angiotensin system and the genetics of human hypertension. In: *Cellular and Molecular Biology of the Renin Angiotensin System* (eds M.K. Raizada, M.I. Philips and C. Summers). CRC Press, Boca Raton, FL, pp. 73–94.

Jeunemaitre, X., Soubrier, F., Kotelevtsev, Y.V., *et al.* (1992) Molecular basis of human hypertension: role of angiotensinogen. *Cell* **71**: 169–180.

Jeunemaitre, X., Charru, A., Chatellier, A., Dumont, C., Sassano, P., Soubrier, F., Ménard, J. and Corvol, P. (1993) M235T variant of the human angiotensinogen gene in unselected hypertensive patients. *J. Hypertension* **11(suppl. 5)**: S80–S81.

Jeunemaitre, X., Inoue, I., Williams, C., *et al.* (1997) Haplotypes of angiotensinogen in essential hypertension. *Am. J. Hum. Genet.* **60**: 1448–1460.

Kageyama, R., Ohkubo, H. and Nakanishi, S. (1984) Primary structure of human preangiotensinogen deduced from the cloned cDNA sequence. *Biochemistry* **23**: 3603–3609.

Kakinuma, Y., Hama, H., Sugiyama, F., Yagami, K., Goto, K., Murakami, K. and Fukamizu, A. (1998) Impaired blood-brain barrier function in angiotensinogen-deficient mice. *Nature Med.* **4**: 1078–1080.

Kamide, K., Hori, M.T., Zhu, J.H., Barrett, J.D., Eggena, P. and Tuck, M.L. (1998) Insulin-mediated growth in aortic smooth muscle and the vascular renin–angiotensin system. *Hypertension* **32**: 482–487.

Kamitani, A., Rakugi, H., Higaki, J., Yi, Z., Mikami, H., Miki, T. and Ogihara, T. (1994) Association analysis of a polymorphism of the angiotensinogen gene with essential hypertension in Japanese. *J. Hum. Hypertension* **8**: 521–524.

Kanai, H., Matsuzawa, Y., Kotani, K., Keno, Y., Kobatake, T., Nagai, Y., Fujioka, S., Tokunaga, K. and Tarui, S. (1990) Close correlation of intra-abdominal fat accumulation to hypertension in obese women. *Hypertension* **16**: 484–490.

Kew, M.C., Leckie, B.J. and Greeff, M.C. (1989) Arterial hypertension as a paraneoplastic phenomenon in hepatocellular carcinoma. *Arch. Intern. Med.* **149**: 2111–2113.

Kiema, T.R., Kauma, H., Rantala, A.O., Lilja, M., Reunanen, A., Kesaniemi, Y.A. and Savolainen, M.J. (1996) Variation at the angiotensin-converting enzyme gene and angiotensinogen gene loci in relation to blood pressure. *Hypertension* **28**: 1070–1075.

Kim, H.S., Krege, J.H., Kluckman, K.D., *et al.* (1995) Genetic control of blood pressure and the angiotensinogen locus. *Proc. Natl Acad. Sci. U.S.A.* **92**: 2735–2739.

Kimura, S., Mullins, J.J., Bunnemann, B., *et al.* (1992) High blood pressure in transgenic mice carrying the rat angiotensinogen gene. *EMBO J.* **11**: 821–827.

Klett, C., Hellmann, W., Müller, F., Suzuki, F., Nakanishi, S., Ohkubo, H., Ganten, D. and Hackenthal, E. (1988) Angiotensin II controls angiotensinogen secretion at a pretranslational level. *J. Hypertension* **6(suppl. 4)**: S442–S445.

Kobashi, G. (1995) A case-control study of pregnancy-induced hypertension with a genetic predisposition: association of a molecular variant of angiotensinogen in the Japanese women. *Hokaido Igaku Zaosshi* **70**: 649–657.

Kotelevtsev, Y.V., Clauser, E., Corvol, P. and Soubrier, F. (1991) Dinucleotide repeat polymorphism in the human angiotensinogen gene. *Nucl. Acids Res.* **19**: 6978.

Kunz, R., Kreutz, R., Beige, J., Distler, A. and Sharma, A. M. (1997) Association between the angiotensinogen 235T-variant and essential hypertension in whites: a systematic review and methodological appraisal. *Hypertension* **30**: 1331–1337.

Lindpaintner, K., Jin, M., Niedermaier, N., Wilhelm, M. J. and Ganten, D. (1990) Cardiac angiotensinogen and its local activation in the isolated perfused beating heart. *Circ. Res.* **67**: 564–573.

Lodwick, D., Kaiser, M.A., Harris, J., Cumin, F., Vincent, M. and Samani, N.J. (1995) Analysis of the role of angiotensinogen in spontaneous hypertension. *Hypertension* **25**: 1245–1251.

Lynch, K.R. and Peach, M.J. (1991) Molecular biology of angiotensinogen. *Hypertension* **17**: 263–269.

Lynch, K.R., Hawelu-Johnson, C.L. and Guyenet, P.G. (1987) Localization of brain angiotensinogen mRNA by hybridization histochemistry. *Brain Res.* **388**: 149–158.

Makino, N., Sugano, M., Ohtsuka, S. and Sawada, S. (1998) Intravenous injection with antisense oligodeoxynucleotides against angiotensinogen decreases blood pressure in spontaneously hypertensive rats. *Hypertension* **31**: 1166–1170.

Mast, A.E., Enghild, J.J., Pizzo, S.V. and Salvesen, G. (1991) Analysis of the plasma elimination kinetics and conformational stabilities of native, proteinase complexed and reactive site cleaved serpins: comparison of a1-proteinase inhibitor, a1-antichymotrypsin, antithrombin III, a2-antiplasmin, angiotensinogen and ovalbumin. *Biochemistry* **30**: 1723–1730.

McGehee, R.E., Ron, D., Brasier, A.R. and Habener, J.F. (1993) Differential-specific element: a cis-acting developmental switch required for the sustained transcriptional expression of the angiotensinogen gene during hormonal-induced differentiation of 3T3-L1 fibroblasts to adipocytes. *Mol. Endocrinol.* **7**: 551–560.

Ménard, J., El-Amrani, A.I.K., Savoie, F. and Bouhnik, J. (1991) Angiotensinogen: an attractive and underrated participant in hypertension and inflammation. *Hypertension* **18**: 705–706.

Merrill, D.C., Thompson, M.W., Carney, C.L., Granwehr, B.P., Schlager, G., Robillard, J.E. and Sigmund, C.D. (1996) Chronic hypertension and altered baroreflex responses in transgenic mice containing the human renin and human angiotensinogen genes. *J. Clin. Invest.* **97**: 1047–1055.

Misumi, J., Gardes, J., Gonzales, M.-F., Corvol, P. and Ménard, J. (1989) Angiotensinogen's role in angiotensin formation, renin release, and renal hemodynamics in isolated perfused kidney. *Am. J. Physiol.* **256**: F719–F727.

Morgan, L., Baker, P., Pipkin, F.B. and Kalsheker, N. (1995) Pre-eclampsia and the angiotensinogen gene. *Br. J. Obstet. Gynaecol.* **102**: 489–490.

Morgan, T., Craven, C., Nelson, L., Lalouel, J.M. and Ward, K. (1997) Angiotensinogen T235 expression is elevated in decidual spiral arteries. *J. Clin. Invest.* **100**: 1406–1415.

Morise, T., Katenchi, Y. and Takeda, R. (1995) Rapid detection and prevalence of the variants of the angiotensinogen gene in patients with essential hypertension. *J. Intern. Med.* **235**: 175–180.

Nabika, T., Bonnardeaux, A., James, M., Julier, C., Jeunemaitre, X., Corvol, P., Lathrop, M. and Soubrier, F. (1995) Evaluation of the SA locus in human hypertension. *Hypertension* **25**: 6–13.

Naftilan, A.J., Zou, W.M., Ingelfinger, J.R., Ryan, T.J., Pratt, R.E. and Dzau, V.J. (1991) Localization and differential regulation of angiotensinogen mRNA expression in the vessel wall. *J. Clin. Invest.* **87**: 1300–1311.

Nibu, Y., Takahashi, S., Tanimoto, K., Murakami, K. and Fukamizu, A. (1994) Identification of cell type-dependent enhancer core element located in the 3′-downstream region of the human angiotensinogen gene. *J. Biol. Chem.* **269**: 28598–28605.

Niimura, F., Labosky, P.A., Kakuchi, J., *et al.* (1995) Gene targeting in mice reveals a requirement for angiotensin in the development and maintenance of kidney morphology and growth factor regulation. *J. Clin. Invest.* **96**: 2947–2954.

Nishimura, S., Kario, K., Kayaba, K., Nagio, N., Shimada, K., Matsuo, T. and Matsuo, M. (1995) Effect of the angiotensinogen gene M235T variant on blood pressure and other cardio-vascular risk factors in two Japanese populations. *J. Hypertension* **13**: 717–722.

Niu, T., Xu, X., Rogus, J., *et al.* (1998) Angiotensinogen gene and hypertension in Chinese. *J. Clin. Invest.* **101**: 188–194.

O'Byrne, S. and Caulfield, M. (1998) Genetics of hypertension. Therapeutic implications. *Drugs* **56**: 203–214.

Ohkubo, H., Nakayama, K., Tanaka, T. and Nakanishi, S. (1986) Tissue distribution of rat angiotensinogen mRNA and structural analysis of its heterogeneity. *J. Biol. Chem.* **261**: 319–323.

Ohkubo, H., Kawakami, H., Kakehi, Y., *et al.* (1990) Generation of transgenic mice with elevated blood pressure by introduction of the rat renin and angiotensinogen genes. *Proc. Natl Acad. Sci. USA* **87**: 5153–5157.

Oxvig, C., Haaning, J., Kristensen, L., Wagner, J.M., Rubin, I., Stigbrand, T., Gleich, G.J. and Sottrup-Jensen, L. (1995) Identification of angiotensinogen and complement C3dg as novel proteins binding the proform of eosinophil major basic protein in human pregnancy serum and plasma. *J. Biol. Chem.* **270**: 13645–13651.

Peng, J.F., Kimura, B., Fregly, M.J. and Phillips, M.I. (1998) Reduction of cold-induced hypertension by antisense oligodeoxynucleotides to angiotensinogen mRNA and AT1-receptor mRNA in brain and blood. *Hypertension* **31**: 1317–1323.

Phillips, M.I. (1997) Antisense inhibition and adeno-associated viral vector delivery for reducing hypertension. *Hypertension* **29**: 177–187.

Phillips, M.I., Speakman, E.A. and Kimura, B. (1993) Tissue renin–angiotensin system. In: *Cellular and Molecular Biology of the Renin–Angiotensin System* (eds M.K. Raidaza, M.I. Phillips and C. Summers). CRC Press, Boca Raton, FL, pp. 97–130.

Pratt, J.H., Ambrosius, W.T., Tewksbury, D.A., Wagner, M.A., Zhou, L. and Hanna, M.P. (1998) Serum angiotensinogen concentration in relation to gonadal hormones, body size, and genotype in growing young people. *Hypertension* **32**: 875–879.

Pravenec, M., Kren, V., Klir, P., Simonet, L. and Kurtz, T. (1991) Identification of genes determining spontaneous hypertension. *Sb Lek* **93**: 136–141.

Racz, K., Pinet, F., Guyene, T.-T., Gasc, J.-M. and Corvol, P. (1992) Coexpression of renin, angiotensinogen and their messenger in human normal and pathologic adrenal tissue. *J. Clin. Endocrinol. Metab.* **75**: 730–737.

Rakugi, H., Jacob, H.Z., Krieger, J.E., Ingelfinger, J.R. and Pratt, R.E. (1993) Vascular injury induces angiotensinogen gene expression in the media and neointima. *Circulation* **87**: 283–290.

Robertson, J.I.S. (1993) Angiotensinogen measurement. In: *The Renin Angiotensin System*, Vol. 1 (eds J.I.S. Roberston and M.G. Nicholls). Gower Medical Publishing, London, pp. 14.1–14.8.

Rotimi, C., Puras, A., Cooper, R., McFarlane-Anderson, N., Forrester, T., Ogunbiyi, O., Morrison, L. and Ward, R. (1996) Polymorphisms of renin angiotensinogen genes among Nigerians, Jamaicans and African Americans. *Hypertension* **27**: 558–563.

Rutledge, D.R., Browe, C.S., Kubilis, P.S. and Ross, E.A. (1994) Analysis of two variants of the angiotensinogen gene in essential hypertensive African-Americans. *Am. J. Hypertension* **7**: 651–654.

Sato, N., Katsuya, T., Rakugi, H., Takami, S., Nakata, Y., Miki, T., Higaki, J. and Ogihara, T. (1997) Association of variants in critical core promoter element of angiotensinogen gene with increased risk of essential hypertension in Japanese. *Hypertension* **30**: 321–325.

Saye, J.A., Lynch, K.R. and Peach, M.J. (1990) Changes in angiotensinogen messenger RNA in differentiating 3T3-F442A adipocytes. *Hypertension* **15**: 867–871.

Schmidt, S., Sharma, A. M., Zilch, O., Beige, J., Walla-Friedel, M., Ganten, D., Distler, A. and Ritz, E. (1995) Association of M235T variant of the angiotensinogen gene with familial hypertension of early onset. *Nephrol. Dial. Transplant* **10**: 1145–1148.

Schorr, U., Blaschke, K., Turan, S., Distler, A. and Sharma, A.M. (1998) Relationship between angiotensinogen, leptin and blood pressur levels in young normotensive men. *J Hypertens* **16**: 1475–1480.

Schunkert, H., Dzau, V. J., Tang, S.S., Hirsch, A.T., Apstein, C.S. and Lorell, B.H. (1990) Increased rat cardiac angiotensin converting enzyme activity and mRNA expression in pressure overload left ventricular hypertrophy. Effects on coronary resistance, contractility and relaxation. *J. Clin. Invest.* **86**: 1913–1930.

Schunkert, H., Ingelfinger, J., Hirsch, A.T., Tang, S.S., Litwin, S.E., Talsness, C. and Dzau, V.J. (1992) Evidence for tissue specific activation of renal angiotensinogen mRNA expression in chronic stable experimental heart failure. *J. Clin. Invest.* **90**: 1523–1529.

Schunkert, H., Hense, H.W., Gimenez-Roqueplo, A.P., Stieber, J., Keil, U., Riegger, G.A. and Jeunemaitre, X. (1997) The angiotensinogen T235 variant and the use of antihypertensive drugs in a population-based cohort. *Hypertension* **29**: 628–633.

Sheu, W.H., Lee, W.J., Jeng, C.Y., Young, M.S., Ding, Y.A. and Chen, Y.T. (1998) Angiotensinogen gene polymorphism is associated with insulin resistance in nondiabetic men with or without coronary heart disease. *Am. Heart. J.* **136**: 125–131.

Skinner, S.L., Lumbers, E.R. and Symonds, E.M. (1972) Analysis of changes in the renin angiotensin system during pregnancy. *Clin. Sci.* **42**: 479–488.

Smithies, O. and Kim, H.S. (1994) Targeted gene duplication and disruption for analyzing quantitative genetic traits in mice. *Proc. Natl Acad. Sci. USA* **91**: 3612–3615.

Stein, P.E., Tewkesbury, D.A. and Carrell, R.W. (1989) Ovalbumin and angiotensinogen lack serpin S-R conformational change. *Biochem J* **262**: 103–107.

Stornetta, R.L., Hawelu-Johnson, C.L., Guyenet, P.G., Lynch, K.R. (1988) Astrocytes synthesize angiotensinogen in brain. *Science* **242**: 1444–1446

Streatfeild-James, R.M., Williamson, D., Pike, R.N., Tewksbury, D., Carrell, R.W. and Coughlin, P.B. (1998) Angiotensinogen cleavage by renin: importance of a structurally constrained N-terminus. *FEBS Lett* **436**: 267–270.

Sutherland, A., Cooper, D., Howie, P., Liston, W. and MacGillivray, I. (1981) The incidence of severe pre-eclampsia amongst mothers and mothers-in-law of pre-eclamptics and controls. *Br. J. Obst. Gynaecol.* **88**: 785–789.

Takahashi, D., Tamura, K., Ushikubo, T., *et al.* (1997) Relationship between hepatic angiotensinogen mRNA expression and plasma angiotensinogen in patients with chronic hepatitis. *Life Sci.* **60**: 1623–1633.

Takimoto, E., Ishida, J., Sugiyama, F., Horiguchi, H., Murakami, K. and Fukamizu, A. (1996) Hypertension induced in pregnant mice by placental renin and maternal angiotensinogen. *Science* **274**: 995–998.

Tamura, K., Umemura, S., Iwamoto, T., *et al.* (1994) Molecular mechanism of adipogenic activation of the angiotensinogen gene. *Hypertension* **23**: 364–368.

Tamura, K., Umemura, S., Nyui, N., *et al.* (1996) Tissue specific regulation of angiotensinogen gene expression in spontaneously hypertensive rat. *Hypertension* **27**: 1216–1223.

Tanaka, T., Ohkubo, H. and Nakanishi, S. (1984) Common structural organization of the angiotensinogen and α1-antitrypsin genes. *J. Biol. Chem.* **259**: 8063–8065.

Tewksbury, D.A. (1990) Angiotensinogen. Biochemistry and molecular biology. In: *Hypertension, Pathophysiology, Diagnosis and Management* (eds J.H. Laragh and B.M. Brenner). Raven Press, New York, pp. 1197–1216.

Tewksbury, D.A. and Dart, R.A. (1982) High molecular weight angiotensinogen levels in hypertensive pregnant women. *Hypertension* **4**: 729–734.

Thompson, M.W., Smith, S.B. and Sigmund, C.D. (1996) Regulation of human renin mRNA expression and protein release in transgenic mice. *Hypertension* **28**: 290–296.

Tiret, L., Ricard, S., Poirier, O., Arveiler, D., Cambou, J.-P., Lue, G., Evans, A., Nicaud, V. and Cambien, F. (1995) Genetic variation in relation to high blood pressure and myocardial infarction: the ECTIM study. *J. Hypertension* **13**: 311–317.

Tiret, L., Blanc, H., Ruidavets, J.B., *et al.* (1998) Gene polymorphisms of the renin-angiotensin system in relation to hypertension and parental history of myocardial infarction and stroke: the PEGASE study. Projet d'Etude des Genes de l'Hypertension Arterielle Severe a moderee Essentielle. *J. Hypertension* **16**: 37–44.

Tomita, N., Morishita, R., Higaki, J., Tomita, S., Aoki, M., Kaneda, Y. and Ogihara, T. (1995) Effect of angiotensinogen on blood pressure regulation in normotensive rats: application of a loss of function approach. *J. Hypertension* **13**: 1767–1774.

Ueno, N., Yoshida, K., Hirose, S., Yokoyama, H., Uehara, H. and Murakami, K. (1984) Angiotensinogen-producing hepato-cellular carcinoma. *Hypertension* **6**: 931–933.

Vasku, A., Soucek, M., Znojil, V., *et al.* (1998) Angiotensin I-converting enzyme and angiotensinogen gene interaction and prediction of essential hypertension. *Kidney Int.* **53**: 1479–1482.

Walker, W.G., Whelton, P.K., Saito, H., Patterson Russel, R. and Hermann, J. (1979) Relation between blood pressure and renin, renin substrate, angiotensin II, aldosterone and urinary sodium and potassium in 574 ambulatory subjects. *Hypertension* **1**: 287–291.

Ward, K., Hata, A., Jeunemaitre, X., *et al.* (1993) A molecular variant of angiotensinogen associated with pre-eclampsia. *Nature Genet.* **4**: 59–61.

Watt, G.C.M., Harrap, S.B., Foy, C.J.W., Holton, D.W., Edwards, H.V., Davidson, R., Connor, J.M., Lever, A.F. and Fraser, R. (1992) Abnormalities of glucocorticoid metabolism and the renin angiotensin system: a four-corners approach to the identification of genetic determinants of blood pressure. *J. Hypertension* **10**: 473–482.

Williams, G.H., Dluhy, R.G., Lifton, R.L., Moore, T.J., Gleason, R., Williams, R.R., Hunt, S.C., Hopkins, P. and Hollenberg, N.K. (1992) Non-modulation is an intermediate phenotype in essential hypertension. *Hypertension* **20**: 788–796.

Williams, G.H. and Hollenberg, N.K. (1991) Non-modulating hypertension: a subset of sodium sensitive hypertension. *Hypertension* **17(suppl. I)**: 181–185.

Wilton, A.N., Kaye, J.A., Guo, G., Brennecke, P. and Cooper, D.W. (1995) Is angiotensinogen a good candidate gene for preeclampsia? *Clin. Exp. Hypertension* (Hypertension in Pregnancy) **14**: 251–260.

Yanai, K., Nibu, Y., Murakami, K. and Fukamizu, A. (1996) A *cis*-acting DNA element located between TATA box and transcription initiation site is critical in response to regulatory sequences in human angiotensinogen gene. *J. Biol. Chem.* **271**: 15981–15986.

11

Pharmacogenomics in arterial hypertension

Patrizia Ferrari and Giuseppe Bianchi

1. Introduction

Pharmacogenomics of arterial hypertension is aimed at explaining individual variation of blood pressure response to drugs with genetic polymorphisms (Ferrari, 1998; Pratt and Dzau, 1999). The published papers on this topic are very scant and not sufficient to evaluate its real impact on therapeutic trends in the future. However, with the increase in our knowledge about the genomic abnormalities associated with the different pathophysiological mechanisms of disease and the genetics of hypertension (Dominiczak *et al.*, 1998), an unavoidable question will be raised both from patients and physicians: what is the relationship between the new achievements on genetics of hypertension and the drugs we are using to cure it?

Primary hypertension is very likely caused by several major genes acting with or against a very wide polygenic background (Lifton, 1996; Soubrier and Lathrop, 1995). Whatever the number of these genes, in order to increase blood pressure they have to reset the function of some key cells, namely to increase the rate of Na transport across the renal tubular cells or tone and contractility of vascular cells. All known mechanisms (nervous and hormonal), which have been shown thus far to increase blood pressure, must in one way or the other change the function of these two types of cells.

The need for a novel pharmacogenomic approach to the therapy of hypertension and its complications also depends upon the efficacy, safety and tolerability of the available drugs to cure this disease. In fact one may argue that some of the existing antihypertensive drugs, such as ACE inhibitors or angiotensin II receptor antagonists already have a good level of tolerability with reduced side-effects, therefore the medical need for having better drugs is less.

Why should we embark on a very complex approach to discover new drugs which selectively interfere with a specific genetic mechanism? Below we list a number of reasons that justify such a novel approach:

(i) ACE inhibitors or angiotensin II receptor antagonists, as all other antihypertensive compounds, when given alone, are active only in a portion of patients

Molecular Genetics of Hypertension, edited by A.F. Dominiczak, J.M.C. Connell and F. Soubrier.

(roughly 50%) (Laragh *et al.*, 1988; Menard *et al.*, 1990; Niutta *et al.*, 1990). We do not know the causes of this high variability in the individual drug response. This ignorance is very disturbing because it decreases the trust of patients towards their physicians and consequently their compliance.

(ii) ACE inhibitors also prevent organ damage because of the involvement of RAS in the mechanisms of the end-organ complications (Butler *et al.*, 1997; Cambien *et al.*, 1992). But, again, this is true only for some patients. Why? Probably a correct pharmacogenomic approach may produce some useful data to answer this question.

(iii) The duration of the treatment of hypertension may last for decades. The best follow-up studies addressed to measure organ damage protection lasted only few years. Therefore, we are left with long-term effects (either positive or negative) of treatments of which we have little knowledge. Certainly these long-term studies are expensive and difficult to carry out particularly with drugs whose life-span on the market is relatively short. This limit may not be so important for a drug that selectively interferes with the genetic mechanisms of hypertension and related organ damage. These drugs may be expected to produce less side-effects, and to stay on the market for a much longer time.

(iv) The increasing number of available antihypertensive drugs may discourage pharmaceutical companies to invest again in this field. The drawback of this policy, however, is that in the absence of a rationale that enables a given drug to be prescribed selectively to individual patients for whom it will be effective and safe, it will be more and more difficult to make the right choice among dozens of drugs. Conversely, by applying pharmacogenomics to the already available antihypertensive compounds, they could be better targetted to the individual patients (Ferrari, 1998; Marshall, 1997; Pratt and Dzau, 1999).

It must be clearly stated that only when the complexity of the genetic mechanisms underlying the regulation of the cellular functions mentioned above will be properly understood, can pharmacogenomics be successfully applied to the treatment of essential hypertension. Since this is a very long-term goal, it is possible, in the meantime, to use this approach to dissect the genetic complexity of this disease throughout the selective correction of well defined cellular effects produced by the various genetic molecular abnormalities associated with hypertension.

Three types of approaches are possible:

(i) to use DNA markers or polymorphisms of appropriate candidate genes, to identify the genetic characteristics of responders to the known antihypertensive drugs or to novel ones (Housman and Ledley, 1998; Marshall, 1997);

(ii) genetic polymorphisms may also be applied to define the characteristics of those patients who can suffer adverse effects from a specific treatment (Jick *et al.*, 1998);

(iii) if, at preclinical level, a novel antihypertensive drug has been developed as a selective blocker of a given genetic polymorphism, this polymorphism may be used to select patients for the clinical development of such a compound

(Marshall, 1997). This last approach is certainly the most stimulating since it may be considered an important tool to assess the contribution of that particular polymorphism to the hypertension of individual patients.

As discussed elsewhere in this book, the genetic dissection of a polygenic multifactorial and quantitative disease, such as primary arterial hypertension, is hampered by both the genetic heterogeneity (Horan and Mockrin, 1992) and the very large component of environmental variance (Hamet *et al.*, 1998). These difficulties also apply to pharmacogenomics and are certainly involved in the individual variation to drug response. Such variation may depend upon a variety of causes, namely: (i) genetic differences in the molecular mechanism underlying hypertension (genetic heterogeneity of hypertension); (ii) genetic difference in drug metabolism (pharmacogenetics); (iii) stage of hypertension; (iv) previous or concomitant therapies; (v) environmental factors. Among all these causes the last three should be better understood and controlled in order to obtain proper information on the relationship between individual genetic differences and therapeutic efficacy. Since the need to minimize the noise due to these causes and to the environment is not really apparent from the majority of the studies of genetics of hypertension, we will briefly discuss the relevance of these causes before dealing with pharmacogenomics.

2. Stage of hypertension

The most clear cut demonstration of the variation in drug response according to the stage of hypertension stems from the results on renovascular hypertension. In fact, the pathophysiological mechanisms that produce the raise in blood pressure change with time (Bianchi *et al.*, 1970). That is, soon after renal artery constriction, the rise in plasma renin and in the vascular peripheral resistance can account for the increase of pressure; later sodium and water renal retention occur and the increase of pressure is also sustained by a rise in cardiac output (Bianchi *et al.*, 1970). Finally all the abnormalities in renin, body fluids and cardiac output normalize and hypertension stabilizes in association with enhanced peripheral resistance caused by mechanisms that are still open to discussion (Bianchi *et al.*, 1970; Guan *et al.*, 1992; Morishita *et al.*, 1993). During the first phase, drugs that interfere with the pressor effect of renin are able to normalize blood pressure (Watkins *et al.*, 1978). Conversely, in the later phases, the effect of these drugs is much less (Carretero and Gulati, 1978; Fernandez *et al.*, 1978) and the normalization of blood pressure (or the prevention of hypertension) is obtained only with a combination of maneuvers which prevent renal sodium retention (Freeman *et al.*, 1979). Also the blood pressure response to the removal of the primary cause of hypertension (ischemic kidney) may vary with time. In fact, this maneuver is able to normalize blood pressure until a certain length of time from renal artery constriction. If the duration and severity of hypertension exceed a critical level the damage to the controlateral kidney or peripheral vessels may prevent the normalization of blood pressure (Edmunds *et al.*, 1991). The same time-related variability of pressure mechanisms has also been demonstrated for other forms of hypertension both in animals (Bianchi *et al.*, 1975) and humans (Lund-Johansen and Omvik, 1990).

Clearly, when trying to establish the involvement of a primary pressor mechanism (either genetic or experimental in nature) and the blood pressure fall after a given drug, the above information must be properly taken into account. In the past, the time-dependent variation of pressor mechanisms has been discussed in detail by Guyton and others (1995); it is due to the progressive re-adjustment of the feedback mechanisms connecting all the physiological pathways that control blood pressure. Of course, each of these pathways is controlled by a peculiar gene polymorphism which modulates its response. Whatever is the genetic major cause(s) of hypertension, such a feedback mechanism must be involved in one way or another to reset the final level of blood pressure.

3. Previous treatment

In all studies aimed at evaluating the antihypertensive activity of a given drug, the patients are given the drug after a washout period of no longer than 1 month. Very valid ethical reasons limit the duration of this period. However, it must be recognized that the blood pressure measurements after therapy withdrawal have clearly shown a very wide range of time intervals (from one to several months) needed for blood pressure to return to the pretreatment levels (MRC-WPMH, 1986; VA-CSGAA, 1975). Moreover, the carryover effect of a given treatment not only applies to blood pressure level but also involves other pressure mechanisms. For example, previous treatment with diuretics which lasted for 3 years, was discontinued for 4 weeks; however, a substantial effect on the renin response was demonstrated when rechallenged with the same diuretic (Swart *et al.*, 1982). This persistance of abnormalities in blood pressure control mechanisms after therapy withdrawal has also been shown by others using different approaches (Boyle *et al.*, 1979; Lowder and Liddle, 1974). Clearly the blood pressure response to a given drug may differ according to the alteration of the underlying pressure mechanism elicited by the previous treatment. This phenomenon may certainly disturb any attempt to establish a possible relationship between the underlying primary genetic mechanism of hypertension and the blood pressure response to a given drug.

4. Environment

The variability of blood pressure levels in a population is due to genetic (VG) and environmental (VE) components (Ward, 1995). In turn, for a quantitative phenotype such as blood pressure, the genetic (VG) may be split into a variance due to additive genetic effects (VA) and one due to dominant effects (VD). It is generally assumed that the VG contribution to the overall blood pressure variance is around 20–40% in humans (Ward, 1995). This rather low level of VG may discourage the genetic approach to primary hypertension. However, in genetic animal models of hypertension, the value of VG is around 40–50%, in spite of the fact that these animals develop under very controlled experimental conditions and are selected with appropriate genetic crosses to enhance VG (Rapp, 1995).

In order to maximize the likelihood of detecting possible associations between a given genetic polymorphism and blood pressure response to a drug interfering with it (see above, approach (iii)), we must minimize the noise due to these three causes of blood pressure individual variations, particularity the environmental factors. It should be realised that this noise is much more disturbing in assessing the effect of this type of drug than that of more 'traditional' drugs that block a physiological mechanism of blood pressure regulation like sympathetic tone, RAS or renal excretion of Na. In fact, 'traditional' drugs, because of their effect on physiological mechanisms, may also interfere with other genetic or environmental factors. As a consequence of these 'nonspecific' activities, 'traditional' therapies may also block physiological pathways necessary for the proper function of organs or cells not necessarily involved in causing hypertension or organ damage. This fact may account for a significant proportion of side-effects on the long run. Conversely, drugs developed according to pharmacogenomics should be active only on the 'genetic' portion of blood pressure variation (that is around 20–40%) particularly when their effects are measured within a relatively short period of time, for example during Phase II Clinical Trials.

For all the reasons given above, the type (iii) pharmacogenomic approach should be carried out out in newly diagnosed and never-treated patients with a relatively recent onset of hypertension. The environment should be properly controlled and kept constant, before and during the treatment period which should last no less than 4 weeks. This length of time is necessary to normalize blood pressure, for instance, in primary hyperaldosteronism treated with spironolactone, which may be taken as a paradigm of a 'causal' drug acting on underlying molecular pressor mechanisms.

It is likely that the different polymorphisms associated with the phenotype of interest may be grouped according to the cellular or biochemical functions regulated by them. Therefore we may organize the available results according to the biological levels of organization, namely molecular, biochemical, cellular, organ or whole-body level. In this way the appropriate genetic pathophysiological link may be established in individual patients.

Even though we can hypothesize that in the future several polymorphisms associated with hypertension and end-organ damage may be discovered, it is likely that only very few will be at work in an individual pedigree. Therefore the proper combination of drugs active on these genetic mechanisms may be used in individual patients.

5. Present status of pharmacogenomics of hypertension

There are only a few published studies which have addressed the problem of an association between a given locus and the published blood pressure response to a specific antihypertensive agent. Dudley *et al.* (1996) investigated whether the M235T polymorphism of the AGT gene and the insertion/deletion (I/D) polymorphism of the ACE gene can predict blood pressure responses to a β-blocker, ACE inhibitor and Ca antagonist in EH patients. They recorded a large variability between individuals in blood pressure responses to these agents which were not associated with the analyzed polymorphisms. Other studies aimed at verifying an association between treatments and the ACE/angiotensinogen genetic variants gave conflicting results. In particular in the study of Hingorani *et al.* (1995) angiotensinogen variants were associated with the magnitude of the decrease in blood pressure after ACE inhibition, whereas in that of Dudley *et al.* (1996) the same variants were not related to the hypotensive effect. More recently, O' Toole *et al.* (1998) studied the effect of ACE I/D genotype on the response to ACE inhibitor therapy in patients with heart failure, carrying out a double-blind, crossover study, and comparing captopril with lisinopril. They found a significant relationship between the ACE genotype and the decrease in blood pressure on captopril, with the II genotype being more sensitive to treatment than the I/D and DD. This relationship was not detected with lisinopril (O'Toole *et al.*, 1998). Although this study has many limitations, it is of particular interest in view of the protective role of ACE inhibitors on mortality, recurrent myocardial infarctions and progression of heart failure in patient with left ventricular dysfunction (Pfeffer, 1992; SOLVD, 1991). However, all the previously discussed criteria for a proper assessment of this type of association have not been applied to these studies. Therefore it is not possible to draw any definitive conclusions.

As will be further discussed in detail below, our group demonstrated, first in rats (Bianchi *et al.*, 1994) and more recently in humans (Casan *et al.*, 1995; Cusi *et al.*, 1987), a genetic association between hypertension and the α-adducin gene polymorphism. In humans, a polymorphism of α-adducin at the position 460, consisting of the substitution of tryptophan (Trp) for glycine (Gly), was detected (Cusi *et al.*, 1997). When compared to patients with the 460 Gly variant, those with the 460 Trp variant show a significantly greater change in blood pressure between high and low sodium balance and a larger fall in blood pressure on chronic diuretic treatment (Cusi *et al.*, 1997).

A different pharmacogenomic approach was used by Vincent *et al.* (1997) to investigate, in genetic hypertensive rats of the Lyon strain, the cosegregation of genetic loci with acute blood pressure responses to drugs acting on the renin–angiotensin system (Losartan), the sympathetic system (Trimetaphan) and calcium metabolism (L-type Ca antagonist, PY 108–068). With the use of microsatellite markers they identified a quantitative trait locus (QTL) on rat chromosome 2 that specifically influenced blood pressure responses to dihydropiridine Ca antagonist, while the same locus has no effect on either basal blood pressure or on responses to the ganglion blocking agent and ATII receptor antagonist (Vincent *et al.*, 1997). These findings represent one of the first examples in which a specific locus has been demonstrated to influence the response to a given antihypertensive therapy.

6. Adducin polymorphism as a tool for pharmacogenomics of hypertension

For many years our group has been pursuing the objective of understanding the molecular basis of human essential hypertension by studying a genetic animal model of spontaneous hypertension, the Milan hypertensive strain of rats (MHS)

(Barber *et al.*, 1994). The strategy adopted to reach this goal and, consequently, to try to develop a pharmacogenomic project aimed at discovering a 'selective' or 'causal' antihypertensive compound, included studies of renal function, cellular, biochemical and molecular characteristics of MHS rats in comparison with those of human subjects in order to identify common genetic-molecular mechanisms underlying the disease in both species.

7. The Milan rat model

Genetic hypertension in MHS rats is primarily due to a renal abnormality in the ability to excrete salt (Bianchi *et al.*, 1975). This alteration is genetically determined, since hypertension can be transplanted with MHS kidney in normotensive (MNS) controls, even when the kidney derives from a young prehypertensive animal (Ferrari and Bianchi, 1995).

Studies on renal function and body fluid metabolism showed that this pressor role is linked to an intrinsic defect in tubular Na reabsorption. GFR and urinary flow are greater (Ferrari and Bianchi, 1995) while plasma and renal renin are lower in young MHS than in MNS of the same age (Bianchi *et al.*, 1975) and these differences tend to disappear as hypertension is fully developed in MHS (Bianchi *et al.*, 1975; Ferrari and Bianchi, 1995). Balance studies showed renal Na retention in MHS during the development of hypertension (Bianchi *et al.*, 1975). Micropuncture studies showed that the single nephron GFR, the whole kidney GFR, renal interstitial hydrostatic pressure and tubular Na reabsorption are greater in young prehypertensive MHS than in MNS, while the MHS tubuloglomerular feedback mechanism is blunted (Ferrari and Bianchi, 1995). All these differences tend to disappear as hypertension develops in MHS (Ferrari and Bianchi, 1995). Similarly, the maintenance of a normal Na excretion in MHS in the presence of a faster GFR and increased tubular sodium reabsorption was also demonstrated on isolated perfused kidneys (Ferrari and Bianchi, 1995).

Looking at the cell functions that can underline such renal dysfunction, peculiar alterations of tubular cell ion transports have been detected: it was shown that Na/H countertransport in brush border vesicles, Na–K cotransport in luminal membrane of the thick ascending limb (Ferrari and Bianchi, 1995), and the activity and mRNA expression of the basolateral Na–K ATPase both in proximal and ascending limb tubules (Ferrari *et al.*, 1996) are all increased in MHS as compared with MNS. Therefore, it is possible to find ion transport abnormalities at the renal cell level that are in keeping with the overall organ dysfunction. Since these differences in ion transport disappear after removal of cytoskeleton (Ferrari and Bianchi, 1995), cross immunization studies were set up to detect possible differences between the cytoskeletal components of the two strains. These studies revealed a difference in a protein subsequently identified as adducin.

Adducin is an α/β heterodimeric protein which participates in the assembly of the spectrin–actin cytoskeleton, modulates the actin polymerization (Hughes and Bennett, 1995), binds calmodulin, is phospholylated by PKC and tyrosin kinase and regulates cell-signal transduction (Matsuoka *et al.*, 1996). Point mutations both in the α- and β-adducin subunits account for around 50% of the difference in blood pressure between MHS and MNS rats (Bianchi *et al.*, 1994). Transfection of 'hypertensive' and 'normotensive' α-adducin variants in rat kidney cells (NRK) showed that the former increases the surface expression and maximal rate of the Na–K pump (Tripodi *et al.*, 1996). Moreover, in a cell-free system, the 'hypertensive' variant of α-adducin directly stimulates the isolated Na–K pump at significantly lower concentrations than that of the 'normotensive' (Ferrandi *et al.*, 1997a). This finding provides therefore the genetic molecular basis for a constitutive increase of tubular sodium reabsorption in MHS rats. The cellular and biochemical alteration caused by adducin polymorphism in MHS rats are also associated with an increased production of an endogenous Na–K pump inhibitor, the so called ouabain-like factor (OLF) (Ferrandi *et al.*, 1992; 1997b), which is involved in the Na–K pump modulation and consequently in the cell ion handling.

8. Comparison between rat and man

As extensively discussed elsewhere (Ferrari and Bianchi, 1995), many similarities have been shown between MHS rats and a subgroup of human hypertensive subjects or subjects prone to develop hypertension. These similarities include renal function (Bianchi *et al.*, 1979), cellular ion transport (Cusi *et al.*, 1991) and plasma levels of renin (Bianchi, *et al.*, 1979) and OLF (Ferrandi *et al.*, 1997b; Rossi *et al.*, 1995). These data support the notion that a primary increase of tubular sodium reabsorption may explain hypertension also in this subgroup of patients.

There is close homology (94%) for the α-adducin gene between rat and man (Bianchi *et al.*, 1994) and this is particularly relevant from the evolutionary and functional point of view since it supports the notion that adducin has important cellular functions in both rat and man, despite more than 40 million years of divergence between these two species. Moreover, a functional mutation in the α-adducin gene (Gly460Trp) has been found to be associated with human essential hypertension, in at least three independent studies (Castellano *et al.*, 1987; Cusi *et al.*, 1987; Iwai *et al.*, 1997). Others, however, failed to confirm these findings (Ishikawa *et al.*, 1998; Kato *et al.*, 1998). Moreover, irrespectively of the presence (Cusi *et al.*, 1987) or absence (Glorioso, N. *et al.*, 1999) of the association between the 460 Trp variant and hypertension, the blood pressure fall after a diuretic in patients carrying the 460 Trp variant was greater than that of patients with 460 Gly. The former group of patients also displayed a greater change of blood pressure after an acute variation of sodium balance (Cusi *et al.*, 1997). Therefore, also in humans the α-adducin polymorphism may be used to identify patients with a salt-sensitive form of essential hypertension which might be treated with drugs able to increase renal sodium excretion. In this respect, it must be noted that increased levels of OLF have been demonstrated in essential hypertensive patients (Ferrandi *et al.*, 1997b; Rossi *et al.*, 1995) and may vary according to salt and water homeostasis. All these findings indicate that a common molecular mechanism, supported by adducin polymorphism, is operating both in rats and in a subgroup of hypertensive subjects.

The link among the individual steps connecting adducin polymorphism to the increase in Na–K pump activity, tubular Na^+ re-absorption, OLF levels and hypertension have not yet been fully elucidated. However, the following hypotheses can be proposed: a primary genetic molecular defect of adducin may affect the actin-cytoskeleton structure and function leading to an increase of Na–K pump units on the cell membrane with a consequent enhancement of the overall Na^+ transport across renal tubular cells; this, in turn, causes an increase of renal Na^+ re-absorption, blood pressure and OLF levels. Moreover, increased OLF levels seems to affect *per se* the expression and activity of the renal Na–K pump. In fact, beside the observed relationship between renal Na–K pump over-expression and OLF in MHS, chronic infusion in normal rats of very low doses of ouabain (Ferrari *et al.*, 1998), which is considered very similar if not identical to endogenous OLF, or incubation of cultured renal cells with ouabain, at nanomolar concentrations, lead to an upregulation of the maximal rate of the Na–K pump. This suggests that chronic exposure to low concentrations of OLF may have a synergistic effect with adducin favoring the pump overexpression (Ferrari *et al.*, 1999a).

9. A new pharmacological target

From these findings, it can be postulated that any therapeutic maneuver able to interfere with the sequence of mechanisms linking adducin polymorphism with renal Na–K pump overexpression, OLF and hypertension might be able to lower blood pressure in those subjects where this is operating. In this respect, the upregulation of the Na–K pump activity and OLF levels could represent a new pharmacological target for the treatment of forms of hypertension sustained by these mechanisms. This implies that not all forms of hypertension may respond to a selective therapy addressed to normalize the expression and activity of the renal Na–K pump altered as a consequence of the adducin mutations. However, it cannot be excluded that, independently from this specific genetic cause, an upregulation of the renal Na–K pump sustained by other molecular mechanisms may be the target for a new therapeutic approach to hypertension.

In line with these findings, a new compound, named PST 2238, has been developed (Ferrari *et al.*, 1998; 1999a, 1999b; Quadri *et al.*, 1997). This compound is able to interfere with the Na–K pump receptor, and to reduce, at nanomolar concentration, the Na–K pump increase induced in renal cell cultures either by cell transfection with the 'hypertensive' variant of adducin or by incubation with very low concentrations of ouabain. Conversely, no effect on Na–KATPase has been observed in normal, wild type cells. When given to MHS rats, PST 2238 lowers blood pressure and normalizes the increased renal Na–K pump at oral doses of 1–10 $\mu g\ kg^{-1}$. Similarly, less than 1 $\mu g\ kg^{-1}$ of PST 2238 is able to completely normalize both blood pressure and the increased renal Na–K pump also in rats made experimentally hypertensive by a chronic infusion of low doses of ouabain. Two aspects must be noted: firstly, PST 2238 in MHS rats does not completely normalize blood pressure, but it does normalize renal Na–KATPase; this can be explained by the polygenic nature of genetic hypertension, in fact this compound is able to interfere with only one of the genetic mechanisms causing hypertension in MHS (i.e. the alteration of the renal Na–K pump consequent on the adducin mutation), while other not yet identified mechanisms still exert their pressor effects. Secondly, from the pharmacogenomic point of view, it is interesting to note that this compound does not lower blood pressure in SHR rats (Ferrari *et al.*, 1999b), in which adducin polymorphism (Tripodi *et al.*, 1997), renal Na–KATPase (Nguyen *et al.*, 1998) and OLF levels (Doris, 1994) do not seem to be involved in causing hypertension. This compound is devoid of cardiac and hormonal effects typical of digitalis or antimineralocorticoid drugs (Ferrari *et al.*, 1999a), it does not bind to other receptors involved in blood pressure regulation or hormonal homeostasis (Ferrari *et al.*, 1998), it has a safe toxicological profile and it is well tolerated in healthy volunteers who underwent phase 1 clinical studies (Ferrari et al., 1999a).

10. Conclusions

In conclusion, we now have the theoretical background and the practical tools to address the problem of improving our approach to the individual therapy of essential hypertension. We may hypothesize that a better understanding of the genetic mechanisms of hypertension, will not only favor the development of new pharmacological approaches and the discovery of novel antihypertensive drugs, but also will furnish powerful tools for a better and more appropiate use of the available ones. As stressed above, it is very likely that two or more major genes are involved in hypertension in the individual patient within a given pedigree. In complex, multifactorial diseases like hypertension, the pressor role of an individual gene polymorphism must be evaluated in the context of its overall genetic and environmental background. Therefore an understanding of the appropriate interactions among genes and between genes and environment, which underline a given pathophysiological alteration, are of paramount importance in the implementation of any pharmacogenomic approach to the therapy of hypertension. Along these lines, we are studying the interaction between adducin and ACE polymorphism on the blood pressure response to acute changes in sodium balance. The result of this study (C. Barlassina and D. Cusi, unpublished data) showed that these two polymorphisms interact epistatically in determining this response. Therefore it is likely that similar interactions are also relevant for dissection of the genetic basis of the individual blood pressure response to drugs.

References

Barber, B.R., Ferrari, P. and Bianchi, G. (1994) The Milan hypertensive strain: a description of the model. In: *Handbook of Hypertension* (eds. D. Ganten and W. de Jong). Elsevier, Amsterdam, pp. 316–345.

Bianchi, G., Tenconi, T. and Lucca, R. (1970) Effect in the conscious dog of constriction of the renal artery to a sole remaining kidney on haemodynamics, sodium balance, body fluid volumes, plasma renin concentration and pressor responsiveness to angiotensin. *Clin. Sci.* **38**: 741–766.

Bianchi, G., Baer, P.G., Fox, U., Duzzi, L., Pagetti, D. and Giovanetti, A.M. (1975) Changes in renin, water balance and sodium balance during development of high blood pressure in genetically hypertensive rats. *Circ. Res.* **36/37(suppl. 1)**: 153–161.

Bianchi, G., Cusi, D., Gatti, M., *et al.* (1979) A renal abnormality as a possible cause of 'essential' hypertension. *Lancet* **1**: 173–177.

Bianchi, G., Tripodi, G. and Casari, G., (1994) Two point mutations within the adducin genes are involved in blood pressure variation. *Proc. Natl Acad. Sci.* **91**: 3999–4003.

Boyle, R.M., Price, M.L. and Hamilton, M. (1979) Thiazide withdrawal in hypertension. *J. R. Coll. Physicians* **13**: 172–173.

Butler, R., Morris, A.D. and Struthers, A.D. (1997) Angiotensin-converting enzyme gene polymorphism and cardioascular disease. *Clin. Sci.* **93**: 391–400.

Cambien, F., Poirier, O., Lecerf, L., *et al.* (1992) Deletion polymorphism in the gene for angiotensin-converting enzyme is a potent risk factor for myocardial infarction. *Nature* **359**: 641–644.

Carretero, O.A. and Gulati, O.P. (1978) Effects of angiotensin antagonist in rats with acute, subacute and chronic two-kidney renal hypertension. *J. Lab. Clin. Med.* **91**: 264–271.

Casari, G., Barlassina, C. and Cusi, D. (1995) Association of the α-adducin locus with essential hypertension. *Hypertension* **25**: 320–326.

Castellano, M., Barlassina, C., Muiesan, M.L., Beschi, M., Cinelli, A., Rossi, F., Rizzoni, D., Cusi, D. and Agabiti-Rosei, E. (1997) Alpha-adducin gene polymorphism and cardiovascular phenotypes in a general population. *J. Hypertension* **15**: 1707–1710.

Cusi, D., Fossali, E., Piazza, A., *et al.* (1991) Hereditability estimate of erythrocyte Na–K–Cl cotransport in normotensive and hypertensive families. *Am. J. Hypertension* **4**: 725–734.

Cusi, D., Barlassina, C., Azzani, T., *et al.* (1997) Polymorphism of α-adducin and salt sensitivity in patients with essential hypertension. *Lancet* **349**: 1353–1357.

Dominiczak, A.F., Clark, J.S., Jeffs, B., Anderson, N.H., Negrin, C.D., Lee, W.K. and Brosnan, M.J. (1998) Genetics of experimental hypertension. *J. Hypertension* **16**: 1859–1869.

Doris, PA. (1994) Ouabain in plasma from spontaneously hypertensive rats. *Am. J. Physiol.* **266**: H360–H364.

Dudley, C., Keavney, B., Casadei, B., Conway, J., Bird, R. and Ratcliff, P. (1996) Prediction of patient responses to antihypertensive drugs using genetic polymorphism: investigation of renin–angiotensin system genes. *J. Hypertens*ion **14**: 259–262.

Edmunds, M.E., Russel, G.I. and Bing, R.F. (1991) Reversal of experimental renovascular hypertension. *J. Hypertension* **9**: 289–301.

Fernandez, M., Fiorentini, R., Onesti, G., Bellini, G., Gould, A.B., Hessan, H., Kim, K.E. and Swartz, C. (1978) Effect of administration of Sar1-Ala8-angiotensin II during the development and maintenance of renal hypertension in the rat. *Clin. Sci. Mol.* Med. **54**: 633–637.

Ferrandi, M., Minotti, E., Salardi, S., Florio, M., Bianchi, G. and Ferrari, P. (1992); Ouabainlike factor in Milan hypertensive rats. *Am. J. Physiol* **263**: F739–F748.

Ferrandi, M., Tripodi, G., Salardi, S., *et al.* (1996) Renal Na–KATPase in genetic hypertension. *Hypertension* **28**: 1018–1025.

Ferrandi, M., Ferrari, P., Salardi, S., Barassi, P., Rivera, R., Manunta, P. and Bianchi, G. (1997a) Influence of the hypertension related adducin polymorphism on the Na–K ATpase activity. *J. Hypertension* 5**(suppl. 4)**: S60. Abs.

Ferrandi, M., Manunta, P., Balzan, S., Hamlyn, J.M., Bianchi, G. and Ferrari, P. (1997b) Ouabain-like factor quantification in mammalian tissues and plasma: comparison of two independent assays. *Hypertension* **30**: 886–896.

Ferrari, P. (1998) Pharmacogenomics: a new approach to individual therapy of hypertension? *Curr. Opin. Nephrol. Hypertension* **7**: 217–222.

Ferrari, P. and Bianchi, G. (1995) Lessons from experimental genetic hypertension. In: *Hypertension: Pathophysiology, Diagnosis and Management* (eds J.H. Laragh and B.M. Brenner). Raven Press, New York, pp. 1261–1280.

Ferrari, P., Torielli, L., Ferrandi, M., *et al.* (1998) PST 228: a new antihypertensive compound that antagonizes the long-term pressor effect of ouabain. *J. Pharm. Exp. Ther.* **285**: 83–94.

Ferrari, P., Ferrandi, M., Torielli, L., Tripodi, G., Melloni, P. and Bianchi, G. (1999a) PST 2238: a new antihypertensive compound that modulates Na-KATPase and antagonizes the pressor effect of OLF. *Cardiovasc. Drug Rev.,* in press.

Ferrari, P., Ferrandi, M., Tripodi, G., Torielli, L., Padoani, G., Minotti, P., Melloni, P. and Bianchi, G. (1999b) PST 2238: a new antihypertensive compound that modulates Na-K ATPase in genetic hypertension. *J. Pharm. Exp. Ther.*, **288**: 1074–1083.

Freeman, R.H., Davis, J.O., Watkins, B.E., Stephens, G.A. and De Forrest, J.M. (1979) Effects of continuous converting enzyme blockade on renovascular hypertension in the rat. *Am. J. Physiol.* **236**: F21–24.

Glorioso, N., Manunta, P., Filigheddu, F., *et al.* (1999) The role of α-adducin polymorphism in blood pressure and sodium handling regulation may not be excluded by a negative association study. *Hypertension* (in press).

Guan, S., Fox, J., Mitchell, K.D. and Navar, L.G. (1992) Angiotensin and angiotensin converting enzyme tissue levels in two-kidney, one clip hypertensive rats. *Hypertension* **20**: 763–767.

Guyton, A.C., Hall, J.E., Coleman, T.G., Manning, R.D. and Norman, R.A. (1995) The dominant role of the kidneys in long-term arterial pressure regulation in normal and hypertensive states. In: *Hypertension: Pathophysiology, Diagnosis and Management* (eds J.H. Laragh and B.M. Brenner). Raven Press, New York, Chap. 78, pp. 1311–1326.

Hamet, P., Pausova, Z., Adarichev, V., Adaricheva, K. and Tremblay, J. (1998) Hypertension: genes and enviroment. *J. Hypertension* **16**: 397–418.

Hingorani, A.D., Jia, H., Stevens, P.A., Hopper, R., Dickerson., J.E. and Brown, M.J. (1995) Renin–angiotensin system gene polymorphisms influence blood pressure and the response to angiotensin converting enzyme inhibition. *J. Hypertension* **13**: 1602–1609.

Horan, M.J. and Mockrin, S.C. (1992) Heterogeneity of hypertension. *Am. J. Hypertension* **5**: 110S–113S.

Housman, D. and Ledley, F.D. (1998) Why pharmacogenomics? Why now? *Nature Biotech.* **16**: 492–493.

Hughes, C.A. and Bennett, V. (1995) Adducin: a physical model with implications for function in assembly of spectrin–actin complexes. *J. Biol. Chem.* **270**: 18990–18996.

Ishikawa, K., Katsuya, T., Sato, N., Nakata, Y., Takami, S., Takiuchi, S., Fu, Y., Higaki, J. and Ogihara, T. (1998) No association between α-adducin 460 polymorphism and essential hypertension in a Japanese population. *Am. J. Hypertension* **11**: 502–506.

Iwai, N., Tamaki, S., Nakamura, Y. and Kinoshita, M. (1997) Polymorphism of α-adducin and hypertension. *Lancet* **350**: 369.

Jick, H., Garcia Rodriguez, L.A. and Perez-Gutthan, S. (1998) Principles of epidemiological research on adverse and beneficial drug effects. *Lancet* **352**: 1767–1770.

Kato, N., Sugiyama, T., Nabika, T., Morita, H., Kurihara, H., Yazaki, Y. and Yamori, Y. (1998) Lack of association between the alpha-adducin locus and essential hypertension in the Japanese population. *Hypertension* **31**: 730–733.

Laragh, J.H., Lamport, B., Sealey, J. and Alderman, M.H. (1988) Diagnosis ex juvantibus. Individual response patterns to drugs reveal hypertension mechanisms and simplify treatment. *Hypertension* **12**: 223–226.

Lifton, R.P. (1996) Molecular genetics of human blood pressure variation. *Science* **272**: 676–680.

Lowder, S.C. and Liddle, G.W. (1974) Prolonged alteration of renin responsiveness after spironolactone therapy. *N. Engl. J. Med.* **291**: 1243–1244.

Lund-Johansen, P. and Omvik, P. (1990) Hemodynamic patterns of untreated hypertensive disease. In: *Hypertension: Pathophysiology, Diagnosis and Management* (eds J.H. Laragh and B.M. Brenner). Raven Press, New York, Chap. 22, pp. 305–327.

Marshall, A. (1997) Getting the right drug into the right patient. *Nature Biotech.* **15**: 1249–1252.

Matsuoka, Y., Hughes, C.A. and Bennet, V. (1996) Definition of the calmodulin-binding domain and sites of phosphorylation by protein kinase A and C. *J. Biol. Chem.* **271**: 25157–25166.

Medical Research Council Working Party on Mild Hypertension (MRGWPMH) (1986) Course of blood pressure in mild hypertensives after withdrawal of long-term antihypertensive treatment. *Br. Med. J.* **293**: 988–992.

Menard, J., Bellet, M. and Serrurier, D. (1990) From the parallel group study to the crossover design, and from the group approach to the individual approach. *Am. J. Hypertension* **3**: 815–819.

Morishita, R., Higaki, J., Okunishi, H., Nakamura, F., Nagano, M., Mikami, H., Ishii, K., Miyazaki, M. and Ogihara, T. (1993) Role of tissue renin angiotensin system on two-kidney, one-clip hypertensive rats. *Am. J. Physiol.* **264**: F510–F514.

Nguyen, A.T., Hayward-Lester, A., Sabatini, S. and Doris, P.A. (1998) Renal Na–K ATPase in SHR: studies of activity and gene expression. *Clin. Exp. Hypertension* **20**: 641–656.

Niutta, E., Cusi, D., Colombo, R., Pellizzoni, M., Cesana, B., Barlassina, C., Soldati, L. and Bianchi, G. (1990) Predicting interindividual variations in anti-hypertensive therapy: the role of sodium transport systems and renin. *J. Hypertension* **6(suppl. 4)**: S53–S58.

O' Toole, L., Stewar, M., Padfield, P. and Canner, K. (1998) Effect of the insertion/deletion polymorphism of the angiotensin-converting enzyme gene on response to angiotensin-converting enzyme inhibitors in patients with heart failure. *J. Cardiovasc. Pharmacol.* **32**: 988–994.

Pfeffer, M.A., Braunwald, E., Moye, L.A., *et al.* (1992) Effect of captopril on mortality and morbidity in patients with left ventricular dysfunction after myocardial infarction: result of the Survival and Left Ventricular Enlargement Trial. *N. Engl. J. Med.* **327**: 669–677.

Pratt, R.E. and Dzau, V.J. (1999) Genomics and hypertension. Concepts, potentials and opportunities. *Hypertension* **33**: 238–247.

Quadri, L., Bianchi, G., Cerri, A., Fedrizzi, G., Ferrari, P., Gobbini, M., Melloni, P., Sputore, S., and Torri, M. (1997) 17b-(3-Furyl)–5b-androstane–3b, 14b, 17a-triol (PST 2238). A very potent antihypertensive agent with a novel mechanism of action. *J. Med. Chem.* **40**: 1561–1564.

Rapp, J.P. (1995) The search for the genetic basis of blood pressure variation in rats. In: *Hypertension: Pathophysiology, Diagnosis and Management.* (eds J.H. Laragh and B.M. Brenner). Raven Press, New York, Chap. 76, pp. 1289–1300.

Rossi, G.P., Manunta, P., Hamlyn, J., Pavan, E., De Toni, R., Semplicini, A. and Pessina, A.C. (1995) Immunoreactive endogenous ouabain in primary aldosteronism and essential hypertension: ralationship with plasma renin, aldosterone and blood pressure levels. *J. Hypertension* **13**: 1181–1191.

Soubrier, F. and Lathrop, G.M. (1995) The gentic basis of hypertension. *Curr. Opin. Nephrol. Hypertension* **4**: 177–181.

The SOLVD Investigators (1991) Effect of enalapril on survival in patients with reduced left ventricular ejection fractions and congestive heart failure. *N. Engl. J. Med.* **325**: 293–302.

Swart, S., Bing, R.F., Swales, J.D. and Thurston, H. (1982) Plasma renin in long-term diuretic treatment of hypertension: effect of discontinuation and restarting therapy. *Clin. Sci.* **63**: 121–125.

Tripodi, G., Valtorta, F., Torielli, L., *et al.* (1996) Hypertension-associated point mutations in the adducin α and β subunits affect actin cytoskeleton and ion transport. *J. Clin. Invest.* **97**: 2815–2822.

Tripodi, G., Szpirer, C., Reina, C., Szpirer, J. and Bianchi, G. (1997) Polymophism of γ-adducin gene in genetic hypertension and mapping of the gene to rat chromosome 1q55. *Biochem. Biophys. Res. Comm.* **237**: 685–689.

Veterans Administration Cooperative Study Group on Antihypertensive Agents (VA-CSGAA (1975) Return of elevated blood pressure after withdrawal of antihypertensive drugs. *Circulation* **51**: 1107–1113.

Vincent, M., Samani, N.J., Gauguier, D., Thompson, J.R., Lathrop, M.G. and Sassard, J.A. (1997) pharmacogentic approach to blood pressure in Lyon hypertensive rats. A chromosome 2 locus influences the response to a calcium antagonist. *J. Clin. Invest.* **100**: 2000–2006.

Ward, R. (1995) Familial aggregation and genetic epidemiology of blood pressure. In: *Hypertension: Pathophysiology, Diagnosis and Management* (eds J.H. Laragh and B.M. Brenner). Raven Press, New York, Chap. 5, pp. 67–88.

Watkins, B.E., Davis, J.M., Freeman, R.H., De Forrest, J.M. and Stephens, G.A. (1978) Continuous angiotensin II blockade throughout the acute phase of one-kidney hypertension in dog. *Circ. Res.* **42**: 813–820.

12

Molecular genetics of hypertension: future directions and impact on clinical management

Friedrich C. Luft

1. Introduction

Predicting the future is not only a risky, but also a thankless task particularly when relegated to pessimists. Over a decade ago, with the introduction of restriction fragment-length polymorphism technology as a relatively routine laboratory procedure, the National Heart Lung and Blood Institute of the National Institutes of Health, U.S.A. decided to place major emphasis on molecular genetics in terms of supporting research grant applications (Luft, 1998). Similar emphasis was placed on this area of research in France, U.K. and elsewhere. Nongeneticists (such as yours truly) shifted their research emphasis to the new area in a process termed 'retooling'. It is hard to pick up an issue of any hypertension-related journal without encountering reports on this or that polymorphism, the confirmation of this association study or the refutation of that association study. Nevertheless, the results of this heady research area are a little more sobering. Exactly how many genes have been found that are important to primary hypertension? How have these findings facilitated diagnosis? More importantly, what new therapeutic insights have been developed from these findings? Funding for this area of research has been generous to the detriment of other topics. Sooner or later, we shall be called to account. Perhaps we should start thinking of some answers, if for no other reason than to formulate better questions. After all, we do not want molecular genetics to be regarded as 'the god that failed'. About 2000 papers were published in the last 18 months that Medline identified with the key words 'hypertension' and 'genetics'. One safely made prediction is that the literature on this topic will undoubtedly continue to expand exponentially, at least as long as the funding agencies continue to be generous. My colleagues had the easy and pleasant chapters in this volume. It remains for me to be the grinch that stole Christmas.

Molecular Genetics of Hypertension, edited by A.F. Dominiczak, J.M.C. Connell and F. Soubrier.

2. Monogenic hypertension

Monogenic hypertension is the bright spot in the area of molecular genetics of human hypertension. The notion here is that by elucidating rare monogenic diseases, we shall understand mechanisms applicable to primary hypertension (Luft *et al.*, 1995). This promise has been kept largely through the efforts and successes of Lifton and colleagues, as well as work from the laboratory of White (see Chapter 8). Therefore, I will confine my comments on possible applications of this work.

2.1 Glucocorticoid-remediable aldosteronism

Are there future directions for GRA? Richard Lifton's group will kindly exclude GRA in your patients if you send them a few drops of blood on filter paper. They have found many families with this condition and undoubtedly more will be found. For these families, identification of their genetic condition is very important. Affected members can be identified, nonaffected members firmly excluded and definitive therapy can be administered. However, in the large frame of things, GRA is not terribly important to the hypertensive population at large.

2.2 Apparent mineralocorticoid excess

The facinating possibility that AME might be relevant in the heterozygous state has been raised by Li *et al.* (1997), who observed a patient with apparent mineralocorticoid hypertension at age 38 years, who had a daughter with homozygous AME. The patient had low renin and aldosterone concentrations and was found to have a mutation in the gene for 11β-hydroxysteroid dehydrogenase. These are fascinating findings; however, the importance of AME does not appear great and the condition will probably remain a curiosity and a topic for medical students examinations. Mineralocorticoid receptor antagonists, spironolactone and similar compounds, are already available and continue to be refined.

2.3 Liddle syndrome

Liddle syndrome responds nicely to diuretics. However, it is unlikely that discoveries linked to Liddle syndrome will increase the therapeutic armamentarium. Conceivably, a new loop-type diuretic could be produced by developing inhibitors to CLCNKB; however, loop diuretics are inferior to thiazide diuretics in terms of lowering blood pressure. Currently, we still have difficulties in convincing physicians to apply readily available, inexpensive, diuretic therapy in the first place (Joint National Committee, 1997). It is unlikely in my view that the pharmaceutical industry will invest heavily in the development of future diuretics on the basis of genetic research.

2.4 Autosomal-dominant hypertension with brachydactyly

Thus far, this syndrome has only been described in this Turkish kindred and a similar family in Canada. Recently, we encountered another such family in the

U.S.A. In these three families, the hypertension also follows an autosomal-dominant mode of inheritance and cosegregates 100% with short stature and type E brachydactyly (Toka *et al.*, 1998). A deletion syndrome in a Japanese child with type E brachydactyly, as well as the additional families, has enabled us sharply to decrease the area on 12p containing the gene (Bähring *et al.*, 1997); however, a 4 million base-pair segment remains and we have not yet cloned the gene. Thus, importance of this syndrome to primary hypertension is totally speculative. A putative mechanism explaining hypertension in these families is the presence of a vascular anomaly involving the posterior inferior cerebellar artery at the ventrolateral medulla. A tortuous, looping vessel was found with magnetic resonance imaging in 15 affected subjects with autosomal-dominant hypertension and brachydactyly and in none of 11 nonaffected siblings (Naraghi *et al.*, 1997). Recently, Geiger *et al.* (1998) reported on the favorable results of surgical decompression in this syndrome. However, no detailed neurophysiological studies have been done in subjects with autosomal-dominant hypertension and brachydactyly. Thus, we do not yet know whether or not this form of hypertension has a neurovascular basis. However, the possibility that essential hypertension has in part a neurovascular basis is being actively pursued by several groups (Chalmers, 1998). If autosomal-dominant hypertension with brachydactyly is a suitable model of this condition, the impact for diagnosis and treatment could be considerable.

3. Primary hypertension

A brief overview of papers in 1997 and 1998 on patients with primary hypertension revealed research on the following genes: angiotensin converting enzyme (ACE), angiotensinogen, AT1 receptor, β-2 adrenergic receptor, α-adducin, angiotensinase C, renin binding protein, G-protein beta3 subunit, atrial natriuretic peptide, insulin receptor and eNOS in hypertension of pregnancy. Even telomere length has been raised as being important to primary hypertension (Aviv and Aviv, 1998). In addition to gene variants influencing blood pressure as such, other variants may exist that influence the likelihood to develop complications. Such genetic attributes might be even more important to hypertension-related sequellae. An example is the nitric oxide synthase gene polymorphism predisposing to acute myocardial infarction (Hibi *et al.*, 1998) and stroke (Markus *et al.*, 1998). Despite their interest, discussion of all these genes is beyond the scope of this commentary; however, in my view six genes – ACE, angiotensinogen, α-adducin, the β-2 adrenergic receptor, G-protein beta3 subunit and the T594M mutation in the β subunit of the epithelial sodium channel – are of particular promise.

3.1 Angiotensin converting enzyme

The ACE gene locus was linked to blood pressure in spontaneously hypertensive rats in 1991, and although the ACE gene insertion/deletion allelic variant has been implicated in arteriosclerotic cardiovascular disease, cardiac hypertrophy, restenosis, progression of diabetic renal disease and progression of IgA nephropathy, hanging a guilty verdict in terms of hypertension onto the ACE gene has been difficult

(Soubrier, 1998). O'Donnell *et al.* (1998) found evidence for association and genetic linkage of the ACE gene with hypertension and blood pressure in men, but not in women, when they analyzed over 3000 participants from the Framingham Heart Study. The data were significant, but not robust. Fornage *et al.* (1998) studied 583 three-generation pedigrees from Rochester, MN, U.S.A., and were able to show that variations in a microsatellite marker within the growth hormone gene, which is close to the ACE gene locus, influenced interindividual blood pressure differences in young white men, but not in women. What these findings will mean in terms of complications is not clear. Recently, Girerd *et al.* (1998) could find no association between renin–angiotensin system gene polymorphisms, and wall thickness of the radial and carotid arteries in a sizable middle-aged cohort. Could genetic variations in the ACE gene have any therapeutic inference in terms of ACE inhibitor therapy? The answer is possibly yes, but not much (Ueda *et al.*, 1998).

3.2 Angiotensinogen

Jeunemaitre *et al.* (1992) first reported linkage of the angiotensinogen (AGT) gene locus to hypertension in hypertensive siblings from France and Utah. Subsequent screening identified the so-called AGT 235T variant in hypertensive cases, as being more frequent than in controls. The variant is associated with higher AGT levels and appears to be in tight linkage disequilibrium with a promoter mutation -6 bp (G-6A) upstream of the initiation site of transcription (Inone *et al.*, 1997). This mutation may result in a higher basal transcription rate. The haplotype combining the AGT 235T and G-6A polymorphisms appears as the ancestral allele of the human AGT gene and is the haplotype associated with hypertension (Jeunemaitre *et al.*, 1997).

Caulfield *et al.* (1996) have investigated AGT extensively and reported linkage of the AGT locus to blood pressure in 77 European families ($P<0.001$). Their studies in African Caribbeans also supported the notion that the AGT locus is linked to hypertension. Since the initial reports, many studies have been published on the association between allelic variants in AGT and hypertension. Hunt *et al.* (1998) genotyped large numbers of subjects from the Trials of Hypertension Prevention study. Patients with the -6 bp AA genotype seemed to do better with salt restriction than other genotypes. Kunz *et al.* (1997) have recently reviewed the evidence on AGT 235T from 11 studies of 14 populations. Data on 5493 patients showed that the AGT 235T allele was significantly associated with hypertension (odds ratio, OR 1.2, confidence interval, CI 1.11–1.29). These data were significant statistically; however, their clinical significance is another matter. The authors concluded that much more than AGT 235T had to be genetically responsible for primary hypertension. The AGT gene has been the most scrutinized and the most promising finding of the primary hypertension genes thus far; however, the AGT 235T variant explains only a relatively small part of blood pressure variance.

3.3 α-Adducin

To my knowledge, α-adducin is the only example of rat molecular genetic research contributing pertinent information to the molecular genetics of human hypertension. A mutation in rat α-adducin was found to be responsible for 50% of the hypertension in the Milan hypertensive rat. The mutation was shown to be responsible for an increase in Na–K pump activity in renal cell transfection experiments. Linkage and association studies were subsequently performed in hypertensive patients and controls and a point mutation (G460W) was found in the human α-adducin gene. The 460W variant was shown to be more frequent in hypertensive patients than in controls. The pressure–natriuresis relationship was subsequently studied in 108 hypertensive patients. The relationships suggested a shifted, reduced-slope, salt-sensitive pressure–natriuresis curve in persons bearing the W variant. The α-adducin studies combine molecular genetics and physiology in rats and patients and present a truly remarkable story of careful observations, patience, and scholarship, which has been summarized elsewhere (Manunta *et al.*, 1998). However, the importance of α-adducin to other hypertensive populations and to salt-sensitive hypertension must await additional studies. Recently, a Japanese group (Kato *et al.*, 1998) and an Australian group (Kaminati *et al.*, 1998) were unable to find an association between α-adducin allelic variants and essential hypertension.

3.4 β-2 Adrenergic receptor

A restriction fragment length polymorphism in the β-2 adrenergic receptor gene was associated with and linked to salt sensitive hypertensive persons of African origin in earlier studies (Svetkey *et al.*, 1997). An amino terminal variant in the β-2 adrenoceptor, which encodes glycine instead of arginine at basepair position 16 (Arg16→Gly), has been described which appears to have functional significance (Yang-Feng *et al.*, 1990). The variants showed equal affinity for epinephrine or isoproterenol; however, the Gly16 variant exhibited increased downregulation in response to isoproterenol, compared with the Arg16 variant (Green *et al.*, 1994). Such a downregulation pattern could lead to impaired vasodilatory responses to circulating β-2 adrenergic agonists. This hypothesis is supported by *in vivo* studies showing that pulmonary β-2 adrenoceptors with the Gly16 variant also exhibit increased down-regulation in response to salbutamol, compared to the Arg16 variant. Furthermore, a recent report indicating that the Gly16 variant in the β-2 adrenoceptor is associated with nocturnal asthma renders further support to the notion that this polymorphism may have major functional importance (Turki *et al.*, 1995). Finally, increased β-2 adrenoceptor downregulation might serve to explain the decreased β-2 adrenoreceptor expression on the fibroblasts of salt-sensitive, compared with salt-resistant normotensive Europeans (Kotanko et al., 1992). Recently, Kotanko *et al.* (1997) performed an association study in 136 African Caribbeans with hypertension and 81 unrelated control persons from the island of St Vincent. They found significant support for an association of the pro-downregulatory Gly variant with hypertension. Hypertensive persons of African origin would be expected to be salt sensitive, although they were not tested. Data from the Bergen Blood Pressure study support the idea that the Arg16→Gly allelic variant is important to increased blood pressure (Timmermann *et al.*, 1998). However, in that study, the Gly variant was associated with lower blood pressures in a dose-dependent fashion. Recently, Liggett *et al.* (1998) found that another polymorphism in this same gene predicted which patients with congestive heart failure were more likely to die.

Obviously, much remains to be done to elucidate the role of the β-2 adrenergic receptor gene in primary hypertension and cardiovascular disease.

3.5 G-protein beta3 subunit

The notion that G-proteins might be involved in primary hypertension stems from observations that pertussis toxin-sensitive G proteins in lymphoblasts and fibroblasts from selected patients with primary hypertension engaged in enhanced signal transduction (Siffert, 1998). Siffert *et al.* (1998) detected a novel polymorphism (C825T) in exon 19 of the gene encoding the G-protein beta3 subunit (GNB3). The T allele is associated with the occurrence of a splice variant, which causes a loss of 41 amino acids and one WD repeat domain of the G beta subunit. The splice variant was shown to be active in expression studies. A genotypic analysis of 427 normotensive and 426 hypertensive subjects suggested a significant association of the T allele with essential hypertension. Schunkert *et al.* (1998) have since identified an association between this polymorphism and lower renin values, as well as elevated diastolic blood pressure levels in a German cohort. Additional support comes from an association study in the Canadian Oji-Cree indians (Hegele *et al.*, 1998). The relevance of these findings will require additional confirmatory studies.

3.6 T594M mutation in the β subunit of the epithelial sodium channel

A variant of the β-subunit of the amiloride-sensitive sodium channel was described by Su *et al.* (1996) who also observed increased channel activity in lymphocytes in African Americans. Baker *et al.* (1998) recently studied 206 hypertensive black patients and 142 normotensive black control subjects in London. Seventeen (8%) of the hypertensive blacks had the T594M mutation, compared to 2% of normotensive blacks. Persons with the mutation had lower plasma renin activity, supporting the notion of increased sodium reabsorption. Thus, the T594M mutation may serve to explain some degree of salt-sensitivity and hypertension in blacks. The elucidation of Liddle syndrome led to the discovery of this allelic variant. The finding underscores the potential relevance of rare monogenic diseases to complex genetic disease. Nevertheless, flies surround the ointment. Munroe *et al.* (1998) were unable to find linkage of the epithelial sodium channel to hypertension in black Caribbeans. Recently, Melander and colleagues (1998) described the βArg564X mutation as the cause for Liddle syndrome in a Swedish kindred. However, they were unable to identify polymorphisms in the gene, which were associated with hypertension or diabetic nephropathy. Persu *et al.* (1998) found seven polymorphisms in this gene, which were more common in subjects of African origin, compared to Caucasian subjects. Functionally significant properties could not be shown for these genetic variants.

4. Genetic and environmental factors: Pickering's work revisited

Although major efforts have been expended, excellent experiments have been performed, and exciting stories have been told, the results in the area of human molecular genetics of hypertension are modest. In terms of genetically explaining blood pressure variance for specific genes, we have a long way to go. The above six genes and their allelic variants are worthy of special discussion, in my view, because of the thought processes involved in their evaluation. Linkage analysis was successful in the case of three of the genes – AGT, α-adducin, and β-2 adrenergic receptor. However, we cannot conclude that these genes were found by linkage analysis. They were candidate genes which were selected by investigators who then sought linkage of their loci with the phenotype, hypertension. That the genes of the renin–angiotensin system and the genes for catecholamine receptors, and genes for the enzymes involved in their production and degradation might be involved in hypertension, would have occurred to students of hypertension 50 years ago. α-Adducin and the G-protein beta3 subunit were identified as candidates by whole animal and cell physiology approaches, which resulted in their being selected as candidate genes. A linkage analysis was subsequently performed in hypertensive sibling pairs and the α-adducin gene locus was indeed linked to hypertension.

To my knowledge, thus far no gene for a complex disease has been discovered and cloned on the basis of a linkage analysis. This state of affairs is also true for rat genetics. Despite huge breeding colonies, countless numbers of transgenic strains, major financial support, and boundless enthusiasm, the rat cardiovascular geneticists have not cloned a single new gene responsible for elevated blood pressure. I hasten to point out that I too am heavily involved in rodent research and some of my best friends are rat geneticists. Recently, Aitman *et al.* (1999) identified Cd36 (Fat) as an insulin-resistance gene causing defective fatty acid and glucose metabolism in spontaneously hypertensive rats. They used an impressive, high-tech, micro-array linkage approach. Their findings are indeed interesting; however, the relevance to hypertension is not sufficient to warrant my eating a crow dinner! Cd36 in man is known; a molecular basis of Cd36 deficiency was elucidated by Kashigwagi *et al.* (1995) earlier, and no connection with hypertension has been shown.

One possible explanation for this result is that the genes we seek have relatively small effects. If there are many genes with small effects (such as the gene variants above), the sample size necessary for linkage studies will be prohibitive. Perhaps the right linkage studies have not been done and indeed cohorts of hypertensive sibling pairs exceeding 1000 pairs are being subject to total genome scans in the United States and Europe; however, I am not optimistic.

A review of Pickering's work is enlightening in my view (reviewed in Pickering, 1982). His group obtained a sample of the population at large believed to be representative; first degree relatives of patients with essential hypertension; and first degree relatives of patients without essential hypertension. The data acquisition and data analysis took 4 years. Pickering found that the frequency distribution curves for blood pressure in the relatives of subjects without hypertension were indistinguishable from those of the population sample. Those for relatives of subjects with essential hypertension were similar in shape but were shifted upwards, by about the same amount at all ages. The increase in blood pressure with age was the same in the relatives of subjects with hypertension as in the rest of the population, but the relatives tended to have higher pressures at all ages.

Miall and Oldham (1955) performed a similar study in a Welsh mining valley and measured blood pressure in a sample of the population and their first degree relatives. The regression coefficient of blood pressure of relatives and propositi was about 0.2, similar to that observed by Pickering's group. Thus, blood pressure appeared to be inherited as a graded characteristic over the whole range of blood pressure, irrespective of the classifications: hypotension, normotension or hypertension. Pickering concluded that the inheritance of blood pressure was quite analogous to the inheritance of height. Were we to consider heights in excess of 170 cm as abnormal, we might as well be looking for 'tallness' genes. In all likelihood, we are facing many genes with small effects.

What are our options? The relative power of linkage and association studies for the detection of genes involved in hypertension have been reviewed by Jones (1998). He performed power calculations according the methods developed by Risch and Merikangas (1996), and showed that if a single major locus causing susceptibility to hypertension were present, nonparametric linkage strategies using affected sibling pairs may prove effective. However, if as suggested by the experiences of the last decade the number of genes is large and their effect is small, the sample size for such linkage analyses will be massive. In that case, a systematic search for allelic association may be more appropriate because of the dramatic reduction in the excess allele sharing for genes of small effect. The transmission disequilibrium test is an example (Ewens and Spielman, 1995). This test requires the collection of trios of two parents and an affected child. The frequency at which alleles are transmitted and not transmitted to the affected offspring is compared to the Mendelian expectation of 50:50. Importantly, this test determines allelic association requiring the presence of both linkage and linkage disequilibrium in order to yield a significant result. By using nontransmitted alleles as the control population, problems of population admixture and mismatched controls are avoided. Spielman and Ewens (1998) have recently expanded the transmission disequilibrium test to permit accruing information from units in which the parents are already dead, as is often the case in cardiovascular diseases. They describe a method termed the sib transmission disequilibrium test that overcomes this problem by using marker data from unaffected siblings instead of from parents, thus allowing application of the transmission disequilibrium test to sibships without parental data.

An alternative approach might be to identify quantitative trait loci (QTLs) for blood pressure in normotensive individuals. In my view, Pickering would have liked that approach, since he was not at all convinced that 'hypertension' existed as a discrete entity. That such QTLs might be relevant to primary hypertension is highly likely. We have employed studies in monozygotic and dizygotic twins and the parents of the latter. An analysis of monozygotic and dizygotic twins allows heritability estimates to be made. The dizygotic twins and their parents then lend themselves to a linkage analysis. We have linked the insulin-like growth factor (IGF)-1 gene locus to systolic blood pressure and heart size with this approach (Nagy *et al.*, 1997). In that study, we also found linkage to the loci for Liddle syndrome, and AT_1 receptor gene for systolic blood pressure. Linkage for diastolic blood pressure was found at the autosomal-dominant hypertension with brachydactyly locus. Both systolic and diastolic blood pressure were linked to the renin gene locus. However, the linkage was most consistent for the IGF-1 gene locus and systolic blood pressure. We also found linkage between the IGF-1 gene locus and posterior cardiac wall thickness, septal thickness and left ventricular mass index. We suggest that these QTLs may be important for the subsequent detection of allelic variants for elevated blood pressure. In retrospect, it is not surprising that these genes, with important blood pressure-regulatory effects, are linked to blood pressure in normal individuals. Recruiting families with multiple children is an alternative approach. The children can then be studied in terms of concordance and discordance for blood pressure and other variables. Excellent cohorts are available for such analyses in the United States. The Rochester Family Heart Study and the San Antonio Heart Study are two examples.

QTLs for blood pressure in normal individuals lend themselves to powerful approaches. Normal persons do not stay that way; they become ill and develop hypertension and heart disease at a later date. Identifying the ones that will do so may be possible from QTLs. We recently identified the β-2 adrenergic receptor gene locus as a QTL for blood pressure in normal twin subjects (unpublished work). We relied on the powerful technique of multiplex sequencing developed by Church and Kieffer-Higgins (1988) and sequenced the entire gene in normotensive dizygotic twin subjects. We did the same with the subjects from the Bergen Blood Pressure study mentioned earlier. We identified the known polymorphisms, plus a series of other variants not yet described. We found significant associations of certain polymorphisms with blood pressure spanning 10 mm Hg in these normal subjects. By sequencing QTL genes, we should find more variants in candidate genes affecting blood pressure in normal persons. Performing association studies at a later date in hypertensive individuals or conducting prospective long-term evaluation studies examining the influence of these genetic variants would then be much easier.

We should perhaps reconsider still another approach. A powerful tool for the fine structure localization of disease genes in a complex condition such as hypertension is linkage disequilibrium mapping in isolated populations. This novel approach adapts Luria and Delbrück's classical methods for analysing bacterial cultures to the study of human isolated founder populations with several goals in mind; namely, the estimation of the recombination fraction between a disease locus and a marker; the determination of the expected degree of allelic homogeneity in a population, and the mutation rate of marker loci. Linkage disequilibrium mapping is based on the observation that affected chromosomes descended from a common ancestral mutation should show a distinctive haplotype of the ancestral chromosome. The technique offers increased resolution because it exploits recombination events occurring over the entire history of a population. The best setting in which to apply the method would be a population in which there is a single disease-causing allele with a high frequency, so that the excess of an ancestral haplotype can be detected easily. Furthermore, this allele should have been introduced into the population sufficiently long ago that recombination has made the region of strongest linkage disequilibrium confined, but not too small. The theoretical basis for this 'haplotype sharing' method are described in detail elsewhere (de Vries *et al.*, 1996; Donelly, 1983; Lander and Botstein, 1986). The combination of focused sampling and the method of mass parallel genotyping (genome scanning) represent a new strategy to identify new genes for complex diseases. Isolated populations are available for study. Finland represents an ideal population for linkage disequilibrium mapping (de la Chapelle, 1993). However, there are other relatively isolated populations in many countries on the north American continent, other countries in Europe, the middle east, and elsewhere.

Finally, the discussion should not be ended without some comments as to the phenotype. I love Mozart's A major piano concerto, a remarkable genotype for a given piece of music. However, my performance is a laughable phenotype compared to the phenotype rendered by any capable musician, even though the genotype remains the same. Gene–environment interactions exert their obvious effects. Pickering indicated that blood pressure is a continuous variable and that 'hypertension' as such, does not exist. Since his work, we have been busy assuming the contrary and have divided the nonexistent entity 'hypertension' into low renin, normal renin and high renin hypertension, salt-sensitive and salt-resistant hypertension, modulating and nonmodulating hypertension, white coat hypertension, dipping and nondipping hypertension, neurogenic hypertension and whatever else have you. These intermediate phenotypes may help us; however, to use them a massive additional effort will be necessary. Controlled diets, provocative maneuvers, 24 h ambulatory blood pressure measurements, complex hormonal determinations and much more would be required. If 700 affected sibling pairs is a hindrance, imagine that same number all removed from their medication for a month and phenotyped in terms of salt sensitive or modulator status! However, such an effort may be warranted. Or alternatively, we might concentrate more on the 'normotensive' individuals. As discussed above, we may be helped by searching for quantitative trait loci for blood pressure in putatively healthy persons, 'hypertension' being arbitrary. Another avenue might be to genotype normotensive or hypertensive individuals prospectively for the candidate variation in question (angiotensinogen M235T for example). Thereafter, age- and gender-matched homogenous groups could be phenotyped in terms of intermediate or blood pressure-relevant variables and the genotypic groups compared. This approach would represent a novel association study from the genotype to the phenotype (bottom-up association).

Will molecular genetics eventually be important to hypertension management? The complexities are far greater than we could foresee. As previously mentioned, there are four polymorphisms described in the β-2 adrenergic receptor gene which result in amino acid substitutions and therefore are likely to be of functional significance (Liggett, 1995). Setting up a haplotype analysis of this information is a daunting prospect. Let us assume that we have found all 30 (or more) of Pickering's genes responsible for variation in blood pressure. Technically, it is quite conceivable that all variants in these genes could be placed on a single chip for hybridization and genotyping. For the β-2 adrenergic receptor alone, I know that the number approaches a dozen. Will this huge amount of data, even if we can analyze it, help us pick the patient best treated with a beta blocker or hydrochlorothiazide or an ACE inhibitor? Will we develop new drugs on the basis of this information with different modes of action? Possibly, transcription modulating drugs and drugs interfering with signal transduction are examples;

however, thus far such drugs have not been developed because of genetic inferences (Kurtz and Gardner, 1998). This work is currently emanating from cell biology and molecular pharmacology, not from genetics.

In summary, while our successes with monogenic diseases have been phenomenal, our search for genes causing primary hypertension have been more modest. Interesting finds have been made; however, the surface has barely been scratched. Novel approaches in terms of study design, analyses and populations will be necessary. Future success will depend less on molecular genetic technology and more on investigator ingenuity. In terms of the latter, Pickering provides a brilliant example for us to follow.

References

Aitman, T.J., Glazier, A.M., Wallace, C.A., ***et al.*** (1999) Identification of Cd36 (Fat) as an insulin-resistance gene causing defective fatty acid and glucose metabolism in hypertensive rats. *Nature Genet.* **21**: 76–83.

Aviv, A. and Aviv, H. (1998) Telomeres, hidden mosaicism, loss of heterozygosity, and complex genetic traits. *Hum. Genet.* **103**: 2–4.

Bähring, S., Nagai, T., Toka, H.R., Nitz, C., Toka, O., Aydin, A., Wienker, T.F., Schuster, H. and Luft, F.C. (1997) Deletion at 12p in a Japanese child with brachydactyly overlaps the assigned locus of brachydactyly with hypertension in a Turkish family. *Am. J. Hum. Genet.* **60**: 732–735.

Baker, E.H., Dong, Y.B., Sagnella, G.A., ***et al.*** (1998) Association of hypertension with T594M mutation in β subunit of epithelial sodium channels in black people resident in London. *Lancet* **351**: 1388–1392.

Caulfield, M., Lavender, P., Newell-Price, J., Kamdar, S., Farrall, M. and Clark, A.J.L. (1996) Angiotensinogen in human essential hypertension. *Hypertension* **28**: 1123–1125.

Chalmers, J. (1998) Volhard Lecture: brain, blood pressure and stroke. *J. Hypertension* **16**: 1849–1858.

Church, G.M. and Kieffer-Higgins, S. (1988) Multiplex DNA sequencing. *Science* **240**: 185–188.

de la Chapelle, A. (1993) Disease gene mapping in isolated human populations: the example of Finland. *J. Med. Genet.* **30**: 857–865.

de Vries, H., van der Meulen, M.A., Rozen, R., Halley, J.J.D., Scheffer, H., ten Kate, L.P., Buys, C.H.C.M. and te Meerman, G.J. (1996) Haplotype identity between individuals who share a CFTR mutation allele 'identical by descent': demonstration of the usefulness of the haplotype-sharing concept for gene mapping in real populations. *Hum. Genet.* **98**: 304–309.

Donelly, K.P. (1983) The probability that related individuals share some section of genome identical by descent. *Theor. Pop. Biol.* **23**: 43–63.

Ewens, W.J. and Spielman, R.S. (1995) The transmission/disequilibrium test: history, subdivision, and admixture. *Am. J. Hum. Genet.* **57**: 455–464.

Fornage, M., Amos, C.I., Kardia, S., Sing, C.F., Turner, S.T. and Boerwinkle, E. (1998) Variation in the region of the angiotensin-converting enzyme gene influences interindividual differences in blood pressure levels in young white males. *Circulation* **97**: 1773–1779.

Geiger, H., Naraghi, R., Schobel, H.P., Frank, H., Sterzel, R.B. and Fahlbusch, R. (1998) Decrease of blood pressure by ventrolateral medullary decompression in essential hypertension. *Lancet* **352**: 446–449.

Girerd, X., Hanon, O., Mourad, J.J., Boutouyrie, P., Laurent, S. and Jeuenemaitre, X. (1998) Lack of association between renin–angiotensin system, gene polymorphisms, and wall thickness of the radial and carotid arteries. *Hypertension* **32**: 579–583.

Green, S.A., Turki, J., Innis, M. and Liggett, S.B. (1994) Amino-terminal polymorphisms of the human β-2 adrenergic receptor impart distinct agonist-promoted regulatory properties. *Biochemistry* **33**: 9414–9419.

Hegele, R.A., Harris, S.B., Hanley, A.J.G., Cao, H. and Zinman, B. (1998) G protein b3 subunit gene variant and blood pressure variation in Canadian Oji-Cree. *Hypertension* **32**: 688–692.

Hibi, K., Ishigami, T., Tamura, K., *et al.* (1998) Endothelial nitric oxide synthase gene polymorphism and acute myocardial infarction. *Hypertension* **32**: 521–526.

Hunt, S.C., Cook, N.R., Oberman, A., Cutler, J.A., Hennekens, C.H., Allender, P.S., Walker, W.G., Whelton, P.K. and Williams, R.R. (1998) The angiotensinogen genotype, sodium reduction, weight loss, and prevention of hypertension: trials of hypertension prevention, phase II. *Hypertension* **32**: 393–401.

Inoue, I., Nakajima, T., Williams, C.S., *et al.* (1997) A nucleotide substitution in the promoter of human angiotensinogen is associated with essential hypertenson and affects basal transcription. *J. Clin. Invest.* **99**: 1786–1797.

Jeunemaitre, X., Soubrier, F., Kotelevtsev, Y.V., *et al.* (1992) Molecular basis of human hypertension: role of angiotensinogen. *Cell* **71**: 169–180.

Jeunemaitre, X., Inoue, I., Williams, C., *et al.* (1997) Haplotypes of angiotensin in essential hypertension. *Am. J. Hum. Genet.* **60**: 1448–1460.

Joint National Committee (1997) The sixth report of the Joint National Committee on prevention, detection, evaluation, and treatment of high blood pressure. *Arch. Intern. Med.* **157**: 2413–2446.

Jones, H.B. (1998) The relative power of linkage and asscociation studies for the detection of genes involved in hypertension. *Kidney Int.* **53**: 1446–1448.

Kamitani, A., Wong, Z.Y., Fraser, R., Davies, D.L., Connor, J.M., Foy, C.J., Watt, G.C. and Harrap, S.B. (1998) Human alpha-adducin gene, blood pressure, and sodium metabolism. *Hypertension* **32**: 138–143.

Kashiwagi, H., Tomiyama, Y., Honda, S., *et al.* (1995) Molecular basis of CD36 deficiency. *J. Clin. Invest.* **95**: 1040–1046.

Kato, N., Sugiyama, T., Nabika, T., Morita, H., Kurihara, H., Yazaki, Y. and Yamori, Y. (1998) Lack of association between the α-adducin locus and essential hypertension in the Japanese population. *Hypertension* **31**: 730–733.

Kotanko, P., Höglinger, O. and Skrabal, F. (1992) β-2 adrenoceptor density in fibroblast culture correlates with human NaCl sensitivity. *Am. J. Physiol.* **263**: C623–C627.

Kotanko, P., Binder, A., Tasker, J., DeFreitas, P., Kamdar, S., Clark, A.J.L., Skrabal, F. and Caulfield, M. (1997) Essential hypertension in African Caribbeans associates with a variant of the β2-adrenoceptor. *Hypertension* **30**: 773–776.

Kunz, R., Kreutz, R., Beige, J., Distler, A. and Sharma, A.M. (1997) Association between the angiotensinogen 235T-variant and essential hypertension in whites: a systematic review and methodological appraisal. *Hypertension* **30**: 1331–1337.

Kurtz, T.W. and Gardner, D.G. (1998) Transcription-modulating drugs: a new frontier in the treatment of essential hypertension. *Hypertension* **32**: 380–386.

Lander, E.S. and Botstein, D. (1986) Mapping complex genetic traits in humans: new methods using a complete RFLP linkage map. *Cold Spring Harb. Symp. Quant. Biol.* **51**: 49–62.

Li, A., Li, K.X., Marui, S., Krozowski, Z.S., Batista, M.C., Whorwood, C.B., Arnhold, I.J., Shackleton, C.H., Mendonca, B.B. and Stewart, P.M. (1997) Apparent mineralocorticoid excess in a Brazilian kindred: hypertension in the heterozygous state. *J. Hypertens.* **15**: 1397–1402.

Liggett, S. (1995) Functional properties of human b2-adrenergic receptor polymorphisms. *NIPS* **10**: 265–273.

Liggett, S.B., Wagoner, L.E., Craft, L.L., Hornung, R.W., Hoit, B.D., McIntosh, T.C. and Walsh, R.A. (1998) The Ile164β_2-Adrenergic receptor polymorphism adversely affects the outcome of congestive heart failure. *J. Clin. Invest.* **102**: 1534–1539.

Luft, F.C. (1998) Scientific conference on the genome: applications to cardiovascular biology (editorial). *J. Mol. Med.* **76**: 369–371.

Luft, F.C., Schuster, H., Bilginturan, N. and Wienker, T. (1995) 'Treasure your exceptions': what we can learn from autosomal dominant inherited forms of hypertension. *J. Hypertens.* **13**: 1535–1538.

Manunta, P., Cusi, D., Barlassina, C., *et al.* (1998) α-Adducin polymorphisms and renal sodium handling in essential hypertensive patients. *Kidney Int.* **53**: 1471–1478.

Markus, H.S., Ruigrok, Y., Ali, N. and Powell, J.F. (1998) Endothelial nitric oxide synthase exon 7 polymorphism, ischemic cerebrovascular disease, and carotid atheroma. *Stroke* **29**: 1908–1911.

Melander, O., Orho, M., Fagerudd, J., Bengtsson, K., Groop, P.-H., Mattiasson, I., Groop, L. and Hulthen, U.L. (1998) Mutations and variants of the epithelial sodium channel gene in Liddles syndrome and primary hypertension. *Hypertension* **31**: 1118–1124.

Miall, W.E. and Oldham, P.D. (1955) A study of arterial pressure and its inheritance in a sample of the general population. *Clin. Sci.* **14**: 459–461.

Munroe, P.B., Strautnieks, S.S., Farrall, M., Daniel, H.I., Lawson, M., DeFreias, P., Fogarty, P., Gardiner, R.M. and Caulfield, M. (1998) Absence of linkage of the epithelial sodium channel to hypertension in black Caribbeans. *Am. J. Hypertension* **11**: 942–945.

Nagy, Z., Busjahn, A., Bähring, S., Faulhaber, H.-D., Gohlke, H.-R., Knoblauch, H., Schuster, H. and Luft, F.C. (1997) Quantitative trait loci for blood pressure exist near the IGF-1, the Liddle syndrome, and the angiotensin II-receptor gene loci in man. *Hypertension* **30**: 494A.

Naraghi, R., Schuster, H., Toka, H.R., *et al.* (1997) Posterior fossa neurovascular anomalies in autosomal dominant hypertension and brachydactyly. *Stroke* **28**: 1749–1754.

O'Donnell, C.J., Lindpaintner, K., Larson, M.G., Rao, V.S., Ordovas, J.M., Schaefer, E.J., Myers, R.H. and Levy, D. (1998) Evidence for association and genetic linkage of the angiotensin-converting enzyme locus with hypertension and blood pressure in men but not women in the Framingham Heart Study. *Circulation* **97**: 1766–1772.

Persu, A., Bargry, P., Bassilana, F., Houot, A.-M., Mengual, R., Lazdunski, M., Corvol, P. and Jeunemaitre, X. (1998) Genetic analysis of the β subunit of the epithelial Na^+ channel in essential hypertension. *Hypertension* **32**: 129–137.

Pickering, G.W. (1982) Systemic arterial hypertension. In: *Circulation Of The Blood, Men and Ideas.* (eds A.P. Fishman and D.W. Richards). American Physiological Society, Bethesda, MD, pp. 487–541.

Risch, N. and Merikangas, K. (1996) The future of genetic studies of complex diseases. *Science* **273**: 1516–1517.

Schunkert, H., Hense, H.W., Doring, A., Riegger, G.A. and Siffert, W. (1998) Association between a polymorphism in the G protein beta3 subunit gene and lower renin and elevated diastolic blood pressure levels. *Hypertension* **32**: 510–513.

Siffert, W. (1998) G proteins and hypertension: an alternative candidate gene approach. *Kidney Int.* **53**: 1466–1470.

Siffert, W., Rosskopf, D., Siffert, G., *et al.* (1998) Association of a human G-protein beta3 subunit variant with hypertension. *Nature Genet.* **18**: 8–10.

Soubrier, F. (1998) Blood pressure gene at the angiotensin I-converting enzyme locus: chronicle of a gene foretold. *Circulation* **97**: 1763–1765.

Spielman, R.S. and Ewens, W.J. (1998) A sibship test for linkage in the presence of association: the sib transmission/disequilibrium test. *Am. J. Hum. Genet.* **62**: 450–458.

Su, Y.R., Rutkowski, M.P., Klanke, C.A., *et al.* (1996) A novel variant of the β-subunit of the amiloride-sensitive sodium channel in African Americans. *J. Am. Soc. Nephrol.* **7**: 2543–2549.

Svetkey, L.P., Chen, Y.-T., Mckeown, S.P., Preis, L. and Wilson, A.F. (1997) Preliminary evidence of linkage of salt sensitivity in black Americans at the β2-adrenergic receptor locus. *Hypertension* **29**: 918–922.

Timmermann, B., Mo, R., Luft, F.C., *et al.* (1998) β-2 Adrenoceptor genetic variation is associated with genetic propensity to essential hypertension: The Bergen Blood Pressure Study. *Kidney Int.* **53**: 1455–1460.

Toka, H.R., Bähring, S., Chitayat, D., *et al.* (1998) Families with autosomal-dominant brachydactyly type E, short stature, and severe hypertension. *Ann. Intern. Med.* **129**: 204–208.

Turki, J., Pak, J., Green, S.A., Martin, R.J. and Ligget, S.B. (1995) Genetic polymorphisms of the β-2 adrenergic receptor in nocturnal and nonnocturnal asthma: evidence that Gyl6 correlates with the nocturnal phenotype. *J. Clin. Invest.* **95**: 1635–1641.

Ueda, S., Meredith, P.A., Morton, J.J., Connell, J.M.C. and Elliott, H.L. (1998) ACE (I/D) Genotype as a predictor of the magnitude and duration of the response to an ACE inhibitor drug (enalaprilat) in humans. *Circulation* **98**: 2148–2153.

Yang-Feng, T.L., Xue, F.Y., Zhong, W.W., Cotecchia, S., Frielle, T., Caron, M.G., Lefkowitz, R.J. and Francke, U. (1990) Chromosomal organization of adrenergic receptor genes. *Proc. Natl Acad. Sci. USA* **87**: 1516–1520.

Index